THE SOCIAL DIRECTION OF THE PUBLIC SCIENCES

SOCIOLOGY OF THE SCIENCES

A YEARBOOK

VOLUME XI – 1987

THE SOCIAL DIRECTION OF THE PUBLIC SCIENCES

Causes and Consequences of Co-operation between Scientists and Non-scientific Groups

Edited by

STUART BLUME

Department of Science Dynamics, University of Amsterdam, The Netherlands

JOSKE BUNDERS

Department of Biology and Society, The Free University, Amsterdam, The Netherlands

LOET LEYDESDORFF

Department of Science Dynamics, University of Amsterdam, The Netherlands

and

RICHARD WHITLEY

Manchester Business School, University of Manchester

D. REIDEL PUBLISHING COMPANY

A MEMBER OF THE KLUWER ACADEMIC PUBLISHERS GROUP

DORDRECHT / BOSTON / LANCASTER / TOKYO

Library of Congress Cataloging in Publication Data

The Social Direction of the Public Sciences

CIP-data appear on separate card.
ISBN 90-277-2381-8
ISBN 90-277-2382-6 (Pbk)

Published by D. Reidel Publishing Company,
P.O. Box 17, 3300 AA Dordrecht, Holland

Sold and distributed in the U.S.A. and Canada
by Kluwer Academic Publishers,
101, Philip Drive, Norwell, MA 02061, U.S.A.

In all other countries, sold and distributed
by Kluwer Academic Publishers Group,
P.O. Box 322, 3300 AH Dordrecht, Holland

Printed in The Netherlands

TABLE OF CONTENTS

PART III

Collaborations in National Contexts

PART IV

Collaborations and the Emergence of New Scientific Fields

Epilogue

INTRODUCTION

This volume of the *Sociology of the Sciences Yearbooks* stems from our experience that collaborations between non-scientists and scientists, often initiated by scientists seeking greater social relevance for science, can be of major importance for cognitive development. It seemed to us that it would be useful to explore the conditions under which such collaborations affect scientific change and the nature of the processes involved. This book therefore focuses on a number of instances in which scientists and non-scientists were jointly involved in the generation of scientific results at the "interface" of science and society. Despite the considerable variety of cases reported here, a number of questions are central. Under what conditions do such cooperative processes occur? What perceptions of social relevance and what sorts of collaborations with non-scientific groups are involved? How is this collaboration achieved, and through what forums? How can insights into its conditions and mechanisms stabilize such cooperations over a longer period of time? If they are stabilized, do they really affect science, or do they mainly function to shield the rest of the science system against external influences?

These questions are pertinent both to intellectual problems in the sociology of science and to the practical concerns of modern science policies.

The significance of relations between knowledge producers and knowledge consumers and interest in how these relations affect science and society have changed considerably in recent decades. Historically, attention to these questions has one root in the Marxist studies of Bernal, Needham and others in England in the 1930s (1), and the Marxist historiography of which Hessen's 1931 study of Newton is a famous example (2). A second root developed in the post-war period as economists, initially concerned to explain differences in national growth rates, developed a new interest in the inter-relations between science and technology, innovation, and economic growth. By the 1960s in the study of technological innovation a clear focus on a "science and technology push" model had emerged (3). According to this model the

S. Blume, J. Bunders, L. Leydesdorff and R. Whitley (eds.), The Social Direction of the Public Sciences. Sociology of the Sciences Yearbook, Vol. XI, 1987, vii–xii.

development of fundamental knowledge is distinct from and prior to its application in techniques: scientific results are "transferred" into the market place. But increasingly in the 1970s other authors, following Schmookler (4), emphasized the determinant importance of economic "demand" in channelling scientific and technological development in modern society. More recently, the "supply-side" has been put back on the agenda (5) as a result of the growing awareness that scientific results and technological products have to be produced under uncertain market perspectives and over such long periods of time, and that for those involved, research, development and production cannot be related to market developments except in terms of uncertainties and strategies (6). In recent theorizing about technological innovation concepts such as natural trajectories (7) and technological paradigms (8) tend, on the one hand, to link the analysis of technological developments to the well-known analysis of the knowledge production process in terms of Kuhnian paradigms, and on the other hand, to shift the focus of attention from concepts at the system level (e.g. "technology push", "market pull") to a more precise understanding of how the various actors involved socially accomplish the scientific enterprise (9).

Attention to the demand side of the science system makes us, as students of the sociology of science, aware that it is inadequate to regard "interests" and "contexts" as no more than explanatory factors for scientific developments (10). We have to see them as also denoting the channels through which the integration of the science system into the larger social and economic system occurs. If the sociology of science is not to restrict itself to its special domain but also to contribute to the larger body of sociological theorizing, these are matters which demand consideration. The traditional opposition of "internal" versus "external" explanations has proven unfruitful, but also the extension from the science system into ever wider social circles (also introduced as "transepistemic arenas" (11)) becomes problematic. The generation of scientific ideas and their development are not only related to (and to be understood in terms of) social and economic "interests" or "contexts" but are themselves processes through which a culture, a society, an economy, are constituted and transformed. The special character of the "mature" science system — its institutions, but more especially its cognitive structure — seems to make it less accessible to direct political intervention than, for example, the economic system. Any understanding of the possibilities of such intervention as we possess, however, has tended to be based on attempts at government "science policy", "top down" stimulation (12), and is in this way theoretically limited. The properly *sociological* analysis of relations between non-scientific actors and scientists can help to elucidate the problem of intervention, of "steering" (and thus of science policy) by examining these relations not only from the "top down", as the objects of state policies, but also from the "bottom up", as the process of integrating scientific and non-scientific groups.

In this context, the intermediary function of "control structures" is also of importance. Whitley (1984) has noted the relations between external dependencies and the capacities of fields and specialists to rely on more internal reputational control structures (13). As the organized knowledge production system comes to be seen as necessary for social development, managerial control structures should emerge and possibly interfere with the existing older control systems. This perspective requires us to relate these "bottom up" processes, and their organizational formalization, to the aspects of cognitive structure studied by philosophers, historians, and sociologists of science. How, for example, do new organizational forms emerging from the relations between non-scientists and scientists affect the content of science, its problem formulations, the strategies and methods being chosen to solve these problems? Do the new forms of organization in fact relate to cognitive developments in the disciplines involved (14), or only to the institutional settings of these developments?

The present book originated from a theme paper in which we elaborated these and related questions about "bottom up" collaborative efforts, drawing attention to several instances, mainly in Europe, where experimental attempts to integrate scientific and social perspectives had been pursued over several years (15). We proposed to focus on those developments which were primarily directed toward the transformation of the science system and which managed to maintain that direction over time. When this theme was accepted as the subject for Volume XI of the *Yearbook* series, the editors specifically invited papers from persons familiar with the cases cited in the theme paper. In some instances, as the reader will notice, this has been successfully achieved; while in others, for example the women's movement and the development of women's studies, to the editors' regret and despite serious efforts, no contribution to this volume could be secured.

In November 1985, a conference on the theme was held in Amsterdam. This volume contains some of the papers presented at that conference. The first part of the volume, on "Co-operative Processes and the Production of Scientific Knowledge", begins with an essay by *Blume* on the questions for science studies and science policies arising from current interests in university-industry relations. In an article on the meaning of collaborations for the research process in biology, *Bunders* analyzes the specific functions of such collaborations in Dutch university research. *Von Gizycki*'s story of the collaboration between medical researchers and patients, which can be called prototypical for the type of collaborations at issue here, opens the second part of the volume on "Collaborations at the Grassroots". In their article on the development of knowledge interests of the environmental movement, *Cramer, Eyerman & Jamison* extend the analysis of such collaborations to the larger social context. In two following articles, *Benguigui* and *Leydesdorff & Van den Besselaar* focus, respectively, on the conflicts which may arise when groups oppose the "scientification" of their domain, or when they try to organize an alternative.

The larger social context is a more dominant issue in the third part of the volume, on "Collaborations in National Contexts". The urgent needs of a third world country (India) for the problem-solving capacities of the scientific enterprise, as these have been successively defined and redefined, are discussed in *Ramasubban & Singh*'s contribution. This can also be compared with earlier attempts by the Soviet system to breed its own science cadres in the first decade of Bolshevik power, as described in *Balzer*'s contribution. *Amsterdamska* concludes this section with an analysis of the role of intellectuals in the Polish trade-union movement "Solidarity".

In a final part on "Collaborations and the Emergence of Scientific Fields" we take this broader contextualist understanding back into the scientific enterprise itself, and look once more at the meaning of such collaborations for the development of the sciences. *Fridjonsdottir* and *Wagner* explore this meaning for different branches of sociology in their respective contexts. The emergence of new fields under the influence of "contexts" is not only to be found in sociology, as is demonstrated in *Groenewegen*'s account of the emergence of toxicology in the Netherlands. The volume concludes with an epilogue in which *Bunders* and *Leydesdorff* again raise the question of what can be learned of relevance to the sociology of the sciences from the experiences in the case studies described in the volume. Elaborating on the distinction between the socio-political conditions in the various specific national histories and cultures, and the systematic significance of the processes which were brought about in the science system (16), they attempt to develop a framework of the relevant dimensions for comparison between the cases, with a view to the possibilities for further research.

The editors and authors are indebted to the *Directorate for Science Policy of the Ministry of Education and Sciences* of the Netherlands and to the *Royal Dutch Academy of Sciences*, which jointly financed the November 1985 Conference from which some of the papers have served as the basis for this volume. The editors also want to thank the two Amsterdam universities which made it possible to organize both the conference and this volume. Finally, we are grateful to Gene Moore, who corrected the English of most of the papers in this volume.

Amsterdam, July 1986 LOET LEYDESDORFF and JOSKE BUNDERS

Acknowledgement

We would like to thank Stuart Blume for his help with the preparation of this introduction.

Notes and References

1. G. Wersky, *The Visible College*. London: Allen Lane, 1978.
2. B. Hessen, "The social and economic roots of Newton's 'Principia' " in: *Science at the Cross Roads*. (Reprinted London: F. Cass and Co. 1971), 1931.
3. Z. Griliches, "The sources of measured productivity growth: United States agriculture 1940—60", *Journal of Political Economy*, 1963, 331—346; Office of the Director of Defense Research and Engineering, *Project Hindsight Final Report*. Washington, D.C., 1969; Illinois Institute of Technology, *Technology in Retrospect and Critical Events in Science*. Washington, D.C.: National Science Foundation, NSF-C535, 1969.
4. J. Schmookler, *Innovation and Economic Growth*. Harvard University Press, 1966.
5. D. Mowery, N. Rosenberg, "The influence of market demand upon innovation. A critical review of some recent empirical studies", *Research Policy* **8**, 1979, 102—153.
6. M. Teubal, "On user needs and need determinations: Aspects of the theory of technological innovation", in M. J. Baker (ed.), *Industrial Innovation. Technology, Policy, Diffusion*. London: Macmillan Press, 1979, pp. 266—289.

7. R. R. Nelson, S. G. Winter, "In search of a useful theory of innovation", *Research Policy* **6**, 1977, 36—76.
8. G. Dosi, "Technological paradigms and technological trajectories", *Research Policy* **11**, 1982, 147—162.
9. *Ibid.* See also: E. von Hippel, "A customer active paradigm for industrial product idea generation", *Research Policy* **7**, 1978, 240—266; M. Callon, "Some elements of a sociology of translation: domestication of the scallops and the fishermen of St. Brieuc Bay", in J. Law (ed.), *Power, Action and Belief. A New Sociology of Knowledge*, London/Boston: Routledge & Kegan Paul, 1986, 196—233; R. Laudan (ed.), *The Nature of Technological Knowledge: Are Models of Scientific Change Relevant?* Dordrecht: Reidel, 1984.
10. D. Bloor, *Knowledge and Social Imagery*, London/Boston: Routledge & Kegan Paul, 1976. See also: S. Woolgar, "Interests and explanation in the social study of science", *Social Studies of Science* **11**, 1981, 365—394.
11. K. Knorr, *The Manufacture of Knowledge: An Essay on the Constructivist and Contextual Nature of Science*. New York: Pergamon Press, 1981.
12. W. v.d. Daele, W. Krohn, P. Weingart (eds.), *Geplante Forschung*. Frankfurt a.M.: Suhrkamp, 1979.
13. R. D. Whitley, *The Intellectual and Social Organization of the Sciences*. Oxford: Oxford University Press, 1984.
14. G. Böhme, W. v.d. Daele, W. Krohn, "Finalisierung der Wissenschaft", *Zeitschrift für Soziologie* **2**, 1973, 128—144. See also: L. Leydesdorff, "The development of frames of references", *Scientometrics* **9**, 1986, 104ff.
15. J. Bunders, L. Leydesdorff, "The social direction of the public sciences: Causes and consequences of collaboration between scientists and non-scientific groups". Amsterdam internal paper, 1983.
16. *Ibid.*

BIOGRAPHICAL NOTES OF THE CONTRIBUTORS

OLGA AMSTERDAMSKA, born in 1953, received her Ph.D. in sociology from Columbia University in 1983. She is a postdoctoral fellow in the Department of Science Dynamics at the University of Amsterdam. Her book on the development of schools of thought will appear shortly in the Sociology of the Sciences Monographs Series (Reidel). Currently, her research focusses on the relationship between the clinical practice of medicine and the development of the biomedical sciences, particularly bacteriology.

HARLEY BALZER is a Research Associate in History and Russian Area Studies at Georgetown University, Washington, D.C. He received his Ph.D. in History from the University of Pennsylvania in 1980. Building upon his dissertation research on technical training and engineers in Tsarist Russia, he has extended his research into the role of professional groups in Russian and Soviet society, the Soviet education system, and the history of Soviet science and technology. He is currently completing two books on the role of engineers in Russian and Soviet society, and has begun a study of Soviet scientific/technical education and manpower for the U.S. National Science Foundation. His most recent publications are "Is Less More? Soviet Science in the Gorbachev Era", in *Issues in Science and Technology* 1: 4 (Summer 1985), pp. 29—46, and "Soviet Research and Development: Information and Insights from the Third Emigration" (Washington, D.C.: National Council for Soviet and East European Research, July 1986).

GEORGES BENGUIGUI, is Directeur de recherche at the CNRS, and belongs to a team located in the University of Paris VII. His main areas of interest are the sociology of work and sociology of the sciences. Among his recent publications are: "L'Etat et les petits poissons" and "Les physiciens, sont ils de gauche et les chimistes de droite?"

STUART BLUME, born in 1942, graduated from the University of Oxford. In 1966 he joined the Science Policy Research Unit at the University of Sussex, and has since worked in a number of settings, both academic and governmental. The author of *Toward a Political Sociology of Science* (Free Press, 1974) and many articles in the fields of sociology of science, science policy and social policy, he is at present completing a book on technological innovation in diagnostic medicine. Dr. Blume is currently Professor of Science Dynamics at the University of Amsterdam.

JOSKE BUNDERS studied inorganic chemistry and physics. From 1975 till 1980 she was an editor specialising in the popularisation of the natural sciences. Since 1980 she has been working in the new department of Biology and Society at the Free University of Amsterdam. Her reseach is focussed on the analysis of cooperations between university biologists and non-scientific groups. Apart from publications in Dutch journals, she has published "Popularisation within the sciences: The purposes and consequences of inter-specialist communication", T. Shinn and R. Whitley (eds.), *Expository Science: Forms and Functions of Popularisation*, Sociology of the Sciences, Volume IX, 1985, 61—77.

JACQUELINE CRAMER is a tenured faculty member at the Department of Science Dynamics, University of Amsterdam. She has published studies of the dynamics of the field of ecology and the use of ecological knowledge in decision-making.

RON EYERMAN teaches sociology at the University of Lund. Together with co-authors A. Jamison and J. Cramer and the Danish social-psychologist J. Lassöe, he is currently engaged in a comparative study of the environmental movements in Sweden, Denmark, and Holland.

KATRIN FRIDJONSDOTTIR is a researcher at the Swedish Collegium for Advanced Study in the Social Sciences at the University of Uppsala, and teaches sociology at the University of Lund. Her main areas of interest are sociology of science, science policy studies, and studies in the development of the social sciences.

PETER GROENEWEGEN, born in 1952, studied chemistry and sociology at the University of Leiden and sociology at the University of Amsterdam. He has been interested in science and society since the mid-seventies and was involved in science shop work and health and safety activities. He is co-editor of "Macht over Kennis" a compilation of studies relevant to science policy, and his other publications focus on the role of expertise in chemical risk issues. He was previously in the Department of Science Dynamics of the University of Amsterdam, where he taught chemistry and society and studied the development of toxicology, and he is currently a lecturer in science and society in the Department of Chemistry of the University of Groningen.

ANDREW JAMISON is a lecturer and research fellow at the Research Policy Institute, University of Lund. He received his doctorate in theory of science from the University of Gothenburg in 1983, with a thesis entitled "National Components of Scientific Knowledge". He is the author of *The Steam-Power Automobile* (Indiana, 1970), and most recently a co-editor of *Technological Development in China, India and Japan* (Macmillan, 1986) and *The Biotechnological Challenge* (Cambridge, 1986).

LOET LEYDESDORFF is a senior lecturer at the Department of Science Dynamics at the University of Amsterdam. He graduated as a biochemist and as a philosopher, worked with the Science Shop for some time, and wrote his Ph.D. in sociology on the role of employees in technology policy (1984). Currently, his research focusses on the use of scientometric methods for the conceptualization of the sciences as cognitive structures.

RADHIKA RAMASUBBAN and BHANWAR SINGH are social scientists from India. They share a common interest in the relationship between the cognitive and social structures in the newly emerging critical scientific groups in India. Radhika Ramasubban's research and teaching work is directed towards the understanding of the economic and political imperatives of nation building for the formulation of science policy, the emergence of new scientific fields, and the environment of research, particularly in the Indian context.

She is currently engaged in an historical study of the emergence of modern medical science in India under the impact of colonialism. She is on the sociology teaching faculty of the University of Bombay and is actively involved in the research work of the Centre for Social and Technological Change (Bombay) of which Bhanwar Singh is Director. Bhanwar Singh's research is mainly in the field of the social implications of technological changes particularly in the rural Indian context. His book *Agrarian Structure, Technology Change and Poverty* deals with the green revolution technologies. His other recent studies have dealt with energy technologies, the technologies of production, storage, and marketing of horticultural products, and technological changes in dairying.

PETER VAN DEN BESSELAAR studied mathematics at the University of Utrecht and philosophy at the University of Amsterdam. He is now at the University of Amsterdam, Department of Social Science Informatics where he studies and teaches the social aspects of computing.

RAINALD VON GIZYCKI, born in 1942, is senior researcher at the Technology Management and Assessment (TMA) unit of the Battelle Institute in Frankfurt. His main research interests are the dynamics of technological development and its consequences for vocational training, the international division of labour, and changes in the work place. He is the author of a number of articles and books on these topics, e.g., *Mikroprozessoren und Bildungswesen* (1981), and *Microelectronics: A Challenge for Europe's Industrial Survival* (with I. Schubert, 1984).

PETER WAGNER, born in 1956, is a political scientist and a research fellow at the Wissenschaftszentrum Berlin für Sozialforschung. His main research interests are in the relation between science and society and in science and technology policy. His most recent research is part of a larger project on the social and intellectual factors involved in the historical development of policy-oriented social science. He has published a number of articles on this topic.

PART I

CO-OPERATIVE PROCESSES AND THE PRODUCTION OF SCIENTIFIC KNOWLEDGE

THE THEORETICAL SIGNIFICANCE OF CO-OPERATIVE RESEARCH

STUART S. BLUME

Department of Science Dynamics
University of Amsterdam

Introduction

For those sociologists of science who admit some possible demarcation of science from non-science, and who seek some kind of "ordering principles" in the dynamics of science, the question of how we are to understand the influence of "external factors", or society, on science is a central one. Methodologically, historical studies have provided considerable insight as well as a means by which the various theoretical positions have been exemplified. Theoretically, a number of recent formulations have achieved considerable influence and have stimulated considerable debate. These include the "finalization thesis" (Böhme et al. 1973), and the "strong programme in the sociology of knowledge" developed in Edinburgh. According to the first view, the openness of science to external influence, to external goal setting, is to be seen as a function of the theoretical maturity of the field in question. For the authors of the second view, which albeit in attenuated form has had a certain fascination for many "contextualist historians", knowledge is always "constructed out of available cultural resources in ways which are specific to particular times and situations", inspired by interests, and to be understood in terms of its "role in activity" (Barnes 1977). The knowledge constituted around these interests is what people then *agree* to call science. Marxist theoreticians, rejecting the extreme relativism of this position, look at the matter rather differently, by clearly separating between "science" ("direct knowledge" which "does not contain anything that indicates its social origin") and "research" ("tied to social conditions and relations of production") (Farkas 1983).

There are numerous difficulties with all of these positions, some of

S. Blume, J. Bunders, L. Leydesdorff and R. Whitley (eds.), The Social Direction of the Public Sciences. Sociology of the Sciences Yearbook, Vol. XI, 1987, 3–38.

which have been reviewed by Whitley (1983) and by Milic (1984). Among the difficulties which have received rather little attention is the following. How should we try to understand *what* exactly in the "social context" of scientific activity is of cognitive importance for that activity? Purely a posteriori identification (imputation?) of "interests" through historical reconstruction, without theoretical basis, is surely inadequate. Nor is it adequate to assume that social influence is wholly to be identified with those *policies* designed to bring about certain sorts of scientific change. The sociology of science must seek theoretically to account for the way in which a particular set of interests, policies, values comes to be of significance for — to be "constitutive" of — scientific activity. This theoretical requirement is implied in the question posed by this volume: "under what conditions can grass-roots movements, bottom-up processes, influence scientific activity?" Having admitted the importance of this theoretical requirement, a further question arises. In what *way* (or ways) is this relevant social context significant or constitutive?

If we are to make any progress with these questions, I believe we must choose the activities we study with great care. In particular we should attend in the first place to research activities where some kind of simplification, some delineation of what can justifiably be taken as the constitutive social context, is possible (Blume et al. 1985). In this paper I propose to look in very broad terms at an area of activity which may have some interest in this regard, though it is one which is today much discussed in very different terms.

University-Industry Relations: A Problem of Science Policy

In the past five years or so the relations between universities and industry have become a major issue in science policy discussion in almost all industrialised Western countries. In the United States the National Science Board (NSB), the policy making body of the National Science Foundation, devoted its 1982 Annual Report to the subject (NSB 1982). The Report begins:

> Research relationships between American universities and industry have been numerous, constructive and important since the turn of the century. The persistence of this long and fruitful relationship has rested, and continues to rest in good measure upon industry's overriding need for highly qualified new scientists and engineers. Recent

developments suggest that university research programs are becoming of increasing interest to industry in their own right.

A recent report from the OECD, which reviews the situation in a number of OECD Member countries, makes a very similar point (though it puts the beginnings of the transformation further back than I did):

In the last decade, industry-university relations have undergone major shifts and changes. Traditional approaches have turned out to be insufficient in the face of rising expectations stimulated by intense international competition. Although industry still expects the universities to focus their primary efforts on the training (and re-training) of scientists and engineers, academic research is increasingly seen as offering specific opportunities which call for cooperation. (OECD 1984, p. 9)

It is not difficult to identify the symptoms of this new interest: a burgeoning literature (Stankiewicz has provided a 35-page bibliography; Stankiewicz 1984), and a variety of initiatives taken by firms, by universities, and by governments. Whilst these initiatives are to be found throughout the Western world (as the OECD report shows), both quantitatively and qualitatively the underlying transformation is most marked in the United States. The very size of some of the agreements reached between American universities and major corporations in the past few years has been a matter of intense debate. They include Monsanto's "partnership" with Washington University and (largest of all, and by far the most discussed) the German chemical company Hoechst's $50 million award to Massachusetts General Hospital (part of Harvard Medical School) for work in the field of genetic engineering. Of qualitative interest is the recent development of research agreements linking a number of (potentially competing) firms with a single university. These include the "University-Industry Co-operative Research Centers" established under the auspices of the National Science Foundation (see Eveland 1982), and private initiatives such as the Center for Integrated Systems at Stanford University. Some "clarification" of anti-trust regulations (which prevent the formation of cartels) has been necessary for the establishment of these industrial consortia. The implications of this (newly legalised) spirit of co-operation between major American firms are profound.

It is widely believed that the newly emerging connections between universities and industry are in certain important respects *different* from longer established relations. The NSB Report tries to express this sense of newness:

> Experiments with complexes of overlapping non-profit and for-profit organizations such as the arrangements between the University of California at Berkeley, Stanford University, and six corporations (Engenics Inc) aim to develop commercial processes and also to channel significant portions of the profits back into academic research . . . The potential problems generated by the very rapid growth of university-industry research relationships in biotechnology prompted an unprecedented two-day meeting in March 1982 between the presidents of five leading research universities and their counterparts in ten high-technology firms involved in biotechnology to develop suggested guidelines . . .
>
> Research managers from most major chemical companies have held three annual conferences with chairmen of university chemistry and chemical engineering departments which have resulted in the establishment of a Council for Chemical Research . . . The Semiconductor Industry Association has set up a non-profit subsidiary, the Semiconductor Research Cooperative, the intent of which is to encourage increased efforts by manufacturers and universities in long-term semiconductor research . . . (NSB 1982, p. 13)

Moreover there is a particular aspect of this 'newness' which is frequently noted, as in the following extract from an article in *Science* on the Monsanto-Washington University arrangement:

> The second feature of the (. . .) arrangement that sets it somewhat apart is the extent of constant, intimate collaboration it anticipates between researchers at the two institutions . . . (T)his deal provides for what Howard A. Schneiderman, senior vice president of Monsanto, terms a "true partnership". Dozens of company scientists may be working on campus at any one time, once the agreement is in full swing, he notes, adding that Monsanto researchers will not be "token" members of the collaborative team. Indeed, the desire for close collaboration was one of the reasons Monsanto decided to deal with Washington University. Says Schneiderman, not only is it a "major research university", it also has the distinct practical advantage of being "only 15 minutes away" from company headquarters in St. Louis. (Culliton 1982)

Beyond the United States there are many initiatives pushing in the same direction. The Alvey programme in Britain, designed to foster a collaborative approach to the development of advanced information technology, is a clear example, and it would require no great effort to produce a list of other such measures.

What lies behind all this? Why has this precoccupation with university-industry relations emerged? Why is there a sense that they are (and must be) of a new form? Why do they appear principally to be concerned with certain sorts of technologies? There are a number of factors which are certainly relevant to the new state of affairs (or state of mind) pertaining to university-industry relations. The first seems generally agreed to be of major importance in the United States. It may well have wider import. Jerome Wiesner put it most succintly as follows:

> It is becoming clear that . . . competition, rather than occurring primarily between different firms, increasingly pits nation versus nation. (Wiesner 1982)

Roland Schmitt, among whose functions is that of Senior Vice President for corporate R&D of the General Electric Company in the USA, has made a similar point in other words. He writes of the emergence of a new era

> in which the science and engineering base is emerging as the critical component in the US science and technology enterprise. That emergence is driven by heightened international competitiveness, by broadening demands on science and technology, and by sharpening R&D cost structures.
>
> American industry is far more dependent on the strength of that base than it has ever been before. . . . The primary reason is foreign competition. Foreign competition is driving our industry to turn to technology leadership as its primary source of competitive advantage. (Schmitt 1985)

Moreover, Schmitt argues, this attempt to develop technology as the means of advantage is leading industry sharply to increase its interaction with the academic research system. Industry recognises that there are tasks relevant to successful innovation which universities, not firms, are best able to carry out. One of course has to do with education and training. The other suggests something of a complementarity of role. "Industry", writes Dr Schmitt,

> cannot explore all of the exciting ideas that promise potentially large payoffs for the nation, but are as yet too undefined to justify substantial investment by a company seeking to make a profit. So the more emphasis that industry places on technology, the more industry depends upon the new knowledge and talent generated by our universities.

But though the challenge of foreign competition may be an important factor in the aetiology of the present state of affairs, it does not (at least not without much more conceptual refinement) seem to provide any understanding of the form and foci of recent initiatives.

We can proceed a little further when we note the *areas* of technology in which advanced industrial nations believe they can best compete. Reports produced in many countries at the beginning of the eighties show a common focus of interest in biotechnology, computers, microelectronics, materials science and the like (see for example Rothwell and Zegveld 1982, Chapter 5). Schmitt's examples of shifting industrial interest are similar:

> General Motors has moved from automobiles to data systems; General Electric has moved from electricity into engineering plastics, computer-aided design, information services, automation, and many other areas; Monsanto has moved from saccharine to biotechnology; and AT&T . . . is trying to shed its staid Ma Bell image and become a player in the computer game.

These "areas of opportunity" may not only provide the central foci of concern in university-industry relations, but may to some degree also dictate the search for new forms. This may be particularly true with regard to biotechnology:

> The potential industrial applications of biotechnology . . . have emerged directly from publicly funded academic biomedical research. As biotechnology has been moving to the market, universities have been buffers in commercialising the fruits of public funding, because they are virtually the sole source of basic know-how. Many of the new firms in the field of biotechnology have sprung out of academia, whereas in the semiconductor field, ample D&D procurement helped to create industrial know-how and encouraged industrial spin-off. In the area of biotechnology, the traditionally distinct roles of university as a source of research and training and of industry as a source of commercialisation are blurred. (OTA 1984, quoted by Rothwell and Zegveld 1985, p. 236)

Foreign competition and the areas of technology in which advanced industrial nations seek to compete seem to be two of the factors behind the emergence of university-industry relations as an issue in science policy. I believe there is also a third, though its significance is less easily illustrated by quotation. The new perspective on technology, which the various reports analysed by Rothwell and Zegveld introduced to Western governments around 1978—80, is characterised not only by its explicit sense of "technology as strategic opportunity". It is not only that the various new technologies to which I referred are, at last, to provide growth and jobs. Less explicit in formal pronouncements and policy documents, but there nonetheless, is something of a shift in the process by which innovation is thought to take place. This parallels (unsurprisingly but still gratifyingly) the emergence of a new and more complex understanding within the *study* of technological innovation. I can make my point only briefly, for its details lie outside the scope of this paper. In the '60s, it seems to me, much theorising about science, technology and industrial innovation was based upon a sort of "science push" mode. Studies such as Hindsight and TRACES may have differed in the relative importance they attached to basic and applied research, but they shared the assumption that science led to innovation. Joseph Ben-David made the same point in an influential report (Ben-David 1968). By the seventies, by contrast, the theoretical emphasis came to lie on market, demand, factors. Here the spirit of Jacob Schmookler ruled, and empirical studies such as "Project SAPPHO" (SPRU 1972) showed the preminence of market factors in successful innovation. Now it seems to me that much work in the social studies of science and technology is motivated by a much more complex theoretical idea. Neither "science push" nor "demand pull" are paramount, and we have come to believe in some kind of *iterative* process, such as has been spelled out by Nathan Rosenberg (e.g. Rosenberg 1976). The relations between producers and consumers have been transformed for us by Nelson and Winter (e.g. Nelson and Winter 1977), who offer us the notion of a firm developing its innovation strategy in relation to the signals received from the environment in which it operates. In some industrial sectors blessed by highly sophisticated customers (product users), these users through *their* research may play an important role in the process of technological change (see e.g. van Hippel 1976). The result of all this, simplified, is that it is no longer adequate for policy makers to attend solely to the health of the science system (as Ben-David proposed in 1968), nor solely to the expression of effective demand. It is the relations between the two, the "interface", which has become the focus of attention.

Typologies and Variations

In the light of this burgeoning interest, it is scarcely surprising that university-industry relations has become a topic for sustained analysis. Numerous empirical studies have been carried out in the last few years. In approaching their task, in seeking to describe, to offer functional analyses, to prescribe, authors have largely recognised that the phenomenon is a complex one. Relations, initiatives, are not to be ranked on a unitary scale, good-bad or strong-weak, and single indicators (for example the volume of contract funding by industry in the universities) are far from an adequate means of describing. Thus, one of the most comprehensive of the recent empirical investigations (Fusfeld and Peters 1982) — involving interviews in a sample of around one hundred universities and firms — pointed out:

Our field investigation of university/industry research interactions documents their variety and multi-faceted character ... interactions can be formal or informal. They involve not only monetary support of research, but also include donations, transfers, exchanges and sharing of people, equipment, and information. The duration of successful interactions can be for less than an hour or for more than thirty years. An important interaction can be as simple as a telephone call, or as intricate as a ten-year contract. Some require collaborative efforts either among scientists or different disciplines or between university and industry scientists, others the work of only one scientist (Fusfeld and Peters, 1982, Chapter 5)

There have been a number of attempts to develop typologies of these interactions, generally based upon the notion that different sorts of interactions are to be understood in terms of the different *functions* which can be ascribed to them. Stankiewicz (1982), whilst recognising the variety of obstacles to effective relationships between institutions "characterized by strongly divergent value systems, norms, functions and working styles", goes on to distinguish two essentially different kinds:

the *assistance* mode ... in which ... the academic scientist — acting in some more or less clearly defined technical role — assists a non-academic entrepreneur, normally a firm. This mode differs sharply from another one which we will call *entrepreneurial*. The distinction between these two modes is quite fundamental. ... In the entrepreneurial mode ... it is the academic scientist himself who assumes the ultimate responsibility for a technical development project.

A recent report from the OECD does it slightly differently (OECD 1984). This report, based on inquiries carried out in fourteen countries, distinguishes a number of what it calls "networks". These networks of relations, formal and informal, can be categorised in terms of three fundamental purposes (though a given structure or incentive may be intended to serve more than one purpose). The three purposes distinguished are "promotion of long-term linkages between universities and industry"; "promotion of specific areas of scientific and technological activity"; and "liaison systems". Many examples of each of these are provided, some of which are intended to show *how* effective initiatives might be taken. For example, "long-term linkages" ("the most fundamental of all, since they may often require a reassessment of established procedures and values . . .") include the establishment of science parks, of university-based firms; they include the kinds of relationships which large high-technology companies (such as DuPont, IBM, or Philips) traditionally have with certain academic institutions. The second "mode", "promotion of special areas of science and technology," is seen as rather new, and is aimed at developing the academic infrastructure in the new fields of strategic attention. Many of the various biotechnology schemes are classified here, whether initiated by government (as in the Netherlands and Sweden), by an individual university seeking industrial sponsorship (e.g. Cornell University in the USA, with its new Biotechnology Institute), or by a firm (as in the multi-million dollar grants made by Hoechst and Monsanto to which I have already referred). The third mode, liaison systems, includes offices attached to universities designed to channel requests for help to an appropriate point within the university. Aimed specifically at the needs of firms without established academic contacts, they have expanded considerably in the last few years, stimulated in some cases by government initiatives (e.g. the Dutch transfer points, and the schemes operating in the UK). One final typology will suffice. Taken from Brodsky et al. (1980) it is reproduced in Table I. Here two dimensions are distinguished: collaborative research mechanisms versus knowledge transfer mechanisms, and "ongoing" versus "time-limited". I think these distinctions are fairly clear. The first element, "collaborative research mechanisms", corresponds more or less to the first two modes distinguished in the OECD study (though in this US study the examples are all American). Also of interest in this typology is the horizontal axis, wherein the particular kinds of contribution to be expected from each of the various elements are distinguished (and scored). A sequence running from "knowledge" through "products and processes" to development of markets is offered.

I think we can accept without difficulty the general assumption behind these typologies, for which ample evidence is provided, that university-industry relations *are* of different kinds and directed at rather diverse functions. Of particular relevance for the concern of this paper is the distinction which all these typologists make between networks or modes directed at providing knowledge to be used in problem-solving ("assistance", "liaison schemes", "knowledge transfer"), and networks or modes directed at the collaborative development of new knowledge or at the development of strategic fields of science or

TABLE I
Potential contributions to innovation

	Knowledge/ Experience Pool	New Concept	Development of Products and Processes	Market Development	
Collaborative Research Mechanisms					
A. *Ongoing Modes*					*Example*
University-based institutes serving industrial needs	+++	+++	+++	+	Textile Research Institute
Jointly owned or operated laboratories	++	++	++	+	Laboratory for Laser Energetics
Research Consortia	++	++	++	+	Michigan Energy and Resource Research Association
B. *Time-Limited Modes*					
Industry-funded university research (contracted)	++	+++	++	+	Harvard-Monsanto contracted research effort
Government-funded cooperative research programs	++	+++	++	+	MIT Polymer Processing Program
Knowledge Transfer Mechanisms					
A. *Ongoing Modes*					
Industrial liaison programs	++	++	+	+	MIT Industrial Liaison Program
Innovation Centers	+++	++	++	++	Carnegie-Mellon University Center for Entrepreneurial Development
Technology brokerage and licensing	++	++	++	+++	Research Corporation
Continuing education	+++	+	+	+	Various
Co-op study programs	+++	+	+	+	Various
Industrial parks	++	++	++	+	Research Triangle Park
B. *Time-Limited Modes*					
Consulting	++	++	++	+	Various
Personnel exchange	+++	++	+	+	Various
Seminars, Speakers programs, publication exchange	+++	+	+	+	Various

+ = low probability for contribution
++ = moderate probability for contribution
+++ = high probability for contribution

Source: Brodsky et al. (1980). Reprinted with permission from N. H. Brodsky, H. G. Kaufman, J. D. Tooker, *University/Industry Co-operation*: A preliminary analysis of existing mechanisms and their relationship to the innovation process, Copyright © Center for Science & Technology Policy. Reproduced here by kind permission of the Center for Science & Technology Policy, Graduate School of Public Administration, New York University, and the authors.

technology ("collaborative research mechanisms" in the Brodsky et al. report).

Typologies of this kind must of course be regarded as a useful *instrument* in the investigation of university-industry relations, rather than the end-product of such investigation. A further step, which has to my knowledge nowhere been taken, would be to examine the relationships of industry with the various institutions within a national academic system in terms of a typology of this kind. What I think we would expect to find is that the various institutions differ significantly from one another in terms of the intensity and structure of their industrial relationships. The evidence for this proposition is patchy and circumstantial. For example, in European countries having distinctive Technological Universities it is usually a matter of general belief that these Technological Universities have much closer and more complex links with industry than do the old established traditional (multi-faculty) institutions. In Sweden, for example, contract research (much of which was for industry) produced an income at the Technical University in Stockholm of 14.8 million Swedish kroner in 1979 (but only 6.3 million at the University of Stockholm). In Gothenburg the difference was still greater: 16.9 million at the Chalmers Technical University and only 0.9 at the University of Gothenburg (1). For the United States we have some interesting statistical data relating industrial contract research income to total research income/expenditure for the 200 top research performing campuses (NSB 1982). Here an interesting phenomenon emerges. If we rank the 200 institutions in terms of the amount of money they receive from industry for research, we find that at the top of the list come institutions such as MIT ($11.4 million), the University of Rochester ($7.9 million), Penn State ($7.8 million), and the University of Southern California ($7.5 million). Many famous universities come very low indeed: Northwestern ($0.3 million), Chicago, Columbia, Princeton (all around $0.4 million). If we now do the same thing in terms of the *percentage* which industrial research income represents of total research, then a quite different set of institutions heads the list: Brigham Young University (55%), Worcester Polytechnic Institute (44%), Stevens Institute of Technology (29%). On *this* ranking, USC, Penn State and Rochester have modest positions (around 10—11%); MIT still more modest (7%) — Columbia and Chicago remain near the bottom. It is no injustice to these data to suggest that research carried out with or for industry must have a different place in the research activities of, shall we say, MIT, Worcester Polytechnic Institute, and Columbia University. I am well aware that these figures are no more than suggestive in relation to the question which I posed above. Nevertheless, they seem to me to make the assumption that institutional differences exist more reasonable than the counter-assumption that they do not exist.

Why should such institutional differences in the nature and extent of interactions with industry be a characteristic of (many) academic systems? An explanation which will occur immediately to the reader is that they reflect differences in the disciplines emphasized by the various institutions. Technological universities specialise in the technical sciences, and these are the fields in which industry is principally interested. Though this is true, and must be part of any explanation, it is far from sufficient. There are indications (alas no more than that) that a number of other elements must enter into an explanation of this diversity.

There are, in the first place, differences in academic *cultures*, associated with differences in the role which a given institution sees itself as essentially fulfilling. Burton Clark has developed the notion of the "organizational saga" ("a collective understanding of unique accomplishment in a formally established group"), and has suggested that some at least among American academic institutions can be distinguished in terms of the organizational sagas they have developed and the roles these sagas play (Clark 1970, 1972). Statements by the heads of universities can also be indicative of different interpretations of the essential roles and responsibilities of the various institutions. Compare the following remarks:

> such involvement risks putting one's students and research associates in ambiguous circumstances, such that the graduate or postdoctoral student would not know, when working with a professor, for whom he or she was working — the university, the professor, or the company. Of all members of the university community, the student especially ought to be working for himself or herself, and ought to be guided in research and trained in skills and techniques that are designed to produce a first-rate scholar, not profit for a company in the private sector. (Giamatti 1982)

> With these programs came the entrepreneurs: graduate students, instructors, and research assistants who had ideas of their own for a product or service, and who wanted to develop these ideas. We decided to facilitate their efforts through our incubator program . . . we become a resource for them that is the equivalent of a corporate R&D laboratory. We also provide introductions to appropriate financial institutions . . . . (Low 1983)

The first of these quotations is taken from an article by the President of

Yale, the second from a speech by the President of Rensselaer Polytechnic Institute (an American university which has made spectacular progress in developing its links with American industry). The two documents, set against each other, suggest a considerable difference in emphasis. It is not that the two are in direct conflict with each other; rather that there are clear differences in emphasis, and in the resolutions suggested for conflicts of interest (for example over patents, publication, licencing and so on). In articulating the values and responsibilities thought appropriate to their particular institutions, these Presidents and their colleagues are necessarily plotting different *development strategies* for their universities.

The history of university chemistry in the United States shows that it is meaningful to speak of differences in strategy at the level of the individual discipline too. It is not that there exists a "fixed" industrial interest in one field or another such that this interest *dictates* the strategy adopted. Arnold Thackray has described the development of chemical research in three major US universities in the period following the First World War (Thackray 1982). At the University of Illinois, Roger Adams (who moved there as assistant professor of chemistry in 1916) developed that university's department of chemistry "into the world's greatest producer of PhDs" and "forged an unrivalled network of connections in industry". Adams began, very rapidly, by turning a summer course in organic preparations into "Organic Chemical Manufacturers": to compensate for the embargo on German goods, the Illinois department set about manufacturing rare organic chemicals for university and industrial concerns.

> Organic Chemical Manufacturers had many utilities. It provided good contacts for Adams and made him visible in the industrial world. It gave his students practical experience in scaling up laboratory processes to industrial quantities. It led to the creation of two serials, *Organic Syntheses* and *Organic Reactions*, which Adams edited. (Thackray, loc. cit.)

Adams' department trained astonishingly large numbers of PhDs (184 between 1918 and 1958), of whom 65% went directly into industry (compared to 8% at Johns Hopkins). Many attained very high industrial positions.

> Through his PhDs Adams had a rich network of links with industrialists and industrial concerns, while he himself demonstrated by his actions the potentialities of "men with

this training". He consulted on a regular basis. . . . Indeed he was instrumental in helping to shape Du Pont's effort in basic research, both through his advice and through his supply not only of outstanding leaders (Carothers, Cairns) but also of rank-and-file researchers . . .

At the same time Adams was highly successful in persuading industry to support graduate work and research. Thackray concludes of his career at Illinois that he "found a harmonious way of combining the growing interdependence of academic research and chemical manufacturing in the 1920s and 1930s."

Other institutions were not necessarily as convinced of the virtues of such interdependence, nor as imaginative in grappling with problems that inevitably arose when academic and industrial interests found themselves in competition rather than cooperation.

At MIT, irreconcilable differences emerged between two powerful figures with very different ideas of how chemistry should develop within the institution (Servos 1980, Thackray 1982): A. A. Noyes (trained in Leipzig) and W. H. Walker (trained in Göttingen). At the end of the nineteenth century MIT (until 1916 "Boston Tech") was a local engineering school of modest resources. Differences emerged as to how chemistry and chemical engineering (combined in the same department) should be developed in the institution.

Some, under the leadership of Noyes, entertained the ambition of converting MIT from a simple engineering school into a science-based university complete with a graduate school oriented toward basic research. (Servos, loc. cit.)

Pursuing this ambition, Noyes succeeded in 1903 in establishing a Research Laboratory of Physical Chemistry, independent of the chemistry department, to carry out basic work in this rapidly developing field. However, despite the support which Noyes won from colleagues and patrons of MIT, this conception of its future did not go unchallenged.

Shortly after Noyes opened the doors of his research laboratory, a second faction began to lobby vigorously for a very different vision of that future.

According to this view the institute should specialise in the training of applied scientists for industry. "It was only through exposure to problems drawn from industry that the student could learn the uses and limitations of theoretical chemistry".

The institutional vehicle for Walker's ideas on applied chemical research was the Research Laboratory of Applied Chemistry. Organized five years after Noyes' . . . (Laboratory) . . . (this) was intended to serve as a clearing house for problems in applied chemistry . . . the bulk of the expenses were to be met with income drawn from research contracts with industrial firms and trade associations.

The two perspectives could never wholly be reconciled, and with the growing popularity of and support for the applied research laboratory, Noyes found his position increasingly unacceptable. At the end of 1919 Noyes resigned, and moved to Caltech. Gradually, however, opposition built up to the mundane character of much of the applied research carried out, and to the restrictions on publication which sponsors often imposed on work which had any broader interest. Largely speaking, however, it appears from Servos' study that until Karl Compton (who took over as President in 1930) began to set MIT on a new road, its research came more and more to be *dominated* by industry. Servos concludes his study with the following observation:

As academic chemical engineers came to define a set of research priorities that differed from those of the industries they served, they increasingly found themselves allied with their erstwhile adversaries in the basic sciences in demanding greater freedom. . . . Although they did not wish to abrogate all ties with business, they grew to appreciate the need for greater independence. . . . In a sense applied and basic science did converge, but in the process academic applied scientists gradually took on the values of their "purer" but poorer cousins.

At Caltech, where Noyes migrated in 1919, he, Hale and Millikan succeeded in building up the institution "by emphasizing cooperative research, and by drawing on a troika of patrons: the large private foundations (Carnegie, Rockefeller), local Los Angeles and Pasadena wealth, and private industry" (Thackray). Here is a clear contrast, which Thackray brings out. Adams at Illinois found a true co-operation, a style of chemical research which was simultaneously of direct industrial interest and yet of academic standing. This provided the basis for his department. At Caltech, it was a less direct relationship: the translation (as Thackray puts it) of "private industrially-derived wealth . . . into the culturally-applauded triumphs of abstract research". MIT was not able to balance the alternative visions of Noyes and of Walker, and on the departure of the former came under the clear dominance of industrial interests: a "programme" of chemical research dominated by short-term problem-solving.

It might be thought that this sort of analysis, emphasizing the alternative development strategies of charismatic scientists, is of relevance only to the decentralized and competitive academic system of the United States. A study by Bauer and Cohen suggests that similar sorts of differences can be identified in the very different French system. Bauer and Cohen have studied the development of two schools of chemical engineering: that of Nancy and that of Lyon (Bauer and Cohen 1981). From their foundation in the 1880s, the study shows, curriculum development in the two schools followed quite different logics; different choices were being made:

> . . . the two schools developed policies each having its own logic and articulated around unique definitions of chemical knowledge and of the role of the graduate chemical engineer. It is these logics which give meaning to the set of reforms which were initiated, and which cannot be accounted for in terms of a progressive process of innovation: the differences cannot be attributed to "resistance to change" any more than they can to more or less "capacity for adaptation". They appear rather as the product of coherent political choices organised around specific symbolic and social configurations. (Bauer and Cohen, p. 192)

This can be accounted for in terms of the different kinds of university-industry coalitions (*Coalitions industrialo-universitaires*) through which curricula in each case were determined. The details of the study need not be presented here. Essentially, it is argued, the developments in Nancy (based around the notion that "chemical knowledge should restructure our understanding of nature and should inspire major transformations in the industrialisation process which form part of the 'chemical era' . . .") were distinguished from those in Lyon (wherein "the cumulative development of chemical knowledge alone was priviledged, and which presupposed an instrumental conception of the efficacy of that knowledge") in terms of the different "coalitions" which dominated. In Nancy it was a coalition dominated and led by academics which triumphed over a competing industry-led coalition; in Lyon the school sought to respond to the interests articulated by an industry-led grouping.

Academic culture (essentially, the institution's sense of its own particular traditions and responsibilities) and "development strategy" are two factors which shape the relations of a given academic institution with industry. Let me now suggest a third. I have already quoted from a speech by the President of Rensselaer Polytechnic Institute. Elsewhere in that speech Dr Low referred to the *environment* of his university:

mostly because we do not have around us the high technology environment that is so essential for a thriving research university, we decided to develop Rensselaer Technology Park, our own research park. On land owned by RPI, we have installed the roads and utilities for the first phase of this development, and are actively negotiating with potential tenants.

RPI is far from the only university to have tried in this way to develop its industrial environment. Indeed, the notion of a "science park" (a tastefully planned grouping of high-technology companies sharing facilities and — in theory — know-how) is surely the day-dream of many university rectors these days. Many *have* been established, of course, and not only in the United States. (Cambridge, and the Riccarton campus of Heriot-Watt University near Edinburgh are two British examples.) A brochure put out by one of the most famous of them, Research Triangle Park, established in the American state of North Carolina in 1959, extols its virtues in these terms:

Our cluster of three great university campuses is a powerful incentive to new industry. It is their existence within a single close-knit regional community, the challenging intellectual environment they foster, and their receptiveness to innovation and new ideas that have been the most compelling factors in bringing new industry to the area. (Quoted in US Congress 1982)

The third factor to which I refer, as the reader will have inferred, is the *environment* of the university.

In the last two decades, in many countries, universities have been established in development regions on the view that their presence could in some way arrest decline and stimulate industrial growth. Finland has been particularly active in this regard. Unencumbered by inherited notions of "what a university is supposed to be", new universities in Oulu, Kuopio, and Joensuu have sought to define their roles in relation to the needs of the regions in which they are situated. The work of these institutions is in a number of respects quite different from that of "traditional" universities. For example, numbers of university departments have sought *ab initio* to orient their whole research effort around locally relevant problems (for example, geography, local flora and fauna of economic importance) — often in the face of criticism and disapproval from the national scientific community. Also being developed is a *holistic* approach, in which a department orients its research and its education to the whole range of problems faced by (for example) the local fishing industry. In France, too, this kind of regional role has been developed, though less through an institutional redefinition of the role of the university than through regional planning bodies and Regional Chambers of Commerce. Success here has tended to depend upon the presence — in some areas of France — of deeply rooted regional consciousness, and the establishment of a suitable forum in which negotiation between the varies parties can take place. A consequence seems to be that, in some areas, one finds adjacent universities working in very different ways. In the Alsace region of France, for example, the Université Louis-Pasteur at Strasbourg (with its fine tradition in basic research) has links essentially with government research establishments and with national industry. By contrast the nearby Université de Haut-Alsace at Mulhouse was in a sense an emanation of its region, and sought — quite differently — to orient itself rather to the needs of the industry in its region. A major difference from the Finnish examples cited above is that the industrial structure of the Alsace region of France is highly developed, whereas that of Northern and Eastern Finland is largely small-scale and agriculture-based.

It follows from these examples that, under certain circumstances, the development strategy of a university may in some way embody concern with the specific problems of its local or regional environment. Sometimes — as in the cited examples of Oulu, Joensuu, Kuopio and the Université de Haut-Alsace — these problems play a significant part in the research strategy developed by the university. In other cases — such as the Université Louis-Pasteur — this is not so.

Perhaps it would be useful to recapitulate the argument up to this point. It has been that for a variety of reasons (of which the hopes and expectations which have come to be vested in the potentialities of certain fields of technology is an important one) university-industry relations has become a central issue in science policy discussion. As a consequence, a great deal has been written about the matter in the last few years, and a number of studies have been carried out. These agree in suggesting that existing structures of relationships are complex and reflect a variety of functions and objectives. Typologies developed on the basis of classifications of these functions and objectives distinguish (for example) between "collaborative research" and "knowledge transfer" (Brodsky et al.); and "long-term linkages", "promotion of special areas of science and technology" and "liaison systems" (OECD). Moreover, it appears that universities within a given academic system differ from one another in the strength and nature of their interactions with industry. There are a number of reasons why this should be so, including differences in institutional cultures (and interpretations of the roles and responsibilities of the institution); alternative development strategies (which are likely to be in some way an articulation of this "culture"); and (significant for some institutions at least) the particularities of the region or locality in which the university is situated.

Implicit in the discussion thus far is an incipient contradiction, which has significant implications. On the one hand, I argued (from consideration of the development of academic chemistry in the USA) that it seemed to make sense to speak also of the possibility of different development strategies within a given discipline. Adams at Illinois, Noyes and Walker at MIT, and then Noyes at Caltech seemed to develop their departments on the basis of different perceptions of (ideal or desirable) relationships with the developing chemical industry. The French study (Bauer and Cohen) seemed to confirm this point (and went on to interpret it in terms of different kinds of coalitions between academic leadership and that of industry). It seemed to follow that in regard to any particular discipline there was no *determination* (whether by cognitive structure of the field, or industrial structure, or anything else) of the relationships likely to emerge. Yet, in much policy discussion today, much is made of the fact that it is in some senses the exigencies of (notably) biotechnology which have in some way given rise to the particular new forms of co-operation which now attract so much interest. The various studies which I have described tend to reflect such a view. The OECD study, for example, chooses all of its examples of "long-term linkages" and "promotion of special areas of science and technology" from amongst the "new technologies" (biotechnology, materials sciences, membrane technology, microelectronics, telecommunications, robotics, and so on). By contrast, the third element in this typology ("liaison systems") is illustrated by reference to structures and initiatives which are *not* tied to particular areas of technology, but rather to particular sorts of problems (for example professional retraining) or to particular sorts of client-firms.

A number of authors have touched on the question of whether in fact the nature of university-industry relationships within a given area of technology are in some sense a function of the stage of development of that technology. Such a view is implied, for example, in Meyer-Thurow's study of the development of the R&D activities of the

German chemical company Bayer AG between its foundation in 1863 and the First World War (Meyer-Thurow 1982). This author describes a process of, first, a growing dependence of the dyestuffs industry upon academic research from the 1860s. On the one hand economic factors were rendering systematic innovation more important to Bayer; on the other progress in colour chemistry was rendering the field more exploitable by industry. Bayer followed a strategy not only of beginning to hire academically trained chemists, but also of developing connexions with several outstanding research chemists. The 1870s were thus a period of intense interaction: a period of dependence of industrial innovation upon academic research. In the mid 1880s the company began support for an academic laboratory, in which one or two chemists were essentially engaged upon research for Bayer. But in a later phase, from the late 1880s, "the locus of invention" was gradually transferred "from external, mostly academic institutions to industry". The enactment of Germany's first patent law (1877) was one stimulus to this internalisation of the inventive process. But beyond this was a transformation in the nature of the technology and of the manner in which that technology was to be changed. So far as Bayer and the other great German dyestuffs companies were concerned, Meyer-Thurow writes,

> their growing dependence on a continuous supply of inventions made them adapt science to their particular requirements. An irreversible industrialisation of invention was the result. It involved two processes: industrial research became increasingly autonomous as it met industrial and commercial needs, and imitations, inventions and innovations were increasingly reduced to routine.

Maturity of the technology, in other words, implied a research strategy inconsistent with the nature of academic research. Stankiewicz (1984) assembles other empirical work which seems to point in the same direction: to point in other words to a process in which intense university-industry interactions (of the kind now seen as characteristic of biotechnology) represent a given *intermediate* period in the maturing of a technology. But Stankiewicz goes to some pains to try to show that this is a mistaken view: that the maturing of a technology in no *necessary* sense reduces the importance of university research (and connections) for industry. Examples can be adduced in support of this view: for example, I have already alluded to Adams' success in developing a truly symbiotic and collaborative relationship with the chemical industry in his Illinois department in the 1920s and 1930s. Perhaps the matter cannot be resolved by example. How then?

Since the matter is rather central to any attempt whether analytically to understand the development of university-industry relations (and current manifestations of concern) or to contribute to science policy discussion, we cannot leave it thus. Perhaps the question is wrongly posed. I presented it as a disagreement between those who seemed to believe that the state of development of a given technology *determined* the nature of research relations between universities and industry in that field, and those who rejected any such determination. Though it certainly attenuates the policy relevance of my discussion, let me suggest that our difficulties here have significantly to do with the unfortunate word "determine". After all, we know that relations between university research and chemical and pharmaceutical firms in the early history of these industries (see for the dyestuffs industry Van den Belt et al. 1984) differed very significantly between one country and another. And at the same time the "positive" examples — the cases whether of symbiosis (e.g. Illinois, Nancy) or of industrial domination (MIT, Lyon) — also worked in the other direction. The research was very clearly in some sense also a *consequence* of industrial relations. To overcome the (mistaken) contradiction, we have to replace the notion of "determination" with an inquiry into what the possible forms of this (for want of a better term) "mutual shaping" might be. In order to pursue this inquiry, I want to take a rather strange step. Instead of asking "how can we explain (differences in) university-industry relations", I want to go back to the beginning and pose a rather different question.

What Do University-Industry Relations Explain?

As I suggested earlier in this paper, present concern with university-industry relations derives from a greater concern with the promotion of technological innovation. My references to the writings of Rothwell and Zegveld, of Roland Schmitt, and others, suggest that in some sense effective innovation today depends upon the relations that firms enjoy with academic research. In other words, university-industry relations in some sense and to some degree are thought to "explain" success or failure in industrial innovation.

Here is one expression of this view. The authors were on the staff of the Office of Science and Technology Policy in the Carter White House:

For the innovation process to be productive, the generation of new knowledge and the translation of that knowledge into commercial products and services must be linked. Such linkage depends upon close interaction between those who perform basic research and those for whom the results of basic research are the raw materials for product development and commercialization. Because a major share of basic science is done in the universities and colleges, whereas technological development is lodged primarily in industry, strong university-industry relationships can enhance the basic research-innovation linkage. (Prager and Omenn, 1980)

It is not difficult to find studies which bear out this general opinion. Mowery and Rosenberg (1979), in a critical review of a large number of studies of the process of innovation, conclude with some discussion of the policy implications of these studies:

intelligent policies must be directed at institutional aspects of the innovation process, working to encourage the interaction of users and producers, as well as the iterative interactions between more basic and applied research enterprises. We do not yet understand the characteristics of the innovation process sufficiently well, nor do we possess the necessary knowledge base in certain areas of substantial social utility. Useful policies would be those directed at the provision of information, from basic research institutions in the noncommercial sector to private firms and laboratories, as well as from users to producers concerning desired products and characteristics.

More generally, policies directed towards increasing both the frequency and the intimacy of interactions among these separate participant groups may prove to be particularly rewarding.

Let us assume that these highly experienced policy analysts and economists know what they are talking about: productive innovation depends in some way upon effective links between universities and industry. In other words, (effective) university-industry relations in some measure *explain* (productive) innovation processes. Is that all? Perhaps not.

Consider first the process of the invention of the radio described by Aitken (1978). At first sight this might appear a piece of work in the "innovation" tradition, thus adding only one study in support of the conclusion reached above. But the fact that Aitken makes theoretically problematic the *meaning* of the "invention" of radio is indicative of his more complex problematic. Aitken describes a process in which not only does the physics initiated by Clerk Maxwell's electro-magnetic field theory become important for an emerging industrial technology, *but also* this emerging industrial technology becomes increasingly important for the physics. Aitken writes:

there is not the slightest evidence that economic considerations influenced the research programs that Maxwell or Hertz were following . . . decisions as to 'what to tackle next' were taken by criteria other than profitability or the maximisation of program budgets or anything of that sort. As far as we can tell, in this instance the insulation of scientific research from economic pressures or incentives was complete.

And this is precisely why Oliver Lodge is such a pivotal figure in our story. Lodge's status as a scientist was no less than that of Hertz. . . . And in a number of respects . . . his personality characteristics typify the classical stereotype of the true scientist (2).

But Lodge was not *only* a pure scientist, Aitken emphasizes; he wanted science *also* to be useful in finding solutions to practical problems:

His first important research had been on electrostatic precipitation, an industrial problem then as it is today. His work on electromagnetic radiation stemmed partly, it is true, from the same roots as Hertz's: the desire to test Maxwell's model. But it also stemmed from an immediate and urgent practical problem of his own day . . . .

With Lodge, Aitken argues, "the steering impact of market signals becomes visible for the first time". It is perhaps worth recalling that Sir Oliver Lodge became professor of physics at the University of Liverpool at the age of 30 (in 1881), and was elected a Fellow of the Royal Society six years later.

Aitken's analysis derives, of course, from a problematic in which the invention (we might also say the "construction") of a particular artifact stands central. Other authors develop similar analyses — similar at least in one crucial respect — though placing not an artifact but a *research programme* at the focus of their attention. One example is Hanle's study of the development of aeronautics research in Germany, and especially in Göttingen and Aachen (Hanle 1983). Interestingly, the research programme developed at Aachen by Kármán was imported to Caltech at the end of the 1920s, when Robert Millikan persuaded him to leave Germany for California. A further example is Robert Marc Friedman's work on the development of the Bergen school of meteorology (Friedman 1982). Friedman shows how the "new conceptual foundation for atmospheric science", which the group established by Vilhelm Bjerknes succeeded in constructing, was at the same time

part of the process by which they devised an innovative system for weather forecasting

> The special forecasting goals arising from the onset of commercial aviation, the rapid exchanges of weather data and predictions afforded by advances in wireless telegraphy, and the new cyclone model combined to form a single perspective for meteorological discourse. (Friedman 1982)

Friedman's work stresses the attempt of the Bergen group simultaneously to contribute to scientific and to economic goals as the source of this new meteorological discourse:

> the Bergen meteorologists conceived of their 1919 cyclone model as part of a forecasting practice, and the model derived its initial meaning from this practice, sharing its social goals and technological foundation. The surfaces of discontinuity also derived meaning from existing theory . . . and could become objects for theoretical study, but their initial significance for Bjerknes and his assistants derived from their potential use in the detailed and precise short-term forecasts needed for aviation and agriculture.

What these studies seem to show is that the research programmes of Lodge, of Kármán, and of Bjerknes were in some sense shaped by, or embodied, or reflected, their practical concerns. Institutionally, though I have not cited the studies to this effect, these cognitive processes were paralleled by the practical engagement of the scientists in the economic/industrial practices to which the accounts refer. Given the aspirations of this paper, we would of course like to generalise a little beyond the particularities of these three historical accounts: to take them at least as evidence in favour of some theoretical proposition bearing on the importance of university-industry relations. What then might we reasonably induce from these three studies? Each bears on the importance of economic/industrial interests for the development of science. Can we claim them as evidence for any more articulated formulation of this relationship?

Indeed we can. But we must restrain our enthusiasm. The difficulty is, of course, that these studies may be adduced in favour of an indeterminate number of such theoretical formulations. For example, some may wish quite simply to conclude that the sciences of aerodynamics, meteorology and (at a certain point) physics were in some sense constructed around economic interests; others, with a different theoretical bent, that this generalisation has to be related to the stage of development of these sciences: that it was true of this particular stage in the development of each science. Yet others, perhaps with a background in economic history, may note that all of the work analysed was carried out in the period between, say, 1880 and 1920, and that this was the period of the emergence of the communication and aviation industries. Such theoreticians may choose to relate the shaping of these programmes to this specific historical conjuncture. The last, more cautious reading which I want to offer would run as follows. Given the institutional and socio-economic circumstances in which they worked, it *was possible* for Lodge in Liverpool, Kármán in Aachen (and later in California) and Bjerknes in Bergen to develop research programmes having this quality of industrial utility.

The point is, in short, that these exercises in contextualist historiography can be made compatible with many of the theoretical perspectives available in the sociology of science. In other words, the claim that industrial goals or interests can be important in some way for the development of science, without further specification, is compatible with various theoretical positions. Let me elaborate.

For the authors of the "finalization" thesis, for example, the research programmes of Lodge et alia were *rendered historically possible* by the theoretical maturity of physics. According to the "strong programme" the internalisation of economic/industrial interests disclosed by the studies from which I cited is scarcely a matter for surprise. The knowledge constituted around those interests was what people then *agreed* to call "science".

More recently, Knorr (1981, 1982) and Whitley (1983, 1984) have offered more sociologically grounded perspectives which I would like to consider in a little more detail (3). Knorr, probing the laboratory practices through which science is made, argues that these practices have to be understood in terms of a social context, "networks of symbolic relationships which in principle go beyond the boundaries of a scientific community or scientific field" (1981, p. 82):

> the discourse into which the selections of the laboratory are fitted points to *variable transscientific fields.*

It is within these transscientific fields that all the crucial negotiations are conducted, decisions are made, as scientists jockey for position. These processes, the "resource relationships" which they embody and articulate, vital at all stages of the prosecution of a piece of (laboratory) science, are built of local materials.

> . . . a variable transscientific field is not primarily determined by characteristics held in

common by its members. . . . In addition to the scientist in the laboratory, it may include the provost of the university, the research institute's administrative staff, functionaries of the National Science Foundation, government officials, members or representatives of industry. . . . (*Ibid.*)

A central research question for Knorr is then to understand *how* "the contextures we have characterised in terms of resource relationships" relate to the process of knowledge production (p. 88). Knorr stresses the "variability", the locally-constructed nature of these transscientific fields: of which, therefore, relations with industry may be a component.

What happens when we translate the claim that "industrial goals or interests can be important in some way for the development of science" into the notions developed by Knorr? In the first place we are required further to refine our question. We have to ask, "*how* are they important?" More seriously, our focus upon *industrial* relationships comes to appear a crude and one-sided over-simplification. Were we able to investigate the research programmes of Lodge, of Kármán, of Bjerknes in the ethnographic manner required by Knorr, we might well find our view of the importance of their industrial relations confirmed — but we would be certain to identify numerous vital resource relationships. It is the total network, the transscientific field, which is of importance, and we are neither justified in privileging this one element, nor able to generalise as to its significance for other research enterprises.

In Whitley's perspective the organisation of a scientific field stands central: the form of its organisation "constrains and directs" processes of knowledge production and validation. Whitley is seeking a means by which we can *compare* processes of knowledge-production between different sciences:

the modern sciences are particular social organizations for the generation and development of knowledge about environments, and differences between these organisations result in differences between knowledges . . .

Exactly how individual fields develop a particular organization of work and knowledge is the outcome of complex social processes which require sociological analysis. The study of knowledge development and of knowledge producers hence becomes the study of how intellectual enterprises are organized, develop and change in particular circumstances. (Whitley 1983)

Whitley attempts to develop a taxonomy of "social organizations" — the various sciences — in terms of two variables, each of which has two analytically distinct aspects (Whitley 1984). "Mutual dependence" refers to the inter-dependence of scientists in a field, in terms of access to resources, reputation, and evaluations. "As mutual dependence in general increases, competition for reputations and control over the direction of research in that field grows in intensity, as does the strength of its organizational boundaries and identity". The two aspects of mutual dependence are "functional dependence" (which refers to the extent to which the exigencies of reputation demand adherence to common techniques and shared knowledge, to common competence standards and notions of utility) and "strategic dependence" (which refers to the extent to which scientists need to demonstrate the significance of their particular concerns to collective goals) (p. 88). The second major dimension of Whitley's taxonomy is termed "task uncertainty", and refers to the extent to which "work procedures, problem definitions, and theoretical goals are shared between practitioners, and are clearly articulated" (p. 119): in other words to the predictability, the routinization of research tasks. Again two aspects are distinguished: "technical task uncertainty" (the extent to which well-established and understood procedures are available, and give rise to reliable and unambiguous results) (p. 121); and "strategic task uncertainty" (which "encompasses uncertainty about intellectual priorities, the significance of research topics and preferred ways of tackling them, the likely reputational pay-off of different research strategies, and the relevance of task outcomes for collective intellectual goals") (p. 122).

Whitley's discussion of the consequences of allowing each dimension to vary suggests how our concern in this paper can be transposed into his terms. Thus, increasing "mutual dependence", he suggests, implies "a sharper demarcation between scientific fields"; a reduction in the "impact and influence of purely local considerations in setting research strategies"; standarization of research techniques (increasing functional dependence); increasingly agreed and co-ordinated theoretical goals and priorities (increasing strategic dependence). Increasing "technical task uncertainty" implies "greater reliance upon direct and personal control of how research is carried out, considerable local variations in work goals and processes and more informal communication and co-ordination processes" (p. 131). Under high technical task uncertainty, research skills are not standardized, and "local research groups tend to develop distinct ways of conducting and interpreting research". Similarly, increasing "strategic task uncertainty" enhances local variation in research strategies: "choice of problems and topics for research becomes quite diverse across research sites and open to a variety of factors" (p. 137).

What now happens when we translate the claim that "industrial goals or interests can be important in some way for the development of science" into Whitley's terms? In the first place Whitley's problematic requires us to amend our claim. We will now expect the "importance of industrial goals or interests" to be potentially important for *some* fields of science. We are alerted to the importance of inter-field differences. The three historical case studies which I cited (of Lodge, Kármán and Bjerknes) disclosed an internalisation of local economic/industrial goals: the needs of an emerging wireless communication industry in Britain, of a growing aviation industry in Germany and then California, and agriculture and air transport in Norway. Whitley's perspective requires us to postulate that it must have been *possible* for such internalisation of local economic concerns to have taken place. This "possibility" then becomes explicable in terms of the characteristics of the scientific fields discussed. The example of biotechnology at the present time adds a further datum, for there is evidence here for some *convergence* between "public science" and industrial science (or technology) (Narin and Noma 1984). This is likely to be the more possible, according to Whitley's discussion, under conditions of low strategic dependence and high strategic task uncertainty.

Despite the very profound disagreements between the approaches developed by Knorr and by Whitley, the historical evidence as to the importance of industrial objectives for science seems to be compatible with both. I do not think this conclusion would be affected by the introduction of a larger number of such case studies, were they to hand. What is also clear is that the question with which I began this section of the paper — the question of the importance of industrial goals and concerns for scientific change — and which led us to introduce the historical case studies, becomes differently articulated in relation to each of the two approaches. On the one hand we are sensitized to the variation between the sciences: in some sciences, but not in others, scientists will be able to develop their research programmes around (local) economic concerns without compromising their reputational possibilities. In some sciences, but not in others, it will be possible for industrial scientists to contribute significantly to disciplinary goals without compromising their jobs. On the other hand, through Knorr's work we are sensitized to a very different set of concerns. We are led to wonder *in what way* external goals are significant for research, and to speculate as to the social processes through which this significance is established and agreed. Moreover, if we take up the notion of the transscientific field — as the "contexture" within which relevant negotiations take place — we might begin to speculate as to its composition. According to Knorr's formulation, such contextures are constructed by scientists in a "local" and a "strategic" manner. There is nothing to stop us asking ourselves: "under what circumstances would it make sense for scientists to seek to draw representatives of industry into these networks?" We would not then be obliged to answer solely in terms of the dictates of professional reputation as conceived by Whitley. Indeed, though it might be claimed that to negotiate over research goals with external local parties is more meaningful in some fields than in others, we are obliged to look in quite other ways in order to understand *to whom* it makes sense to turn.

I began with the question "what do university-industry relations explain?" and I have tried to inquire into the possibility that in some way they might "explain" scientific change. We can refine the question now, drawing on the two theoretical perspectives. We can say that industrial goals might well be of importance for research in certain fields. We can see that in those fields indicated, we must inquire further into the manner through which their importance is established: into the inherently local and intentional processes through which constitutive networks are established.

Collaborative Research and Common Social Projects

These remarks conclude the digression introduced by the question "what do university-industry relations explain?" Armed with the insights gathered in the course of these theoretical reflexions, we can re-enter the preceding debate. In this final section I shall try to show that not only does theory enrich our contribution to present-day policy discussions, but at the same time this policy discussion (and a reflexion upon it) has major theoretical relevance.

In the earlier part of the paper, I suggested that university-industry relations seemed to be of various kinds, and that classified in terms of their intended functions, these ranged from the provision of ad hoc knowledge to be used in problem-solving, to the collaborative development of fields of science and technology. This latter, in particular, is the focus of concern today, and many have drawn attention to the coincidence of interest between university scientists and industry seemingly manifest in fields such as (notably) biotechnology. Here, as well as to some extent in other new technological fields (materials science, opto-electronics, etc.), it appears to many commentators that there really is a collaborative development of a "scientific field". I argued that despite the financial pressures on universities (which are indeed stimulating a new awareness of the possibilities of industrial finance), there are clear differences — in many academic systems — in the nature and intensity of the relations which individual institutions have developed with industry. I suggested that a number of factors were relevant to this variation, including "academic culture", the "development strategy" of the institution (which in some way is likely to be an articulation of this culture), and (sometimes) the socio-economic environment in which the university is situated. Moreover, on the basis of what we know of the institutionally very varied development of chemistry research in (U.S.) universities, I suggested that it made sense to allow for the existence of varying "strategies" at the level of the individual discipline also.

Under what circumstances does this "collaborative development" take place? This is a question which any science policymaker will admit as of importance today. But now we can put it somewhat differently. In the light of our theoretical reflexions we can ask: "under what circumstances will it make strategic sense for a (university) scientist to develop his or her research programme in such a way that it takes account of, or embodies, industrial goals or interests?" Whitley's analysis suggests part of an answer. The exigencies of professional reputation imply that in some fields of science, but not in others, it makes sense for the scientist to seek to work with externally (locally) set goals.

But this is not the whole of the answer. "Reputational structure" may stimulate or constrain this kind of openness, but it does not in itself say anything about the form or the nature of what results. Do scientists seek to develop an idiosyncratic local theory? Do they develop collaborative research in the way I am trying to discuss? With whom? What determines "to whom it makes sense to turn"? In Knorr's terms, what shapes the "transscientific field"?

Drawing on our earlier discussion of institutional variability in links with the industrial sector (the contrasts which I sketched between Yale and Rensselaer, between Université Louis-Pasteur and Université de Haut-Alsace), I want to argue that institutional characteristics are of great importance here. Within a given field of science, at a given time, the various research programmes developed — goals, concepts, methods of research — will in some sense be consonant with, oriented toward, specific institutional traditions, cultures, development strategies. As Clark has shown clearly enough, universities *are* constructed from the intermingling of two cultures — the international "scientific" culture and the local institutional culture — and though conformity with specifically local norms and values is varyingly enforced, the scientist can rarely neglect them wholly (Clark 1983). There are a number of studies suggestive of the important orienting effect which local institutional values and norms may have upon the development of science. Rupke, for example, discussing the early history of academic geology in the United Kingdom, has shown how the rise to international prominence of the English school — centred in Oxford and Cambridge — in the 1820s and 1830s was shaped by the special characteristics of these ancient universities (Rupke 1983). He writes:

> The central figures of the English school were clerical academics. Buckland, Conybeare, Sedgwick and Whewell were educated in the classical system and took holy orders. At the old English universities they lectured to students many of whom were destined for the Church. *In order to get the new subject of geology accredited they had to align it with the existing educational tradition.* The historical aspect of geology was well suited to this, in contrast to the economic aspect . . . (which was) best left to provincial and metropolitan institutions. (Rupke, 1983, p. 200) (italics added)

Oxford and Cambridge geology took a specific historical form — oriented to natural theology — which clearly differentiated it from Scottish and continental geology. There was little theoretical connexion to the practical problems of mining here. At many German institutions the situation was quite different.

The characteristics of the field of science, in Whitley's terms, and the individual university culture, traditions and values, both have important orienting effects upon the formation of networks of symbolic relationships. But we are still far from a complete account. The deficiency is implicit in the discussion I have already offered of institutional variation in links with industry. I argued earlier that the nature of the socio-economic environment in which a university is situated may also be of importance. In some cases at least, as we have already seen, the economic and social projects of the region or locality may predispose research in one way or another.

To whom does the scientist turn? What makes sense? What can we say about the "fine structure" of the "contextures ... of resource relationships"? Very little, I fear! But we can at least speculate how we might begin to understand this question. Let me offer an heuristic. Let us imagine the scientist, oriented both by reputational organisation and institutional culture to his or her socio-economic environment, casting around. We might imagine him or her considering "which groups or institutions in society appear to regard my kind of science as of significance?" and "which interests appear to me legitimate, or worthy?" We might ourselves add to his speculations. We, the sociologists of science, can further speculate: "with which of these groups is the scientist in contact?" (With whom does he or she have the possibility of negotiating, of establishing a useful resource relationship?) "What is the nature and intensity of this contact?" The notion of "legitimacy" or "worthiness" perhaps needs elucidation. What does it mean to say that a group of scientists accords certain interests a particular *legitimacy*? I would argue that what is involved is, in essence, a sense of *shared social purpose*. This may be based upon shared values (for example, commitment to health, to regional development, to the emancipation of women); or shared interests (in the sense of perceived mutual benefits); or a shared sense of social structure (acceptance of a common authority).

In terms of this heuristic, the formation of scientifically significant links with industry is but a special case of a more general phenomenon. Where the aerodynamicist looks to an emerging industry, the sociologist looks (or looked) to an emerging welfare state, the microbiologist to the cares of his clinical colleagues. Since I would like this point to be retained I would like to propose a term which covers this more general phenomenon of which (I argue) university-industry relations are a special case. Its features are the construction of a research programme in relation to a particular sort of social network: a research programme which consciously embodies the goals, needs, interests or aspirations of actors within this network. I should like to suggest that we use the term *common social project* to refer to research of this kind. According to the perspective I am putting forward, common social projects are constituted when scientific and institutional conditions are favourable; they are not therefore a feature of all science.

In the light of these remarks, we can turn back to science policy and its present concern with the encouragement of collaborative research between universities and industry: the kind of collaborative research which the multi-million dollar biotechnology research agreements exemplify. Under what conditions do these collaborations come into existence? How can they be encouraged? Our analysis suggests that (following Whitley) this phenomenon is feasible in some sciences but much less so in others. It suggests that it is more likely to take place in some universities than in others (though much more would need to be said about "university culture"). It suggests that — in line with experience — a heightened sense of *shared* social purpose (national survival in the face of foreign challenge? regional well-being and development?) as well as more regular and intense contact can play an important role.

The analysis which I have tried to develop also has implications for the sociological understanding of scientific change. It suggests that a sociology of science which seeks "ordering principles" must not limit its search to the internal regulatives of the sciences. In much of science there exists a plurality of theoretically, conceptually and methodologically distinct research programmes which require a sociological analysis going beyond attribution of the "fact" of plurality to the characteristics of the field. In what ways do these various research programmes differ from each other? Such plurality will sometimes be understandable in terms of conflicting social projects: a reflexion of conflict in society at large, wherein some scientists ally themselves with oppositional or underprivileged groups. This situation is not uncommon in the social sciences: in fields such as criminology, urban sociology or medical sociology, and the present volume provides examples from other areas of science too.

There are many questions which need to be asked systematically if the perspective sketched out here is to be developed. We need to inquire into the various significances of conflict for the constitution of common social projects. Scientists may individually be torn, for example, between competing claims on their loyalty, or between moral commitment and self-interest. We need, above all, to inquire into the *various* ways in which shared interests, values, or social structure may be significant for science. My own view is that this difficult question will most usefully be addressed in the first instance to areas of research tightly bound to identifiable social practices themselves open to analysis. (The disciplines of the clinical sciences, and their associated medical practices, are of considerable interest in this regard (Blume et al. 1985).) Scientific advance can, and often does, become embodied in changed practices, thereby leading to a change in the social context constitutive of further scientific advance. It follows, I believe, that sociologists of science would be well advised to develop co-operative studies with sociologists whose concern it is to provide detailed analyses of the practices which will necessarily figure largely in the understanding of scientific change.

Notes and References

1. It must be admitted that other evidence points in the opposite direction. Van Steijn has looked at the growth in the numbers of extraordinary professors in Dutch universities, taking this (as is usually assumed) as an indicator of the strength of connections with external bodies (notably industry). Counter to expectation, the trend in the (multi-faculty) University of Amsterdam precisely paralleled that in the Technological University of Delft (Van Steijn 1985).
2. I understand that recent historical research has called this appreciation of Clerk Maxwell's "insulation" from economic concerns into doubt. I do not think my argument is greatly affected, but I am grateful to Robert Friedman for the information.
3. In commenting on an earlier draft of this paper both Anja Hiddinga and Helga Nowotny wondered whether my discussion of Knorr and Whitley was of value here. On reflection I decided to retain it, for a number of reasons which are perhaps worthy of note. First, it enables me to "locate" the perspective which I am trying to develop in relation to existing and known positions. Second, it seems to me didactically useful to see how "the" problem of university-industry relations translates in these two ways. And third, and most importantly, it enables me to provide some preliminary grounding of two of the issues which I see as of particular importance: inter-field comparisons, and the notion of the "strategic behaviour" of scientists. I recognise, as Nowotny points out, that I might also have tried to do this by introducing more general "structure" and "action" theories from sociology.

4. H. G. J. Aitken, "Science technology and economics: the invention of radio as a case study" in Krohn, Layton and Weingart (eds.), *The Dynamics of Science and Technology*. Dordrecht: Reidel, Sociology of the Sciences Yearbook, volume 2. 1978.
5. B. Barnes, *Interests and the Growth of Knowledge*. London: RKP, 1977.
6. M. Bauer and E. Cohen, "Politiques d'enseigement et coalitions industrialo-universitaires: l'exemple de deux grandes écoles de chimie, 1882—1976", *Revue Française de Sociologie* **12** (2), 1981.
7. H. van den Belt, B. Gremmen, E. Homburg and W. Hornix, *De Ontwikkeling van de kleurstoffenindustrie*. Nijmegen: W&S Programme, 1984.
8. J. Ben-David, *Fundamental Research and the Universities*. Paris: OECD, 1968.
9. S. Blume, O. Amsterdamska, P. Groenewegen and A. Hiddinga, *Dynamics of Biomedical Sciences*. Amsterdam: Department of Science Dynamics, 1985.
10. G. Böhme, W. van den Daele and W. Krohn, "Finalisierung in der Wissenschaft", *Zeit. für Soziologie* **2**, 1973.

11. N. H. Brodsky, H. G. Kaufman and J. D. Tooker, "University-industry cooperation: a preliminary analysis of existing mechanisms and their relationship to the innovation process". New York: NYU Center for Science and Technology Policy, 1980.
12. B. R. Clark, *The Distinctive College*. Chicago: Aldine, 1970.
13. B. R. Clark, "The organizational saga in higher education" *ASQ* **17**, 1972.
14. B. R. Clark, *The Higher Education System: Academic Organisation in Cross National Perspective*. Berkeley: University of California Press, 1983.
15. Barbara J. Culliton, "Monsanto gives Washington U $23.5 million", *Science* **216**, 1982.
16. J. D. Eveland (ed.), *The Development of University-Industry Cooperative Research Centers: Historical Profiles*. Washington D.C.: NSF, 1982.
17. J. Farkas, Chapter 4 in *Science Research Technology: the Hungarian Case*. Budapest: Hungarian Academy of Sciences, 1983.
18. R. M. Friedman, "Constituting the polar front, 1919—1920", *Isis* **73**, 1982.
19. L. Peters and H. Fusfeld, "Current U.S. University-Industry research connections", 1982.
20. A. Bartlett Giamatti, "The university, industry and cooperative research", *Science* **218**, 1982.
21. P. A. Hanle, *Bringing Aerodynamics to America*. Cambridge: MIT Press, 1982.
22. E. A. von Hippel, "The dominant role of users in the scientific instrument innovation process", *Research Policy* **5**, 1976, 212—39.
23. K. D. Knorr-Cetina, *The Manufacture of Knowledge*. Oxford: Pergamon, 1981.
24. K. D. Knorr-Cetina, "Scientific communities or transepistemic arenas of research? A critique of quasi-economic models of science", *Soc. Stud. Science* **12**, 1982, 101—30.
25. J. Liebenau, "Innovation in pharmaceuticals: industrial R&D in the early twentieth century" *Research Policy* **14**, 1985, 4.
26. G. M. Low, "The organisation of industrial relationships" in T. W. Langfitt (ed.) *Partners in the Research Enterprise*. Philadelphia: University of Pennsylvania Press, 1983.
27. V. Milic, "Sociology of knowledge and sociology of science", *Social Science Information* **23**, 1984, 213.
28 D. Mowery and N. Rosenberg, "The influence of market demand on innovation: a critical review of some recent empirical studies", *Research Policy* **8**, 1979, 102—153.
29. F. Narin and E. Noma, "Is technology becoming science?", *Scientometrics* **7**, 1985, 369—381.
30. National Science Board, *University-Industry Research Relationships*. 14th Report of the NSB, Washington D.C.: NSF, 1982.
31. R. Nelson and S. Winter, "In search of useful theory of innovation", *Research Policy* **6,** 1977, 37—76.
32. OECD, *Industry and University: New Forms of Cooperation and Communication*. Paris: OECD, 1984.
33. D. J. Prager and G. S. Omenn, "Research, innovation, and university-industry linkages", *Science* **207**, 1980, 379—384.
34. N. Rosenberg, *Perspectives on Technology*. Cambridge University Press, 1976.

35. R. Rothwell and W. Zegveld, *Industrial Innovation and Public Policy*. London: Frances Pinter, 1982.
36. R. Rothwell and W. Zegveld, *Reindustrialization and Technology*. London: Longman, 1985.
37. N. Rupke, *The Great Chain of History: William Buckland and the English School of Geology (1814—1849)*. Oxford: Clarendon Press, 1983.
38. R. W. Schmitt, "Continuity and change in the U.S. research system". Washington D.C.: School of Public Policy, George Washington University, Occ. Papers No 1, 1985.
39. J. W. Servos, "The industrial relations of science: chemical engineering at MIT 1900—1939", *Isis* **71**, 1980, 531—549.
40. Science Policy Research Unit, *Success and Failure in Industrial Innovation*, Report on Project SAPPHO. Sussex: SPRU, 1972.
41. R. Stankiewicz, "University-industry relations", Report to the Six Countries programme. Delft: TNO, 1984.
42. F. van Steijn, "Part-time professors in the Netherlands", *European Journal of Education* **20**, 1985, 57—65.
43. A. Thackray, "University-industry connections and chemical research: an historical profile", University of Pennsylvania/National Science Foundation, 1982.
44. G. Meyer-Thurow, "The industrialization of invention: a case study from the German chemical industry", *Isis* **73**, 1982, 363—381.
45. Joint Economic Committee, U.S. Congress, *Location of High Technology Firms and Regional Economic Development*, Staff Study. Washington D.C.: U.S. Govt Printing Office, 1982.
46. R. D. Whitley, "From the sociology of scientific communities to the study of scientists' negotiations and beyond", *Soc. Sci. Inform.* **22**, 1983, 681—720.
47. R. D. Whitley, *The Intellectual and Social Organization of the Sciences*. Oxford: Clarendon Press, 1984.
48. J. Wiesner, in *Co-operative Research*, edited by Nam P. Suh and B. M. Kramer. Washington D.C.: National Science Foundation, 1982.

THE PRACTICAL MANAGEMENT OF SCIENTISTS' ACTIONS: THE INFLUENCE OF PATTERNS OF KNOWLEDGE DEVELOPMENT IN BIOLOGY ON COOPERATIONS BETWEEN UNIVERSITY BIOLOGISTS AND NON-SCIENTISTS

JOSKE BUNDERS

Free University, Amsterdam

Introduction

Cooperation between scientists and other groups in the conduct and direction of scientific research has recently become the focus of much political and intellectual concern. This paper reports on the results of a study of how biologists managed such cooperations with a variety of external groups (1). The aim of this study was to acquire insight into the causes and consequences of cooperative relationships and thereby to throw light on the following two related questions.

— How can we understand (or predict) what kinds of externally posed questions scientists in a specific (sub)field can deal with?

— How is the development of knowledge affected by extra-scientific influences exerted in cooperations?

By external groups I mean groups outside the university, such as environmental groups and governmental and commercial institutions. Members of these groups will be referred to here as "non-scientists". The cases studied here were all based on explicit cooperations where the non-scientist formulated the social problem and the scientist the scientific problem, and where largely speaking the research was funded by the non-scientist or the government.

Analysis of these cooperations suggests that the questions of non-scientists which biologists can address are not restricted to the types of questions which colleagues in the field consider interesting. We were confronted with a peculiar situation in which biologists seemed to show

S. Blume, J. Bunders, L. Leydesdorff and R. Whitley (eds.), The Social Direction of the Public Sciences. Sociology of the Sciences Yearbook, Vol. XI, 1987, 39—72.

a high degree of mutual dependence in the choice of their problems, while many cooperations were found in which "non-interesting" problems were addressed.

I will argue that we can understand this contradiction by looking at the roles cooperations can play in the biologists' strategies for carrying out their long-term institutional research programmes. We observed that the resources of non-scientists play important roles in the different strategies designed to reduce several well circumscribed types of risks associated with scientific work.

The frame of analysis here is based on the work of Knorr and Whitley. Although these authors have conducted very different analyses, the work of both seems important for answering our two central questions.

Knorr did an ethnographic study in an institute of biomedical research, from which she concluded that non-specialists strongly influence the production of knowledge: ". . . an external contact, a negotiation about money, or a career, each strategy has immediate technical repercussions" (2). She calls such contacts with non-scientists "resource relationships" because they depend on the mutual exchange of resources. Scientists have many of these resource relationships and are thus involved in a number of what she calls "transepistemic arenas" (3). In her view, these arenas are the loci in which scientific work is organized; the coordination and control of the scientific community need *not* be considered, because "if specialty communities were the locus of the social and cognitive organization of scientific work, we could consider the observable relationships between scientists and non-specialists as irrelevant to the production of knowledge" (4).

This suggests that research carried out on non-scientists' questions is not constrained by the scientific community as a whole. According to Knorr, the development of knowledge in research groups is influenced not by the scientific community but by the transepistemic arenas.

Whitley, on the other hand, has compared patterns of knowledge development in scientific fields, using a unit of analysis quite different from the research groups studied by Knorr. Whitley too looked at the effects which groups other than colleagues can have on the development of knowledge. He suggests that the influence of these groups is strongly related to the particular patterns of organization and control of scientific work in the field involved. These patterns vary from field to field and can be analyzed by looking at two dimensions: the "mutual dependence" among scientists and the "task uncertainty" of their work. He arrived at these two dimensions in the following way:

> Essentially, modern sciences are systems of jointly controlled novelty production in which researchers have to make new contributions to knowledge in order to acquire reputations from particular groups of colleagues. These contributions are assessed in terms of their significance for collective goals and usefulness for others. They therefore have to be different and novel at the same time as being oriented to the work of colleagues and capable of being used by them in their own research. These conflicting demands create particular tensions between scientists and variations in their mutual balance affect the organization of knowledge that was produced in different fields. Many of the major differences between the sciences can be derived from these variations which can be characterized in terms of two distinct dimensions: the degree of mutual dependence between researchers in making competent and significant contributions and the degree of task uncertainty in producing and evaluating knowledge claims (5).

These two dimensions together characterize what Whitley calls the "reputational system", where scientists' reputations are judged in terms of the evaluation criteria applied by their colleagues to their research results. His analysis suggests that the ability of scientists to deal with questions posed by non-scientists will be strongly related to their degree of mutual dependence and task uncertainty. If there is a low degree of mutual dependence, extra-scientific interests may more easily affect the research (6); and when there is a high degree of task uncertainty, scientists will be more free to adjust their tasks to the demands of non-scientists (7).

Knorr and Whitley's analyses can give rise to conflicting predictions concerning possible extra-scientific influences on the development of knowledge. For fields characterised by a high degree of mutual dependence, Whitley suggests that scientists can deal only with those questions posed by outside groups which involve goals and standards similar to those found within the scientific field. The influence which audiences outside these scientific fields can exert on knowledge production is thus likely to be ineffective or very limited. Knorr, on the other hand, suggests that it is non-specialists and not scientific colleagues who are crucial to the development of knowledge. An example can elucidate these different positions.

Whitley says that physics is a field characterized by a high degree of mutual dependence, because access to resources and reputations is controlled by the physics community. In his view the resource-relationships described by Knorr are found typically in biomedical research and fields with comparable patterns of coordination and control of scientific work, such as artificial intelligence (8). Knorr, however, does not restrict her analysis to bio-medical research. She could argue that even in the field of physics, in-depth studies of the development of scientific knowledge show the strong influence of non-specialists. Pinch, for example, has described a problem dominating solar neutrino science, a subfield of physics which emerged from a resource-relationship between a chemist and a physicist (9). This example suggests that even in fields characterized by high mutual dependence the influence of non-specialists can be considerable.

Knorr's analysis, however, like Whitley's does not appear sufficient to explain knowledge production. Knorr does not want to consider the constraints imposed by the scientific community, on the grounds that "if we cannot assume that the 'cognitive' or 'technical' selections of scientific work are exclusively determined by scientists' specialty membership groups, it makes no sense to search for a 'special community' as the relevant setting for knowledge production" (9). In my view, the demonstration that non-specialists influence knowledge production is not sufficient evidence for concluding that scientific communities do not organise and also influence this production.

I suggest that an analysis of the organization and control mechanisms within the field *and* the scientific group is necessary to throw light on our two questions. To understand the types of questions scientists can deal with, we need to examine how the macroscopic mechanisms involving organization and control within the field interact with the microscopic mechanisms involving organization and control in groups, transepistemic arenas, and cooperations. This interaction seems to me to be crucial to an understanding of the process leading to the results of knowledge development published in biological journals.

We shall explore this interaction by examining the research strategies of university biologists. These strategies can be analysed by studying how decisions in research groups are influenced both by the demands of the reputational system and by the questions posed and resources offered by non-scientists. We studied a number of decisions taken by scientific groups who were cooperating or thinking about cooperating with non-scientists. These decisions primarily concerned issues such as the choice of (sub)problems, the approach to these problems, the concepts and techniques to be used, the object of study, and the consequences of these choices with regard to the standards of biological results.

We chose to study the decisions of groups instead of individual scientists because in most biological groups the decisions of individual scientists are heavily constrained by the decisions their groups take as a whole. In our study we found that the group's subproblems, ideas, techniques, and contacts were continually changing. In trying to find the pattern underlying this chaos, we asked the biologists involved why the observed changes did not give rise to problems, and to explain what types of changes would. With this information we tried to understand the roles played by the non-scientists' questions and resources in producing a result publishable in a biological journal. With the answers to these questions we were able to formulate more precise questions for other biologists. We continued this process until we could predict general patterns; then we tried to analyse the patterns of knowledge development within the field, using primary and secondary literature and interviews with biologists who were involved in controlling access to resources or in evaluating biological research groups in the Netherlands.

This particle is organized as follows.

The next section offers an analysis of the pattern of knowledge development in biology, and a discussion of the criteria in terms of which biologists evaluate each other's results. We shall distinguish among demands regarding the choice of problems, goals, approaches, techniques, or objects of study, and we shall also analyse the uncertainty biologists encounter in their work when they try to deal with these criteria. These results are then used to answer the question: what constraints and possibilities do biologists encounter when they cooperate with non-scientists?

In section three the unit of analysis changes from the field to the research group cooperating with non-scientists. Types of cooperations within the field of biology will be distinguished and characterized by examining the ways in which cooperation can influence a group's research programme. The contextual factors affecting various types of cooperation are also considered, including the resources offered by the non-scientist and the demands of the reputational system. As we shall see, various types of cooperation can be related to types of biological research strategies. With this analysis we will answer the question: can resources solve all, none, or only certain types of problems arising from conflicting demands from the reputational system and the non-scientist? Finally, I shall consider the direct and indirect consequences of the various types of cooperation on the development of knowledge. The information used in this section is based on 3 detailed and 17 more general case studies of cooperations conducted between 1982 and 1985, including interviews with the scientists involved and reading their reports and articles (11).

Although I am focussing here on scientists' research strategies, the practical aim of this study is to contribute to the improvement of the intermediary work of biology "shopkeepers" who try to motivate university biologists to carry out research on questions posed by external groups unable to finance the research themselves (12).

2. Patterns of Knowledge Development in the Field of Biology

What constraints and possibilities does the reputational system impose on biologists who cooperate with non-scientists? To answer this question I have analysed the patterns of knowledge development in biology in terms of Whitley's two dimensions: the mutual dependence of scientists and their task uncertainty.

Variations in the degree of mutual dependence and task uncertainty can serve as indicators of the constraints and possibilities facing scientists attempting to deal with a specific non-scientist's question. (For example, a high degree of mutual dependence among researchers with regard to a given aspect of a field means that for this aspect there will be a strong demand for what is considered "good research", and therefore little possibility for scientists in this area to orient themselves to meet the demands of a non-scientist.)

Whitley has described these dimensions on both the "strategic" and the "technical" or "functional" level. Strategic mutual dependence and strategic task uncertainty concern choices of problems, goals and approaches. He explains these two subdimensions as follows:

> The strategic task uncertainty encompasses uncertainty about intellectual priorities, the significance of research topics and preferred ways of tackling them, the likely reputational pay-off of different research strategies and the relevance of task outcomes for collective intellectual goals.
>
> The strategic mutual dependence refers to the extent to which scientists need to demonstrate the significance of their particular concerns to collective goals (13).

Whitley's other subdimensions of technical task uncertainty and functional mutual dependence concern the use of techniques, concepts, and objects of study. He defines these two subdimensions as follows: technical task uncertainty is "the extent to which work techniques are well understood and produce reliable results in various scientific fields"; while functional mutual dependence is "the dimension which indicates to which degree scientific contributions have to fit in with existing knowledge and rely on similar techniques, methods and materials of specialist colleagues" (14).

The following section discusses these dimensions as they apply to biology at the strategic level, followed by an examination of the subdimensions of technical task uncertainty and functional mutual dependence. The analysis is focussed at the level of biologists in general. Because although the discussion applies mostly to the subfield of biology of organisms, we found that the levels of mutual dependence and task uncertainly were virtually the same as in the other biological subfields. This does not imply that all the characteristics of the subfields are similar, as I will argue elsewhere.

2.1. *The choices of problems, goals and approaches in biology*

First I will discuss biological problems, goals and approaches, then proceed to consider variations in the degree of mutual dependence and task uncertainty concerning these features and their implications for biologists' receptiveness to non-scientists' questions.

Biological problems concern the functioning of organisms, their evolution, or their relation to their environment (15). Most biological problems were identified hundreds of years ago, but naturally the ways in which they can be specified and tackled have been subject to radical changes. As Ilse Jahn wrote in *Geschichte der Biologie* (1985),

> Weiterhin erbigt sich, dass bestimmte Fragen und Probleme bereits sehr frühzeitig in der Menschheitsentwicklung auftauchten und dann entsprechend dem jeweils erreichten Niveau der Wissenschaft oder der Weltanschauung immer wieder neu bearbeitet und beantwortet wurden. Das gilt z.B. auch für eine so moderne Problematik wie die Vererbung (. . .) und die Problematik der Beziehungen zwischen der Organismen und ihrer Umwelt (. . .). Das Photosynthese-Problem, das schon im 18. Jahrhundert eingehend untersucht wurde, füllt noch in der Gegenwart ein volles Forschungsprogramm (16).

One striking aspect of biology in particular is the large number and variety of the research objects involved, the large number of different types of cells, organisms, populations, and ecosystems to be studied. Yet it is not just the large number and great variety of the objects studied that distinguishes biology from other natural sciences; most of all, it is the complexity of the phenomena studied. Pantin, for instance, claims that it is the richness and complexity of their phenomena which distinguishes sciences such as biology and geology from the physical sciences (17). The ways in which researchers approach this complexity seem to have undergone great changes in recent decades.

This change was described clearly by the neurobiologist Lever in his opening lecture to a congress on comparative physiology in 1981. He explained that in contemporary biology a research group cannot focus on one narrowly defined problem, such as the functioning of an animal's heart. Instead, many other problems have to be incorporated into a research programme; and with this he arrives at the most important collective goal of biologists: to get "an insight into all aspects of life":

> We are therefore increasingly faced with the need to study the whole organism: to understand one specific function of an animal we must have an insight into all aspects of its life (18).

One consequence of this perspective is that biological problems cannot be dealt with by one group of specialists but have to tackled by several specialists working together. As Lever says,

> Everybody will agree that many contemporary problems being studied (. . . in biology . . .), are so complicated that they can only be solved by an interdisciplinary approach (19).

This new approach, incorporating techniques and methods from strongly diverging subfields into one research group in order to solve a specific biological problem, is characteristic not only of neurobiology but of practically all biological subfields: "During the last decades the map of biological science has been redrawn and the new terrains are often indicated as 'biologies': molecular biology, cell biology, developmental biology, population biology, neurobiology, sociobiology, etc." (20). The same sentiments about biological research were echoed when we spoke to the chairman of a commission that controls access to resources for fundamental research in neurobiology.

I think that in the past neurophysiologists have limited themselves far too much to neurophysiological techniques. I think it's terribly important to master these techniques, it's absolutely essential. But if you're dealing with biology, you have to approach a problem and then consider which techniques you need to solve that problem. So if you have exhausted, let's say, the electro-physiological techniques, you're going to have to look for a different technique. I think it's typical of recent developments that various different techniques have to be integrated. That's why I prefer the term neurobiology over neurophysiology . . . .

Some think that this new research approach means that biological problems will be reduced exclusively to problems on the cellular and molecular levels. Many biologists wholeheartedly disagree with this view. First of all, they argue that subquestions do not deal exclusively with lower levels of integration; subproblems also concern higher levels, as is the case, for instance, when ecological questions are included in physiological and cellular research. In addition, they emphasize the *demand* that results from research on subquestions should always be integrated into the research on the problem at the level at which it was initially defined.

With this in mind we can describe the standards of biological research as follows. Biologists have to work on problems that their colleagues have long considered important. These problems are intricate and can be dissected into subproblems cooperating on high as well as low levels of integration. They can be approached using various methods and techniques. The results of research into these subproblems have finally to be integrated at the original level of the overall problem (21). These standards for biological research can explain the ways in which biology students are trained, research groups are built up, and articles for biological journals are composed. All biologists in the Netherlands receive a very broad basic training. In a report presented in 1986, the deans of the biological subfaculties deemed this broad basic training very important and recommended that it not be changed by granting students the possibility of early specialisation (22). Many of the research groups we studied consisted of biologists trained in strongly divergent subfields working together on one central problem. The standards of biological research are evident from articles in biological journals and in the ways biologists use to distinguish their results from those of scientists in other fields such as biophysics, chemistry, or the medical sciences. These standards and their consequences for research are clearly illustrated by a professor in phytopathology, the chairman of the committee that evaluated biological research groups in the Netherlands:

If you define a biological problem at a high level of integration, you will often see that the definition is very intricate, and that you have to dissect it into subproblems which often slide down to the cellular or molecular levels. I think it absolutely essential that the answers to these subproblems should be incorporated at the higher level of integration and be fitted into the whole. With biological problems, you can't just highlight one subproblem and say "Look, I've found the answer!!" No way! You've got to fit it into a larger structure, and that's why biologists should have a broad basic training, it's a condition. A biophysicist may ask a question like, what would be the kinetic effect of an enzymatic process of a particular kind where A is turned into B? A biologist would answer, very interesting, but where does A come from, and where does B go? How does it fit into the chain, how does it affect the functioning of that organism? It's typical for a biologist that he should integrate, a biological scientist has to integrate by definition. But we haven't been doing it like this for very long. Biologists thought for a long time that they could handle it all by themselves, up until 20 years ago.

This new approach leads us to conclude that a biological subfield can no longer be characterized by its techniques. How then can we characterise biological subfields? Part of the answer can be found in the way biologists themselves categorize their research. In surveys of biological research, they categorize their subfields according to the "level of integration" at which the biological problem is formulated. These levels of integration are: the molecule, cell, organism, population, and ecosystem. Each level of integration encompasses several central biological problems and thus several comparable biological subfields (23). A subfield in biology can thus be defined by its central question and the level at which this question is formulated.

The approach taken by biologists dealing with the complexities of their research has changed. In the past, research groups in subfields like ecology, physiology, or genetics seem to have employed methods and techniques developed for research within their own subfield. If a scientific problem contained a subproblem outside the scientific subfield of the research team concerned, they would not feel responsible for the analysis of that subproblem. Nowadays, researchers try to tackle the most divergent subproblems even though they may not belong to their own subfield. For this purpose, a research team will try to acquire methods and techniques from other (sub)fields, or it will cooperate with another scientific team better equipped to deal with this subproblem.

What determines the choices biologists can make in their approach to a biological problem? This choice involves the ability of biologists to dissect their central problem into subproblems. Here the previous example can be elucidating. In order to deal with the question about the regulation of the animal's heart biologists want to have "an insight into all aspects of its life". It is obvious that such a goal will lead to the possibility of many alternative dissections of the central problem. In the various approches, different subquestions will be answered. The biologist's choice of an approach will depend on locally available knowledge, skills, techniques, objects of study, and, if possible, cooperations with others outside the (sub)field. The success of an approach is not proven until the answer to a subquestion appears relevant to the overall research problem. Of course there is often conflict among biologists over the question of whether a specific result of a subproblem is really relevant to the overall problem. Nevertheless, the criteria for highly rewarded biological results are clear. There is, however, no single foolproof approach to achieving such results. This means that it is very difficult to determine in advance which approach is most likely to be successful. For this reason, one might qualify biological research to a great extent as high-risk research.

We can now draw the following conclusions about the degree of strategic task uncertainty and mutual dependence regarding the problems, goals, approaches, and standards of research.

The degree of task uncertainty regarding problems, goals, and standards of research seems low because they are clearly characterized. The degree of task uncertainty about how to approach problems is high. No preferred way of tackling the problems is articulated, and the reputational pay-off of different research strategies is unsure. The mutual dependence among biologists concerning the choice of problems, goals, and standards of research seems very high: with regard to the control of resources and the evaluation of results, biologists have to respect the prescribed problems, goals, and standards. Problems and standards for research are more evident than goals, which are always more implicit. This clear identity helps the biologists to distinguish biological research results from results belonging to other fields. This does not mean that biologists never publish in journals outside biology, but they cannot restrict themselves, for example, to publishing only in medical journals. For university biologists there is a hierachy in audiences, and publishing in a medical journal conveys fewer rewards than publishing in a more broadly based biological journal. A biologist who has published in an applied or medical journal will not, as a rule, refer to this publication in a paper submitted to a biological journal. As already noted, mutual dependence concerning the approach to problems is low.

From this analysis we can conclude that non-scientists will encounter problems when trying to persuade a research group to tackle a deviant problem or address a deviant goal or research standard. Non-scientists, however, will encounter fewer problems when they try to motivate a research group to change the way it approaches its problems.

2.2. *The choice of techniques, concepts, and objects of study*

Mutual dependence and task uncertainty also influence the choice of techniques, concepts, and objects of study. The new, more integrative goal of biologists has had considerable repercussions on the way research is carried out. The number of new techniques introduced into biological laboratories over the last few years has increased dramatically. The constant need to update their apparatus confronts the scientists with enormous financial problems. This problem is considered unavoidable, as is illustrated by one participant in the world conference of neurosecretion held in Tokyo in 1984.

> Now we've got all these marvellous new techniques, and I thought I could get along for a while without having to install new laboratories, and the next thing you know is that you have to have a lab for recombining DNA, and, or course, before you know where you are, that will prove insufficient too. But we have to have one just the same, it's unavoidable if you want to stay up-to-date. We'll just have to learn to live with it . . . .

In most cases students and staff learn to use this new equipment within a few weeks, and the data gathered from it can be used directly for solving subproblems. These new techniques are usually well understood and produce reliable results, as is illustrated in many case studies (24). All these methods and techniques are standardized, and were developed in most cases not by biologists but by scientists in other fields. The degree of technical task uncertainty in biology is thus relatively low and exists only by virture of the variability of biological material, which is often not easy to standardize.

In analysing the use of techniques in biology we are confronted with a remarkable problem. This problem is caused by high mutual dependence among biologists regarding problems and goals. Biologists need many new techniques to enable them to solve subproblems, but the lack of a good technique is usually not considered an interesting biological problem. Therefore, a major problem arises when a biologist is unable to import methods or techniques in order to solve a subproblem, since biologists are rarely in a position to develop this method or technique for themselves. A biologist working meticulously to develop a new technique for a long time hardly ever receives any appreciation from his fellow biologists. After all, in most cases these methods and techniques can be borrowed from other fields such as biophysics, biochemistry, or the medical sciences. In biology it is not the development of the technique but the results that count, providing of course that they are relevant to the overall problem. Strange though it may seem to scientists outside the field of biology, one cannot find a biologist in the Netherlands who has been working on the development of a new method for three or four years.

From what I know of biological research, I wouldn't be able to give you an example of someone who has been working for years on the development of a certain technique. Of course we do adjust methods and techniques to suit our own problems, but spending so much time on a new technique? No, the biologist will say this is just an aid, let others develop it. (Quoted from the chairman of the committee that evaluated the biological research groups in the Netherlands)

Mutual dependence and task uncertainty concerning the choice of techniques can be summarized as follows. The techniques biologists use are standardized and well understood. In other words, technical task uncertainty in biological science is relatively low and exists only by virtue of the variability of the biological material. The mutual dependence among biologists concerning the use of particular techniques is low because biologists can use techniques from other (sub)fields. The possibility of developing a new technique is very limited.

From this we can conclude that questions addressed to biologists by non-scientists involving use of a specific known technique in dealing with a biological subproblem will not be constrained by the demands of the reputational system. This analysis also suggests that non-scientsts' questions requiring the development of a new technique are constrained by the demands of the reputational system.

The relatively uncomplicated way in which techniques from other fields are included in fundamental research differs from the way in which new or deviant models and concepts can be used. Bunders and

Whitley have described how experimental results based on deviant models and concepts were publishable only if they could be analysed according to accepted ones (25). Articles based on deviant concepts are thus unlikely to be accepted by prestigious journals. One example of such a deviant concept is "environmental dynamics" (milieudynamiek), which is a concept used in environmental policy but heavily criticized by ecologists (26). This discussion indicates that mutual dependence among biologists concerning the use of specific concepts is high. An essential problem in cooperation will arise when an external group suggests that biologists include socially relevant concepts and models in their research. Whether biological concepts are always clearly articulated and understood is difficult to assess. Biology methodologists often criticize the ambiguous nature of the concepts used and the diverging ways in which concepts are used in the various biological subfields (27). In theory, such an ambiguity could be used by biologists to give the non-scientists influence over the way in which the concept is understood and applied.

Finally we have to discuss the constraints and possibilities biologists encounter in choosing their object of study. Here I deal only with the implications of the degree of mutual dependence with regard to the choice of an object of study. The choice of an object of study is a very important and widely discussed aspect of biological research. Sometimes research on certain objects of study is considered less interesting than the same type of research on other objects of study. Research on lower animals (invertebrates, for example snails) is considered less prestigious than research on higher animals (vertebrates, for example rats) (28). The scientists actually working on these lower animals claim, however, that their approach is better suited for dealing with certain biological problems. They express the general idea, widely held among biologists, that objects of study have to be chosen on the basis of their suitability for experimental research. "They select specific animal forms for study only because these are more suitable than others for the experimental attack on specific general problems" (29). Other criteria also play a role in the choice of an object of study: "We selected these animals because they are small, cheap and easy to keep and to breed" (30). Apart from these considerations, biologists also take into account the criteria of social relevance, although here we are confronted with a familiar controversy: the choice between a "model system" or a "dirty system" (31).

Some species, whether they are particularly suitable for the solution of fundamental questions or not, are correctly chosen because of their direct economic significance or becuase they are pests or parasites. Apart from these, history teaches, as we have seen, that the best species to select are the model animals, which in one or more of their aspects are, so to speak, made for science (32).

Mutual dependence among biologists concerning the choice of a specific object of study is thus neither high nor low but medium, and a non-scientist's request that the biologist choose a specific object of study can be considered seriously. However, there are two reasons why biologists do not easily change their object of study. First, their model organism is mainly chosen for the advantages it has over other possible choices. Secondly, biologists have gathered a lot of useful information about their own objects of study which they may lack on the new objects put forward by an external group. Changing one's object often means that no publishable results can be obtained for a year or so.

We can summarize this section as follows. The ability of biologists to deal with a non-scientist's question depends strongly on the type of question involved. An analysis of the dimensions characterizing the pattern of knowledge development in a field can be used to predict which questions are more likely to give rise to problems than others. Can we therefore expect that certain types of cooperation are less likely to occur than others? Before we can conclude this, we must be sure that biologists have no other ways of dealing with these problems. In the next section we will focus on the decisions biologists take in their research groups and analyse the options scientists have for dealing with the problems caused by the demands of the reputational system.

3. Types of Cooperation, their Causes and Consequences

In this section I will characterize types of cooperation in the field of biology (3.1) and look at their causes (3.2) and consequences (3.3). Our analysis in the previous section suggested that certain types of cooperations are more likely to involve problems than others. Following Whitley's line of thought we can assume that these problems make the occurrence of certain types of cooperation unlikely. Following Knorr's line of thought, however, we can assume that resources can solve the indicated problems. But can resources solve all, none, or only certain types of these problems? This question can be rephrased as follows:

under what conditions can resources solve the problems arising from conflicting demands from the reputational system and the non-scientist? I hope to show that this question can be dealt with effectively by looking at two related subquestions. First, what role do resources play in biologists' strategies for carrying out their long-term institutional research programmes? Secondly, how is the development of knowledge in a group related to knowledge development in a field? To answer these questions we must change the unit of analysis from the scientific field to research groups and cooperations.

The role of non-scientists' resources in biological research strategies will be analysed by studying decisions made within the research groups. Various types of decisions will be distinguished and narrowly correlated with specific types of cooperation. There follows a discussion of the contextual factors affecting the types of cooperation: the resources offered by the non-scientists and the demands of the reputational system. On the basis of this analysis, specific types of cooperation will be related to types of biological research strategy. We identified a number of resources in the case studies, including money, manpower, information, apparatus, access to objects of study, the university biologists' ability to influence the appointment of self-trained biologists, the social legitimation of the research programme, etc. The social legitimation of a research programme is an important resource in the Netherlands because the "social relevance" of fundamental research is appreciated and rewarded. In this paper, however, I will focus on those resources that help us to understand the various roles resources can play in research groups that have problems in dealing with the various demands of the reputational system and the non-scientists.

To analyse the relation between the development of knowledge in the group and the field, we will compare the results of the group's research with the results presented in publications in biological journals.

3.3. Types of cooperations and decisions taken in the knowledge development process within a research group

We can distinguish 5 types of decisions taken in research groups and relate these decisions to 5 types of cooperation. The decisions discussed here vary in the degree to which they influence the long-term research programme of the university group. All university biology groups in the Netherlands have a long-term research programme which must be approved by the scientific committee of the biological sub-faculties. The coherence among the research programmes varies (for example depending on the subfields involved), and thus the ability to address the questions posed by non-scientists varies as well. Yet in spite of this variation, all biologists will encounter constraints in dealing with external questions as a result of their commitment to their own programme. This means that the decision to cooperate with non-scientists is narrowly related to the decisions taken with regard to the research programme. I will therefore characterize both the decisions and the cooperations in terms of the degree to which they influence the long-term research programme. To measure this influence it will be necessary to distinguish not only between a strong and a weak influence, but also between an influence which is compatible and one which is incompatible with the demands of the reputational system. This leads us to the following 5 types of decisions.

The first type of decision is a decision not to change anything in the research process. The second type is a decision to choose one from among two or more different ways of continuing the research process, while still keeping within the planned programme. This kind of decision, which is related to scientists' task uncertainty, is taken regularly and has only a weak influence on the research programme. In all scientific work there is a continuous compulsion to choose from among all the various ways of reaching the desired research results. This second type of decision is compatible with the demands of the reputational system.

The third type of decision is similar to the second type, except that the decision taken is *not* compatible with the demands of the reputational system. In the fourth type of decision we are no longer concerned with a choice within a given research programme, but with a decision to change the programme itself in a manner compatible with the demands of the reputational system. This decision has a very strong influence on the research programme and is rarely taken. The fifth type of decision is similar to the previous one, except that the decision taken is incompatible with the demands of the reputational system.

Looking now at actual instances of cooperation, we can consider whether or not the non-scientists' questions and resources influence these decisions. Our case studies show overwhelming evidence that they often do. For example, a non-scientist's question which cannot be dealt with within the research programme may influence the research group to decide to change the direction of the programme. I will illustrate the five types of cooperation related to the five different types of decision in the research group, where possible with examples taken from our case studies. First, however, we will list the 5 corresponding types of cooperation:

1. Cooperations that do not influence any decision in the research programme of the group.

2. Cooperations that influence decisions within the research programme affecting its direction, where the decisions taken are compatible with the demands of the reputational system.

3. The same as type 2 but with the decisions taken in conflict with the demands of the reputational system.

4. Cooperations that influence decisions to change the research programme, where the decisions taken are compatible with the demands of the reputational system.

5. The same as type 4 but with the decisions taken in conflict with the demands of the reputational system.

We found no direct example of the first type of cooperation. However, in one of the cooperations we studied, the non-scientist exerted only a slight influence on the research programme. This cooperation involved research in medical microbiology. The pharmaceutical firm involved wanted to make a new vaccine against calf diarrhoea using techniques based on genetic engineering. The microbiology group involved was interested in the biomedical mechanism which caused the bacterium to cling to particular cells. The university microbiologist said that the aim of the cooperation was to prevent the unnecessary duplication of experiments. Nevertheless, we found that the pharmaceutical firm clearly influenced decision-making in the research process. The university group's priority with regard to experiments was changed at the firm's request, and the apparatus offered by the firm was used in the research. We have thus to conclude that this cooperation really belongs to type 2.

In the second type of cooperation the group decides to choose a research strategy within a programme which coincides with the interests of the non-scientist and does not conflict with the demands of the reputational system. We found examples of such cooperation in several subfields of biology. An ecological group cooperated with a governmental institution which wanted information about the levels of salt and heavy metals in plants in certain regions. The information gathered in this cooperation was used by the university group to help them understand the physiological mechanism in plants. Because of this cooperation, the ecologists were no longer free to choose where to collect their specimens. In the subfield of molecular genetics cooperations are formed with firms who want to change the characteristics of plants or bacteria, and thus want to know the location of particular genes. The geneticists involved in such cooperations are also interested in the location of genes because they are seeking to answer biological questions about the expression of genes.

In the third type of cooperation the scientific group also decides to choose a research strategy within a programme which coincides with the interests of a non-scientist, but in this case the research strategy is in conflict with the demands of the reputational system. In biology for example, subquestions which involve the development of techniques are not considered to be an interesting or rewarding activity. Techniques are in their view "just aids" to which one should not pay too much attention. Nevertheless we found many cooperations formed in order to develop a technique. One example of this is the cooperation between neurobiologists and an anti-cancer organization. The question of this organization was: how can we test the extent to which cytostatics used for fighting cancerous growths are injurious to healthy cells? The neurobiologists expected that the development of a technique to deal with this question could also be used for their study of neuroendocrine cells. Similar cooperations were found in genetics and ecology.

In the fourth type of cooperation the research decision is influenced by the non-scientist to such an extent that the research programme has to be changed. The decisions taken here are not in conflict with the demands of the reputational system. In one of the cooperations we analysed, the object of study, an aquarium fish, was replaced by a fish of greater economic significance.

The fifth and last type of cooperation influences the decision to change the research programme in a way which conflicts with the demands of the reputational system. Such a change might imply the choice of a central question which would not be considered an interesting biological question. We came across the anti-nuclear energy movement's request to assess the environmental and health consequences of a particular nuclear energy plant. Analysis showed (see first reference note 11) that the multidisciplinary character of this question meant that biologists expected a very limited reward from any results they obtained, and moreover, they lacked many of the skills needed to cope with this question. Thus despite the occurrence of such questions we found no examples of this type of cooperation.

We could find neither type 1 nor type 5 cooperations in our case studies, so we tried to find these cooperations by interviewing other biologists, but still without results. Type 1 cooperations are evidently very unlikely to occur. In studying the cooperations carefully, we always found that scientists had to adjust their research to some extent. Type 5 cooperations are also very unlikely to occur, which is easy to understand. Whatever the results of such a programme might be, they would be highly unlikely to be appreciated and rewarded in the field involved. In retrospect, we can sometimes observe that this type of cooperation has often marked the beginning of the emergence of new subfields (see for example the contributions of *Cramer et al.* and *Groenewegen* in this volume). But the emergence of a new subfield is a relatively rare event, which means that type 5 cooperations are also very unlikely to occur regularly. From here on we will deal only with the remaining three types of cooperation: types 2, 3, and 4.

We found examples of types of cooperation that we might have expected on the basis of the analysis made in the previous section, but we also found some cooperations we did not expect. Obviously, the analysis we made in section 2 was not sufficient to explain precisely what type of questions scientists can deal with. For this, more insight into the research process in university groups is necessary. The decisions taken in these groups not only have to be related to the demands of the reputational system but also to the possibilities which additional resources can provide for dealing with the problems caused by these demands.

3.2. Contextual factors affecting types of cooperations

In this section I will examine the factors determining the types of cooperation within a field by describing the relationships between two contextual factors in the cooperations: the resources offered by the non-scientist and the demands of the reputational system.

In the second type of cooperation we can distinguish between two subtypes of cooperation (A and B), in which resources play completely different roles. In type A, the biologists perform a research task which they also need to do for the sake of their own programme. The influence concerns only the way in which this task is to be carried out.

For example the ecologists, for their experiments, needed to know the levels of salt in plants. Through their cooperation with the non-scientist they were told where to gather the specimens and were given photographs to pinpoint the exact locations. In these cases the resources the scientists received were practically unconditional, and could be used freely to extend their activities.

In type B, the role of resources is to reduce the *risk of wasted research time*. The problems and research standards in biology are well articulated, and the criteria for evaluating and rewarding research results are well defined, if not completely unproblematic. At the same time, it is not always clear which approach to a given problem will be most successful. The success of an approach is not proven until the answer to a subproblem is deemed relevant to the overall research problem. The lack of a dominant cognitive structure to indicate the preferred way of tackling a research problem gives the biologist a wide range of options in his approach. However, this wide range can also lead to wasted research time, since the biologists may obtain results that are not directly relevant to the appointed problems. This combination of the characteristics of standards for research and the high task uncertainty of a given approach to a problem leads researchers to develop strategies to limit the risk of failure. One such strategy, for example, involves finding out whether biologically disappointing results might be interesting to a different audience. As one professor of neurophysiology told me:

> Research is often a bit tricky; so I thought that if something special about these growth cells should come out, in other words, something that might help solving the biological problem of growth, I might be able to get it published in a neuroendocrinological journal. Now if these cells appear to be just as normal as plain neurons, well I'm going to have to twist it towards something interesting to pharmacologists, something that goes on well at the moment. What else can you do with it? If the results don't answer your biological question, you can't publish them in biological journals. But you don't want to be stuck with finished research you have been working on for so long.

Another strategy followed by biologists in order to reduce the risk of failure involves using a variety of approaches at the same time. Here, too, they try to find audiences which might be interested in answers to subquestions, regardless of the overall biological question. (However, publications directed to these audiences confer fewer rewards than publications in biological journals.)

These strategies automatically lead to cooperative relationships with scientists from other (sub)fields, and also with non-scientists. These cooperations are caused by the role resources play in reducing the risk of research which is characterized by uncertainty about the research tasks which should lead to results that meet the demands of the reputational system. In cooperative relationships of this kind, the biologist is looking for information (e.g. the characterization of a gene, the function of an enzyme, the influence of the light climate on the generation of algae, etc.) that would otherwise demand a lot of research time, and the use of which cannot be guaranteed in advance. The risk of wasted research time is reduced appreciably if the researcher receives extra money or manpower or can get some of his results published in journals belonging to other fields, or in applied magazines (if he has been working with non-scientists). It is noteworthy, especially for the biology shop policy makers, that sometimes the researcher's aims allow for a choice among various non-scientific partners whose aims are extremely divergent. For example, a botanic geneticist might benefit from cooperation with an industry selling pesticides: the geneticist is trying to answer biological questions concerning the expression of genes, while the industry is trying to build a gene into certain plants to make them more resistant to pesticides (to destroy the weeds and space the crop). Similar information, however, could be derived from co-operation with a state institution or an environmental group that is trying to reduce the use of pesticides by building a gene into plants that makes them. resistant to certain diseases. These two cooperative relationships would have different social consequences, but the information derived from them would be similar. These type B cooperations are abundant in biology and can be explained by the combination of mutual dependence and task uncertainty in biology.

Another important question is whether these type B cooperations are unique to biology or can also be found in other fields. The combination of a high degree of mutual dependence and a high degree of certain task uncertainties may be specific to biology, but Whitley has noted that all scientific work can be characterized by these conflicting demands for scientific results. The scientist's results must always be new, but in order to be accepted they must also be related to the work done by his or her colleagues. We can thus expect type B cooperations caused by conflicting demands on research results to occur in all sciences although in some more than in others.

Risk-reduction is also important in the third type of cooperation but here we are confronted with *risks caused by conflicting demands affecting choices* regarding research problems, goals, approaches, techniques, concepts, the object of study, or the standards for research. In biology, the specific demands on the choice of problems means that certain types of research, such as the development of a new technique, are not considered interesting. Such constraint against certain types of research problem may retard the development of knowledge because it can delay progress in certain biological subfields. One professor in microbiology described this as follows.

I'm afraid that certain problems in our field are not solved for lack of a method. To give you an example: how do plants and their Rhizobium species recognize each other? What it takes to answer this question, of course, is a method with which one can distinguish the various Rhizobium species. Now I've been attending conferences for seven years, and nothing much has happened in those seven years. I think there's hardly any progress in this field, because we don't have the proper methods.

Biologists, then, are confronted with a kind of prisoner's dilemma. They need new methods to continue their research and to address the standards of biological research, but each individual research group works by the rule: "let others develop the techniques". Therefore scientific developments within the field of biology are easily stranded in situations where biophysics, biochemists, and medical scientists do not develop the equipment biologists need at a certain stage of their research. Cooperation with a non-scientist can sometimes solve this problem. Working together with institutions from outside the university means either that research will be financed from the outside, or that it can be legitimized as socially relevant. This formula permits research into the development of method and techniques. If a cooperative relationship should produce a method or technique which can be used to solve a biological problem, the research has created a relevant and highly rewarded result. If the development of a method or a technique should fail to contribute to solving biological problems, the researcher is still in a position to legitimize his efforts on the strength of their social relevance.

Biological science, especially biotechnology, is involved in cooperative relationships of this kind more and more frequently. Practically the entire state-financed budget for biotechnology is spent on the development of techniques. Cooperative relationships are also formed for the development of techniques in other fields: in neurobiology, for instance, for the development of methods for testing cytostatics and for the registration and interpretation of brain signals; and in ecology for the development of bioassays, etc. These techniques are all developed in the interest of society, but once they have been developed, the researcher can apply them to his own fundamental research.

This does not mean, however, that cooperations of this type can solve all conflicting demands. There is an extra condition which must also be met: the researchers must be free to use the results of the cooperation in their group's research programme. In biology the development of techniques is badly rewarded, but biologists are under no constraints to use any newly developed techniques in their research programme. The low degree of mutual dependence regarding the use of particular techniques is thus essential to an understanding of this type of cooperation. This freedom can be contrasted with the constraints on the use of deviant concepts and models which may result from cooperations. In our survey, requests from non-scientists to biologists that they use the concept of environmental dynamics (milieudynamiek) in their knowledge development were not honoured. This fourth type of cooperation is thus determined by a specific combination of conflicting demands and lack of constraint in the biological research process. Problems which result from this combination can sometimes be solved with the resources of non-scientists.

We can again ask whether these findings can be generalized to other fields. I think we can expect this type of cooperation in all other fields, especially those characterized by a high mutual dependence among colleagues. A cooperation in neutrino science between a chemist and theoretical physicist, for example, might be evaluated in this way. Here we are confronted with a theoretical physicist, Bahcall, heavily involved in the development of a technique. Pinch describes the involvement of Bahcall as follows: "Bahcall was now clearly committed to the experiment and was even giving advice on matters of experimental detail" (33). Involvement in the development of techniques is not considered a highly rewarding activity in theoretical physics; but here again the expected reputational pay-off when the technique was developed seemed very high. Bahcall told Pinch why he accepted the chemist's request to make the many not very rewarding calculations needed for the development of the technique: "The reason I decided to work on it, is the result of a conversation with [a colleague] . . . so we got into a discussion about how unique or fundamental the experiment would be . . . I became convinced that it was really a unique way of testing an otherwise very fundamental theory" (34). Pinch does not think his findings are to be considered as very exceptional:

the special nature of the experiment and the concomitant social relationships are useful for illustrating aspects of scientific activity not usually visible in other cases. In particular the joint investments made have become explicit. There is no reason why in principle the investments documented here should not happen in much more mundane pieces of science (35).

In general, we can say that the resources made available in cooperations of this type give the scientists the opportunity, for a while, to escape from their dependency on their colleagues and ignore a conflicting demand. From our case studies it is clear that we cannot consider these cooperations where brilliant scientists "see" the opportunity for an important breakthrough in knowledge development as exceptions. We have to consider such cooperations as common practice in the research process within a given research programme.

The fourth type of cooperation influences decisions that change the research programme in a way compatible with the demands of the reputational system. Before such a decision can be taken, the advantages of the new resources are weighed against the disadvantages resulting from the loss of accumulated knowledge, skills, scientific credits, etc. These types of decisions and cooperations show that the resources controlled by biologists are very limited. In biology a very limited number of changes can be considered because of the high mutual dependence concerning problems and goals. The simplest way of changing the programme is to change the object of study. We observed this change in two of our 20 case studies.

Cooperations of this type are sometimes of short duration. The main aim of the scientists in these cooperations is to explore new ground. The cooperation gives the scientists the opportunity to do exploratory research. The results of these "quick and dirty" research projects are guaranteed to be socially relevant. Often minimum target results can be defined in advance. If these results do not emerge before a certain deadline, the research is terminated and the cooperation ended. *Reduction in the risk of wasted research time* here is thus also crucial to an understanding of the cooperation.

From these findings we can conclude that the relationships between the demands of the reputational system and the resources offered by the non-scientists exert a very strong influence on the decisions taken in the biological groups. The important role resources play in these decisions means that biologists are often actively involved in finding "interesting" non-scientists. In many cases the researchers can indicate the ways in which the social and scientific aims of a possible cooperation can coincide even before they have been in touch with the non-scientist. The researchers' way of thinking can be described as follows: for the progress of our research programme, we need to develop a certain technique, or obtain certain information. If we offer to find out something for a certain non-scientist over a period of four years, at the end of this time we will not only be able to have articles published in applied journals, but we will also have the technique and/or information we require. This strategy means that biologists evaluate the results of cooperation on two levels: first of all, according to the scientific aim of the cooperation, and secondly in terms of their "hidden agenda", the scientific aim of the researchers' own research programme.

The analysis of this section shows the different roles resources play in a research group's decision to enter into one of the various types of cooperation. Resources can create the opportunity to extend the group's research activities; and they can also serve to reduce the risk of wasted research time caused by the tension between mutual dependence and task uncertainty, or by the exploration of new areas, or to reduce the risk of diminishing reputation through research which does not meet an important demand of the reputational system.

These roles strongly suggest that risk reduction is one of the most important catalysts in the formation of cooperations in biology. Therefore we can answer the first question we posed in the introduction as follows. Scientists can most easily deal with the questions posed by non-scientists if they are already part of their research programme. They can also deal with questions which are not in conflict which any demand from the reputational system, provided that the necessary research elements are available. Scientists can also deal with some questions which are not compatible with the demands from the reputational system. It is these last questions that create opportunities for the scientist to deal with the problems arising from the conflicting demands of the reputational system. Questions which are in conflict with the demands of the field involved, but which do not enable the scientist to deal with the conflicting demands of the reputational system, are very unlikely to be tackled.

3.3. *The consequences of cooperation for the development of knowledge*

Having characterized the types of cooperation which are more likely to occur than others, we can now discuss the second question posed in the introduction: how is the development of knowledge affected by extra-scientific influences exerted in cooperations?

From our analysis it is clear that we have to distinguish several types of knowledge development which can be affected. First there is the influence affecting the development of knowledge within a research group, which can be analyzed by looking at the effects of cooperation on a group's final research result. Secondly there is the influence affecting the development of knowledge within a field, which can be analyzed by looking at the effects of cooperation on results which are published in biological journals. Finally there is the influence on the pattern of knowledge development in the field, which can be analyzed by looking at the effects of cooperation on the degree of mutual dependence and task uncertainty in the field. I will deal with these three levels of influence separately.

From the previous section it is evident that cooperations heavily influence the work of university biologists and their results. The first type of cooperation, which does not influence biological research, seems to be highly unlikely. This finding confirms Knorr's work. In our analysis, however, we were able not only to trace the influences, but also to understand which influences are more likely to occur than others. We found that influence affecting the choices of subproblems, techniques, approaches to problems, or objects of study was more likely in biology than influence on the choice of the central research question or the concepts.

Extra-scientific influences affecting the development of knowledge, as characterized by the content of the university group's publications in biological journals, are less strong.

Many of the group's research results which were influenced by non-scientists were published in biological journals. In this way non-scientists strongly influenced the knowledge published in the formal communication system. We noticed that the third type of cooperation had the greatest potential to influence the development of knowledge. In such cooperations the research that was carried out was not especially highly rewarded by colleagues, but when it did not fail it did eventually reduced important barriers to knowledge development. An example of one such barrier in biology was the lack of methods to distinguish among the various Rhizobium species.

Notwithstanding these influences on biological publications, many other influences are highly unlikely. Many of the group's results cannot be published in biological journals because they do not meet the relevant demands. Biologists try to find other audiences for these results, such as pharmacologists, medical scientists, biochemists, or audiences in the applied sciences. Publication in non-biological journals means that a great deal of the influence of non-scientists on the research group's work cannot be found in the biological journals and thus hardly influences knowledge development here.

Finally, let us consider the non-scientist's influence on the pattern of knowledge development in a field. These patterns can be analyzed by looking at the degree of mutual dependence and task uncertainty. Does cooperation with non-scientists affect these degrees? Whitley predicts that the degree of task uncertainty will become higher and the degree of mutual dependence lower as the number and variety of the scientists' audiences increase (see Notes 6 and 7). We do not have enough information to formulate a clear statement about the influence of non-scientists on task uncertainty. In our studies we found that the degree of mutual dependence was *not* likely to decrease as a consequence of new cooperations with non-scientists. A change in the degree of mutual dependence would imply in biology a change in the demands regarding the choice of the central problems, goals, concepts, or standards of research. We found no evidence of this happening because all the biologists in our case studies only formed partnerships which they hoped would contribute to their present or future research programmes, and thus would bring them high reputations within their own field.

The idea that biologists would adjust their scientific aims in their efforts to comply with the wishes of non-scientists is unlikely given the way things are in the Netherlands at present. A professor of microbiology cooperating regularly with a pharmaceutical firm says:

> The danger of cooperation is the loss of autonomy; you always have to adjust yourself, and if you adjust too much, you may forget some of your own aims. You must always stick to your own field, your own expertise. Within those limits, cooperation is just fine. But if the developments in your own field demand that you should follow a certain course, you have to follow that course regardless of the cooperation you're involved in. In may hurt the cooperation, but there's no other way . . . .

Another important question is: how stable is this situation, and under what circumstances can we expect that the pattern of knowledge development will change? Some biologists are very clear about this: if the financial structure and evaluation procedure of a biological research group should change so that they would expect more scientific credit by moving away from their specific problems, work goals, and research standards, they would undoubtedly take this step. At the moment this seems to be happening to some extent in biological fields contributing to the solving of biotechnological problems. Uncertainty however, as to whether the large financial injections available for the development of biotechnology will be made structural makes most researchers reluctant to direct their research too much to audiences outside the field of biology. Obviously very significant and stable resources are necessary to change the pattern of knowledge development in a field. Cooperations of the types discussed in this article do not contribute to such a change.

In my view one of the most surprising outcomes of our case studies is thus that cooperative relationships with external groups in biology seem to strengthen rather than to weaken the degree of mutual dependence among biologists. These cooperations give the biologists an opportunity to deal with problems caused by conflicting demands from the reputational system and to perform high-risk research. Far from threatening the internal coherence of the knowledge development of the field, they may be considered important conditions for maintaining the knowledge development patterns in all fields characterized by high mutual dependence among scientists.

We can summarize this section as follows. The cooperations described in this paper have very little effect on the pattern of knowledge development in biology, but they exert a considerable influence on the articles published in biological journals. The strongest influence is exerted on the group's knowledge development, despite the constraints imposed by the reputational system on some of the non-scientists' influence.

4. Conclusions

In this paper I have tried to analyze the connection between some of the factors influencing the production of biological knowledge and the end products published in biological journals.

We are confronted in cooperations with the interaction of two factors influencing scientists' work: the demands of the reputational system and the demands of the non-scientists who are offering the resources. By analyzing the relationships between these two factors we can understand what types of questions posed by non-scientists biologists can deal with, and thus indicate which types of cooperations are more likely to occur than others. The resources offered by non-scientists are a very important factor influencing knowledge production in biology. Biologists use these resources to extend their activities but also to reduce the risks associated with scientific work. The idea that scientists pay a great deal of attention to risk-reduction is not new. Knorr, for example, wrote that

> there is no reason to believe that scientists are by nature or necessity devoted to risk taking. To say that research is a high risk enterprise . . . does not reflect the reasoning of scientists in the laboratory . . . (36).

In the analysis presented here, the risks scientists try to reduce are of very specific types, caused mainly by the various conflicting demands of the reputational system. I think that the *kind of analysis* described in this paper can be extended to all fields. Scientific fields have different degrees of mutual dependence and task uncertainty, and also different degrees of conflicting demands, and so the types of cooperation to be found in them will also differ. In fields with a low degree of mutual dependence cooperations will be less constrained or influenced by the reputational system (see also Blume's essay in this volume). Non-scientists can therefore affect the results of research published in journals in such fields rather easily. In such fields cooperations based on the described types of risk reduction are not very likely; while in fields with a high degree of mutual dependence, we can expect cooperations based on the described types of risk reduction. The contents of the same types of cooperation will of course differ from field to field. In experimental physics, for example, it is unthinkable that scientists should need a cooperation just to legitimize the fact that they are developing a technique!

Our analysis of the influence of cooperations on knowledge development shows that the degree of influence is strongly related to the unit of analysis studied. The different units of analysis used by Knorr (transepistemic arenas) and Whitley (scientific fields) partly explain the different predictions we can draw from their work.

The results of our analysis, however, suggest that in order to understand knowledge development within a group, the influence exerted by the demands of the reputational system on the decisions taken by the group must also be considered. I think that Knorr should also have considered this aspect of influence in order to understand fully the decisions reached in the laboratory she studied.

Our analysis also suggests that non-scientists' questions can influence the published results of biological work in certain cases even when these questions are not compatible with the demands of the field. This suggests that the concept of mutual dependence is more complex than Whitley allows, and that an elaboration of the dimensions in terms of which mutual dependence can vary is necessary for a detailed analysis of the process of knowledge production in a scientific field.

In general we can conclude that in order to understand extra-scientific influence on knowledge development, the frameworks of Knorr and Whitley are important but too general and need to be refined considerably. The analysis in this paper can be considered as a step in this direction.

Acknowledgements

Parts of this study were carried out with the help of Jeanine de Bruin, Annelies Stolp and Charlotte van der Woude, whom I would also like to thank for their encouragement and assistance. I would like to thank Loet Leydesdorff, Stuart Blume and Richard Whitley for their valuable comments on an earlier draft of this paper. I would also like to thank Jill Crowson for correcting the English.

Notes and References

1. This study formed a central part of the research programme of the Department of Biology and Society of the Free University in Amsterdam.
2. K. Knorr, "Scientific communities or transepistemic areas of research? A critique of quasi economic models of scientce", *Social Studies of Science* **12**, 1982, 102–130, p. 118.
3. K. Knorr, *op. cit.*, 1982, Note 2.
4. K. Knorr, *op. cit.*, 1982, Note 2, p. 103.
5. R. D. Whitley, *The Intellectual and Social Organization of the Sciences*. Oxford: Oxford University Press, 1984, p. 85.

6. Or formulated the other way round:
"The close links between . . . state agencies and social reform movements have . . . reduced the degree of mutual dependence between researchers in those fields and render the development of a high degree of technical and theoretical coherence rather unlikely." R. D. Whitley, *op. cit.*, 1984, Note 5, p. 111.
7. This relation is also formulated the other way round:
"Audience variety can . . . be a significant factor in determining the level of task uncertainty in scientific fields." R. D. Whitley, *op. cit.*, 1984, Note 5, p. 146.
8. R. D. Whitley, *op. cit.*, 1984, Note 5, p. 191.
9. T. J. Pinch, "Theoreticians and the production of experimental anomaly: The case of solar neutrinos," in K. Knorr et al. (eds.), *The Social Process of Scientific Investigation.* Sociology of the Sciences, Vol. IV, Dordrecht: Reidel, 1980.
10. K. Knorr, *op. cit.*, Note 2, p. 118.
11. The results of these case studies have been presented in reports and a publication. Four reports were prepared by graduate students:
J. W. van Someren, "Samenwerkingsverbanden goed 'aangelegd'?," Amsterdam: *Biologie en Samenleving*, Vrije Universiteit, 1985.
K. Wulffraat, "In de huid van een actiegroep gekropen: Over samenwerking tussen actiegroepen en de universiteit," Amsterdam: *Biologie en Samenleving*, Vrije Universiteit, 1982.
J. de Bruin, "Beschrijving en analyse van samenwerkingsverbanden tussen plantegenetici en externe groepen," Amsterdam: *Biologie en Samenleving*, Vrije Universiteit, 1986.
A. Stolp, "Beschrijving en analyse van samenwerkingsverbanden tussen leden van de onderzoeksgroep plantesystematiek en oecologie van de lagere planten en externe groepen," Amsterdam: *Biologie en Samenleving*, Vrije Universiteit, 1987.
Some results have already been published, J. Bunders and J. de Bruin, "Hoe plantebiotechnologisch onderzoek wordt afgestemd op de belangen van het bedrijfsleven: De afwegingen van een onderzoeksgroep," *Wetenschap en Samenleving* **4**, 1985, 10—14.
12. The biology shop that our department is involved with acts as an intermediary in attempts to solve problems and answer questions in fields including ecology, physiology, and microbiology: a difficult and sometimes disappointing task.

Obviously one major problem confronting biology shops is the lack of money to pay for research. Nevertheless we think this is not the only problem facing the biology shops. Indeed, biologists may even benefit from the work they carry out for the biology shop clientele, since it can be considered of social interest; as such, its usefulness is guaranteed, also in the eyes of the outside world.

A second possible cause of the problems involved in this intermediary task is the way in which research problems are defined, and the way in which the customer would like a given problem to be tackled. The chances of failure of the intermediary efforts of the biology shop can usually be weighed fairly accurately in advance merely by looking at the kind of customer the shop is working for, not because the customer is making insane demands, but simply because the research groups are not equipped to answer certain questions. (K. Wulffraat, "In de huid van een actiegroep gekropen: Over samenwerking tussen actiegroepen en de universiteit," Amsterdam: *Biologie en Samenleving*, Vrije Universiteit, 1982).

Naturally this raises the question of which problems the university biologists are prepared to address. Once this has been established, we can attempt to formulate a client's questions in such a way (not, of course, without their advice and approval) that they stand a better chance of being included in the biologists' research programmes.

13. R. D. Whitley, *op. cit.*, 1982, Note 5, p. 122 and p. 88.
14. R. D. Whitley, *op. cit.*, 1982, Note 5, p. 121 and p. 88.
15. *Biologie, Van Levensbelang: Rapport van de Verkenningscommissie Biologie.* Den Haag: Staatsuitgeverij te 's-Gravenhage, 1983.
16. I. Jahn et al., *Geschichte der Biologie: Theorien, Methoden, Institutionen, Kurzbiographieen.* Jena: VEB Gustav Fischer Verlag, 1985, 618—621.
17. C. F. A. Pantin, *The Relations between the Sciences.* London: Cambridge University Press, 1968, 16—23.
18. J. Lever, "Reflections on comparative physiology" in A. D. F. Addink et al. (eds.), *Exogenous and Endogenous Influences on Metabolic and Neural Control.* Pergamon Press, Oxford, 1982, p. 13.
19. J. Lever, *op. cit.*, Note 18, p. 12.
20. J. Lever, *op. cit.*, Note 18, p. 12
21. This view has also been expressed in a lecture on ecophysiological perspectives in biology: E. N. G. Joosse-van Damme, *Oecofysiologische en Oecotoxicologische perspectieven in de Dieroecologie.* Amsterdam, Uitgeverij Vrije Universiteit, p. 7.
22. *Disciplineplan Biologie 1986.* Amsterdam: Uitgave van de Biologische Raad, pp. 3 and 26—28.
23. *Disciplineplan Biologie 1986*, op. cit., Note 23 pp. 11—12; *Biologie, Van Levensbelang: Rapport van de Verkenningscommissie Biologie.* Den Haag: Staatsuitgeverij te 's-Gravenhage, 1983, pp. 17—19, and *Biologisch Onderzoek voor Mens en Maatschappij.* Amsterdam: Uitgave van de Biologische Raad, pp. 11—14.
24. R. D. Whitley, *op. cit.*, 1984, Note 5, especially the case studies mentioned in chapter 4; and also J. Bunders and Whitley, "Popularisation within the sciences: The purpose and consequences of inter-specialist communication," in T. Shinn and R. Whitley (eds.), *Expository Science: Forms and Functions of Popularisation.* Sociology of the Sciences Yearbook, vol. IX, 1985, pp. 61—77.
25. J. Bunders and R. Whitley, Popularisation within the sciences: The purposes and consequences of inter-specialist communication, in T. Shinn and R. D. Whitley (eds.), *Expository Science: Forms and Functions of Popularisation.* Sociology of the Sciences Yearbook, vol, IX, 1985, pp. 61—77.
26. See P. B. Sloep, "Patronen in denken over vegetaties, een kritische beschouwing over de relatietheorie." Diss. Groningen, 1983.
27. See for example the criticism of the work of Grime, who tried to integrate theoretical elements of physiology, ecology, and evolutionary biology into a new theory (J. P. Grime, *Plant Strategies and Vegetation Processes.* Chichester: Wiley, 1975), voiced by W. J. van der Steen and M. Scholten in "Methodological problems in evolutionary biology. IV. Strees and stress tolerance, an exercise in definitions," *Acta Biother.* **34**, 81—90.
28. Many biologists working on snails and rats told us this.
29. J. Lever, *op. cit.*, 1982. Note 18, p. 10. See also E. Florey, "Die Lage der Zoologie und ihre historische Entwicklungen," in Rathmayer, W. (ed.), *Zoologie Heute.* Stuttgart: Fischer, 1975.

30. J. Lever, *op. cit.*, 1982. Note 18, p. 10.
31. See for example, R. Hohfeld, "Two scientific establishments which shape the pattern of cancer research in Germany: Basic science and medicine," in Norbert Elias et al. (eds.), *Scientific Establishments and Hierarchies*. Sociology of the Sciences Yearbook, Volume VI, 1982, 145–168.
32. J. Lever *op. cit.*, 1982. Note 18, p. 14.
33. T. J. Pinch, 1980, *op. cit.*, Note 9, p. 91.
34. T. J. Pinch, 1980, *op. cit.*, Note 9, p. 87.
35. T. J. Pinch, 1980, *op. cit.*, Note 9, p. 103.
36. R. D. Knorr, "The scientist as an analogical reasoner: A critique of the metaphor theory of innovation," in Karin D. Knorr et al. (eds.), *The Social Process of Scientific Investigation*. Sociology of the Sciences Yearbook, Volume IV, 1980, Dordrecht: Reidel, 1980.

PART II

COLLABORATIONS BETWEEN SCIENTISTS AND NON-SCIENTISTS AT THE GRASSROOTS

COOPERATION BETWEEN MEDICAL RESEARCHERS AND A SELF-HELP MOVEMENT: THE CASE OF THE GERMAN RETINITIS PIGMENTOSA SOCIETY

RAINALD VON GIZYCKI

Battelle-Institut, Am Römerhof 35, Frankfurt a.M., West Germany

In the analysis of how non-scientists can influence the scientific process by promoting cooperation between scientists and non-scientists, the activities of the self-help movement of the German Retinitis Pigmentosa Society is an interesting case. In 1977, when the Society was founded, an inquiry showed that nowhere in Germany was systematic research on the eye disease Retinitis Pigmentosa (RP) being carried out. Eight years after its foundation, however, this situation has changed, as can be illustrated by the close contacts maintained between medical researchers and patients and by the existence of the annual meeting of German Retinitis Pigmentosa researchers, which is attached to the annual German Ophthalmological Society Conference. Even more remarkable than this development is the change which the patients have initiated in the concrete research activities of concerned scientists. The aim of this paper is to analyse the circumstances which can explain these developments.

In my view this paper has special features which distinguish it from the other papers presented in this part of the volume. In this case study the cooperation between scientist and non-scientist is described and analysed from the point of view of the non-scientist rather than the scientist. Therefore we shall look at the conditions prevailing inside the non-scientific group which have made it possible to exert influence on scientists, rather than the other way around. Among the important characteristics of this group is, for example, the fact that it is not an established professional body itself, but rather a young lay organisation, a self-help group of patients starting to formulate and implement their own research interests. Secondly, we are dealing not with a basic scientific discipline, but rather with a very heterogeneous field of the

S. Blume, J. Bunders, L. Leydesdorff and R. Whitley (eds.), The Social Direction of the Public Sciences. Sociology of the Sciences Yearbook, Vol. XI, 1987, 75–88.

medical sciences (ophthalmology), covering basic, applied, and clinical medical research in a variety of different institutions. In summary, what follows deals with the experience of the German Retinitis Pigmentosa Society in promoting interdisciplinary research into the causes of this visual handicap which usually leads to total blindness.

The paper is structured as follows: after a section on the self-help movement, I will briefly describe Retinitis Pigmentosa and the specific objectives of this particular patient organization. Then, I will analyse the Society's activities with respect to the promotion of research in the framework of three rough development phases: the establishment of an internal infrastructure, the establishment of framework conditions conducive to RP research, and the actual exertion of influence upon research projects. Finally I will compare the RP Society's activities with comparable lay societies and draw some conclusions as to the general aspects and conditions of this cooperative experience.

The Self-Help Movement

At present, about 10,000 self-help groups are organized in the Federal Republic of Germany — many of these are concerned with various forms of diseases and disabilities. In principle, the phenomenon of self-help is not new. However, its immense spread during the sixties and particularly during the seventies has forced the public and the state to pay special attention to it.

In the past years, two principal reasons have been identified for the emergence of state-supported research projects in Germany concerned with self-help for the disabled. In the first place, patients are often dissatisfied with the "professionals", particularly with the services of physicians and researchers, with state-financed rehabilitation centres, with manufacturers of technical aids, etc. In the second place, patients demand the right to a degree of self-determination, i.e., the exchange of information and experience with others suffering from the same disability. For this reason, many self-help groups have turned their backs on the professional health care system and have started — sometimes even by means of self-diagnosis and self-therapy — to take their concerns into their own hands.

The foremost concern of many small self-help groups is to provide psychological, socio-therapeutical, and informative support to their members, since they believe that the existing anonymous and technocratic health system cannot offer this. Along with these types of self-help groups who have lost their confidence in the professional system, there are other more cooperative groups who have focussed their efforts on improving the professional health system. Unlike the smaller discussion groups, these cooperative groups are larger, are usually more formally organized, and are more outward-oriented. Most of them have substantial personal and financial resources at their disposal. This puts them in the position of exerting significant influence upon their professional environment. The German Retinitis Pigmentosa Society belongs to this latter cooperative type of self-help group; and its development illustrates the conditions under which a non-professional organization can influence and initiate research.

The German Retinitis Pigmentosa Society (DRPV)

Retinitis Pigmentosa, or retinopathy, is one of the main causes of blindness in industralized societies, and it is still incurable. RP is a form of retinal degeneration which manifests itself in a constellation of different but closely connected symptoms.

At the beginning of the disease, often during childhood, night blindness appears as a first symptom, later followed by light dazzling and colour-blindness. Eventually RP causes an increasing loss of visual field, usually ending in total blindness. This gradual loss of visual field has far-reaching consequences for the affected patient: it results in severe restrictions of his professional prospects and of his private life.

Between the ages of 30 and 40 there is an increased risk of accidents, a risk of job loss, reduced leisure time activities, and an early loss of career prospects. The certainty of becoming blind means, above all, a life-long personal strain, often accompanied by personality crisis, depressions, suicide attempts, and social ostracism. Although ophthalmologists have been familiar with RP and other forms of retinal degeneration since the end of the 19th century, they still cannot provide a scientific explanation for the degeneration of retinal cells causing the loss of sight. As a result, there is no scientifically established therapy. This depressing situation prompted patients in the US and England to form self-help societies during the early seventies, the main purpose of which was the promotion of research into the cause of RP and the discovery of an effective therapy.

Founded in 1977, the DRPV, according to its statutes, has the following aims:

1. Supporting RP research with respect to diagnosis and therapy.
2. Exchanging information and experiences among patients.
3. Informing the public about RP and its social consequences.
4. Exerting influence upon public and private persons, organizations and institutions.

Presently, the German RP Society has more than 1,300 members and represents the interests of more than 20,000 visually handicapped and blind persons in the Federal Republic of Germany. Headed by an executive committee of six affected members, the DRPV is a non-profit registered association. In the Federal Republic of Germany it is organized in more than 30 regional groups.

Infrastructure of the DRPV and Resources for Research Promotion

The ability of the DRPV to exert influence on RP research in Germany has clearly been influenced by the lessons which could be learned from the activities of the American National RP Foundation.

The American National RP Foundation (NRPF), founded in 1971, was rapidly successful in establishing research institutes at several American universities (including the Berman Gund Laboratory at Harvard university) primarily concentrating on research into retinal degenerations. Nowadays, the American organization, with an annual budget of nearly 3 million US$, promotes a large variety of research projects. Approximately 60 full-time RP researchers now work at about 12 institutes. Indeed, research on retinal degenerations has risen to become one of the most important research fields within US ophthalmology. After the foundation of the International RP Association (IRPA) in 1978, the NRPF was able to proceed more systematically with the support and coordination of other national research projects. This was done by financing national and international research congresses and by bearing part of the costs of foreign projects. The DRPV also profited from the Americans when it began to establish its own infrastructure, mainly by translating and adapting US information material.

At the same time, three structural elements were established within the DRPV: the position of research referee was created, a regular information service for RP researchers was set up, and a research foundation trust was established at the German Science Foundation (*Stifterverband*). The research referee, a non-professional and an RP patient himself, soon became a key figure in making known and stimulating German RP research. He not only organized the information service for RP researchers (by ordering, analyzing, and circulating project-specific publications), but he also became personally engaged in the field of genetic research. As a member of a genetic university institute and as the research referee of the DRPV, he plays a double role inside and outside the DRPV aiming at the acceptance of RP research promotion within the scientific community.

With the formation of a foundation trust, a fund was established with DRPV capital and other donations. Access to this fund is subject to scientific selection criteria. To the public, the foundation demonstrated the patients' interest in themselves becoming also financially involved in research promotion.

The Promotion of RP Research

Creating the necessary infrastructure and offering funds are not sufficient conditions for promoting systematic research. Indeed, the DRPV infrastructure is still underdeveloped in this respect. It was necessary to create framework conditions to encourage the scientific community to take an active role in research promotion, and to exert direct influence on research targets. It is indeed difficult to say *a posteriori* how much influence the non-professional organization has actually had on the promotion of research, because the research community soon acted all by itself and because other factors (such as the need for funds) became influential as well. It can be assumed, however, that without the involvement of the DRPV, neither a fundamental interest in retinal degeneration research nor any definite research objectives would have been established. This is apparent if we compare the German research with the research carried out in countries without an RP society.

In 1977 the starting situation was miserable indeed. Soon after its foundation the DRPV initiated an inquiry into the scope of RP research at all ophthalmological clinics within the Federal Republic of Germany. The results showed that nowhere was systematic research being carried out. There were few specialists, such as ophthalmologists at university eye clinics, who were familiar with the disease, but who concentrated primarily on the following types of activities:

- detecting the symptoms of RP and making a (more or less) exact diagnosis;
- improving diagnostic aids (for example, perimeter, electroretinogram, etc);
- providing genetic counseling;
- developing and prescribing technical aids (for example, TV-reading devices, filter glasses, etc.);
- prescribing medicines (Complamin, Difrarel, etc.) without expecting any improvement or stabilization of the visual field.

Biochemically oriented research work on the visual process has been underway at one Max-Planck-Institute and one government research institute, which, however, bore no direct relation to human retinal degeneration. It was obvious that research into RP and other forms of retinal degeneration had no relevant personnel at its disposal. There was not a single research institute concentrating its efforts on finding the cause of the disease, and no connection between clinical and basic research efforts.

After contacting researchers and clinicians in order to prepare the first information brochures for its members, the DRPV started to establish the following framework conditions for German RP research:

- the publication of a status quo report on the international situation of RP research (the *Baltimore Report*);
- the concrete formulation of patient demands vis-à-vis the research community;
- the organization of national and international conferences;
- the establishment of a scientific advisory board;
- granting an RP research prize for the "prevention of blindness";
- the development of an integrated RP research programme.

Also, several requests were made to the Federal Ministry of Research and Technology in order to lobby for public funds for RP research.

In 1981, when the DRPV was invited to the second International RP Congress in Baltimore, it urged the Ministry to use public funds to send two German RP specialists, one ophthalmologist and one biochemist, to attend the conference. Subsequently the two German researchers and the chairman of the DRPV wrote up the *Baltimore Report on the*

International Situation of RP Research, which is based largely on the results of the Congress.

This report soon became a key document for researchers, ophthalmologists, doctors and patients alike: it provided an incentive for attracting additional RP researchers and became the starting point for further reflection on German research projects. Moreover, it was in high demand as a reference work for the further training of eye doctors and made the DRPV known to experts as a cooperative self-help association. Within the DRPV much thought was given to the formulation of patients' needs and research requirements. The following list of research priorities was formulated:

1. Research into the biochemical and genetic causes of retinal degeneration;
2. Research into the efficacy of current and future treatments of RP;
3. Improvement of everyday visual conditions for RP patients;
4. Improvement of conditions for effective genetic counseling.

These requests were published in several places and were finally brought to the attention of the first German RP Congress early in 1984 (which was organized by the DRPV in cooperation with the German Ophthalmology Society).

At this congress, chaired by a well-known Belgian ophthalmologist, a large group of interested German researchers (about 90) met international RP researchers for the first time. Here it became clear that RP research in the Federal Republic of Germany was not only a result of patient demand but also constituted a cognitive challenge and a social opportunity for research workers. New research contacts were made, joint reflections on projects were developed, and the basis for future congresses was established. Since this conference, an annual meeting of German RP researchers is attached to the annual German Ophthalmological Society Conference. These personal contacts also led to a feeling of common interest in RP research, which was the motivation for establishing a scientific advisory board consisting of 8 researchers from various disciplines. Although the idea for this body came from the DRPV, the research community reacted positively and the director of the Ophthalmological Clinic in Munich is the present chairman of the board. The scientific advisory board has several functions. In the first place it develops an integrated RP research programme covering a number of interrelated and inter-disciplinary research projects. Secondly, it evaluates research proposals and research results and decides on the allocation of project funds made available by the RP Research Foundation. Finally, it organizes research conferences and awards the annual "Research Prize for the Prevention of Blindness". In 1985 for the first time, at the suggestion of the scientific advisory board, the prize was given to a young researcher at the Max-Planck-Institute for Clinical Physiology in Bad Nauheim. His work showed experimentally that a certain chemical substance has a negative effect on the production of an enzyme (the so-called "phosphodiesterase") which is required for the visual process. Presumably the RP patients lack this enzyme. (The enzyme phosphodiesterase is activated by light into stimulating sight). At the annual meeting of eye doctors in Essen in 1984, the chairman of the scientific advisory board publicly awarded the research prize, which consists of DM 1,000.- and a free trip to attend an annual RP Research Congress in Florida.

After intensive preparatory work, the scientific advisory board has also recently defined the research programme requested by the DRPV as a basis for funding future RP research work. The presence of an official from the Federal Research Ministry at an RP congress lent support to the decision of the Board to develop such an interdisciplinary programme. Before a scientific audience he stressed the Ministry's willingness to support such a research concept. Already prior to this promise, a member of the German Bundestag had officially asked the Ministry for this support. (He is today an honorary member of the DRPV). On this occasion the Ministry of Research declared its intention to grant RP research the "highest priority".

With the completion of this draft of the research programme, the scientific foundations have now been laid for the start of specific and coordinated research projects.

Specific Research Projects

The framework conditions are only the basis for the feasibility of RP research. The DRPV is even more interested in directing actual research projects to meet the needs and requirements of RP patients (in terms of the above-mentioned priority list). This is, for example, done by inviting hand-picked researchers whose research work is of interest to the DRPV to congresses, seminars, and general assemblies, or by suggesting they be invited to join the scientific board.

In general, the direct influence on RP research is exerted via three types of research projects: external research projects, cooperative research projects, and internal research projects. External research by professional researchers is carried out primarily according to the researchers' own criteria (or those of the scientific advisory board) and cannot be financed by the DRPV alone; instead, public support is required. The influence of the DRPV on the research target is rather small. Cooperative research, however, gives the DRPV the opportunity to express its views on research objectives, as patient data and funds are made available to the researchers. Presently, for instance, the DRPV supports a project on lipoproteinemia. This cooperative project is based on the hypothesis that RP patients suffer from a certain defect in the metabolism of fat. If this proves to be true, therapy might be possible and feasible. The DRPV supported this project by calling upon its members (via its journal "RP Aktuell") to donate blood and by making a personal computer available for data analysis.

Another more important resource the DRPV can offer scientists in cooperative research is retinal tissue. Suitable retinal tissue which can be cultivated and tested for biochemical parameters is an important prerequisite for causal RP research. More than half of the members of the DRPV have declared their willingness to donate their eyes for research purposes after death. In cooperation with the German Ophthalmological Society, with the researchers and with patients, a countrywide communication system is presently being set up which will insure fast removal, transportation and, use of dead patients' retinal tissue.

Another example of cooperative research projects is a graduate project for visual field measurement and for the statistical evaluation of long-term data. Standardized equipment for long-term monitoring of visual field development is essential for the examination of therapy trials. Therefore the DRPV is supporting this project with funds as well as information.

Apart from external research projects and cooperative research projects the DRPV also conducts its own "internal" research work. For example, it distributes questionnaires to its members asking about the development of the disease, about training and job problems, about the use of technical aids, etc. These data are analyzed and made available to members. Also, special questionnaires serve to check and analyse the results of the therapy trials of special patient groups ("the therapy control book"). DRPV internal investigations have also been carried out in order to assess the necessity and possible contents of a technical aids catalogue.

One interesting case in which the DRPV had to intervene in the scientific community in order to achieve its goals was in the field of immunological research. For some years now, RP-related basic research has been carried out in this field, and serious therapy trials (from many patients' point of view) are being conducted in Spain and Argentina. (It should be added that the examination of therapy effects for RP patients is very difficult and takes several years due to the slow progression of the disease). The doctors (and researchers) applying this therapy are not accepted by the established scientific community, but were invited by the DRPV to its annual general assembly and to an international RP congress.

However, in spite of the patient organization's cooperative achievements with the scientific community, it has failed to initiate research outside the state-supported science system. So far, for example, the pharmaceutical industry has not been willing, except for some small donations, to make any important contributions to RP research. Their main argument is that the small patient market does not allow them to recoup the high costs of research efforts which are of unpredictable duration and outcome. From the patient's point of view investment in research on the prevention of blindness appears too risky for the German pharmaceutical industry.

This industry, however, is of special importance for producing and testing pharmaceuticals deemed effective in treatment. Recently, for example, an Italian pharmaceutical company was issued a European patent for a medicine (called Bendazac) allegedly suitable for the treatment of RP. The scientific advisory board of the DRPV was not able to obtain more detailed information on the production, composition, and effects of this medicine; the manufacturer remained silent.

The patients argue that the pharmaceutical industry is already making profits on a large number of ineffective medicines, and that these profits should be channeled back into serious and systematic basic and applied research on RP by the same industry.

Comparison with Other Lay Societies

To understand the circumstances which allowed the DRPV to influence research we cannot restrict ourselves to an analysis of the cooperation between the DRPV and the medical world, but also must compare the activities of this movement with other comparable endeavors. In contrast to what one might first believe, the size of the infrastructure of a patients' self-help organization seems not to be the factor determining the intensity of its cooperation with researchers. As a comparison with other societies reveals, the determining factor is rather the existence or non-existence of a scientific society at the time of the foundation of the patient society and the degree of differentiation of the related scientific discipline.

Although there appears to be a wide variety of forms of cooperation between the patients and the relevant researchers, two basic types of cooperation can be distinguished: the "patient-dominated type" and the "researcher-dominated type". These types describe the institutional forms of cooperation rather than the degree of influence on the contents of research, although more direct influence seems to be exerted by the patient-dominated type; the latter is characterized by the integration of research interests into the patient organization, while the researcher-dominated type is characterized by the integration of patient interests into the research organizations. If, at the time of its foundation, the patient organization is confronted with an existing specialized scientific or medical society, it tends to rely on these research activities. The traditional norms of research endeavours tend in such cases to be accepted, and only at a later stage of its development may the patient organization become sufficiently self-conscious to exert its own influence. However, this can still be done only by subordinating and adapting its own claims to those prevailing among researchers or medical experts.

The German Diabetes Society, for instance, founded in 1950, today has 18,000 patient members and a large number of full-time lay activists. Although its infrastructure is much stronger and more efficient than that of the German RP Society, it has not yet succeeded in forming effective framework conditions for research promotion. Until 1981, the only connection between the Diabetes Society and the relevant medical organization (the Diabetes Association) was the formation of a scientific committee which was (and still is) equivalent to the "lay committee" of the Diabetes Association. In addition, since 1981 there has existed a joint committee consisting of selected members of the executive boards of both organizations, who meet two or three times a year to exchange information of common interest. Only in 1985, on the occasion of a congress organized jointly by the lay and the medical organizations, was a scientific foundation established with the promotion of research into diabetes as its main objective.

To this day, the German Diabetes Society has limited its own activities to providing information, training, and help to improve the quality of everyday life for diabetes patients. According to a letter from the executive director of the German Diabetes Society "The German Diabetes Society has shown great interest in ongoing research activities, but has never attempted to intensify research by itself" (1).

The paradigm of ongoing diabetes research involving looking into the symptoms and consequences of diabetes, classifying patients in terms of these symptoms, and formulating hypotheses about their causes is taken for granted. Systematic research into the causes of diabetes has hitherto been neither undertaken nor promoted. Since most of the research funds are provided by government research programmes, the need for research promotion was never explicitly formulated by the patients. The German Rheuma Society in its (analogous) relationship to the (professional) German Rheumatology Society seems to have undergone a development similar to that of the German Diabetes Society.

In contrast, the German Mucoviscidose and the Multiple Sclerosis societies have been much more successful in laying effective foundations for research into the causes of their respective chronic diseases. At the time of its foundation in 1965 the Mucoviscidose Society (with today only 2,500 members) was not confronted with any specialized research society or any structured research activities. Hence, during the 70s and 80s the society systematically built up the following framework conditions for the promotion of Mucoviscidose research:

- the establishment of an internal group for "Medicine and Research", deciding about applications for research funds;
- the establishment of a research fund (with 160,000.-- DM in assets today);
- the establishment of a scientific committee consisting of representatives of the internal working group and of selected scientists, to make final decisions about research projects;
- the establishment of a research prize for the best German Mucoviscidose researchers (a cash award of 10,000.-- DM provided every two years);

— influencing government research promotion on the federal and state level (e.g. the society is represented on a committee allocating 2.5 million marks in research funds provided by the Land of Baden-Württemberg).

The effects of these promotional activities have been extremely fruitful. Research into the causes of Mucoviscidose is today considered most promising; as in the case of Retinitis Pigmentosa, the defective gene believed to be responsible for the chronic handicap has recently been identified.

The Multiple Sclerosis Society, with its own research fund, two research prizes, and a generously equipped interdisciplinary research institute in Würzburg has also been successful, although one important reason for success in this case has certainly been the patronage of the wife of the German Federal President since 1979 (the society has more than 20,000 members out of about 80,000 affected patients).

The general lesson to be learned from this comparison seems to be that it is easier for a patient organization to formulate and achieve its research goals in an unstructured research environment, even if its resources are small, than vis-à-vis a well-established, state-supported scientific community.

Concluding Remarks

The German RP Society encountered a rather "favourable" research environment at the time of its foundation. In a relatively loose and unstructured ophthalmological community with no previous specialized RP research interests, the DRPV was able to interest researchers in its own research needs and to implement the desired institutional framework conditions.

The older generation of ophthalmologists, rather frustrated by earlier unsuccessful attempts to find a cure for RP, left the initiative to the younger generation. These younger ophthalmologists were attracted to the field not only by the new cognitive problems posed by the disease, in cooperation with other disciplines, but also by their recognition of the severe social and personal problems associated with the irreversible process of going blind. "Researchers help patients — patients help researchers" soon became the motto of the RP society, symbolizing the assertion of patients' interests in a general situation of mutual dependence. Researchers were in need of funds and information as much as patients were in need of better diagnosis and therapy.

Hence, the decisive factors of the DRPV's success in gaining influence among the German RP research community can be located, on the one hand, in the formation of the RP society itself and in its collective demand for interdisciplinary research into the causes of the disease, and on the other, in the willingness and capability of the scientific community to respond positively to these demands. Also, the intense personal involvement of some executive members of the DRPV (themselves affected patients) who can be described as "half-professionals" (i.e., non-medical academics like the research referee and the chairman) have contributed to this success story. On the basis of this experience, it seems essential that non-professionals should strive to "invade" the professional community, and that the professionals should be willing to accept, at least partially, the objectives and values of non-professionals.

Note

1. Letter to the author dated Dec. 12, 1985.

THE KNOWLEDGE INTERESTS OF THE ENVIRONMENTAL MOVEMENT AND ITS POTENTIAL FOR INFLUENCING THE DEVELOPMENT OF SCIENCE

JACQUELINE CRAMER, RON EYERMAN and ANDREW JAMISON

1. Introduction

Analysis of the link between public interest and the production of scientific knowledge has received increasing attention over the past decades. There are a wide range of studies which document how different social groups have influenced the development of science and, vice versa, how society at large has been affected by the development of science. Concern with this relationship has come both from science policy analysts and from researchers in the sociology of science. That science and technology can possibly have negative effects on society is a relatively new idea which is perhaps best symbolized by the destructive potential of the atom bomb and of nuclear weapons in general. Such negative effects of scientific knowledge production have made more people aware of the necessity of aligning scientific and technical developments with general social needs. However, the idea of broadening discussions about the specific goals of knowledge production and attempting to influence the development of scientific knowledge through social and political means has not always been openly embraced by professional scientists. Scientists, often motivated by a fear of losing their professional autonomy, frequently try to shield their research from "outside" influence.

One exception to this general rule of professionalization occurs among environmental researchers, a group of scientists who, at least on first sight, appear more sympathetic to a closer relation between science and the solving of social problems. Banding together with environmental activists, they have called for the development of a new kind of science which can contribute to the solving of environmental problems.

S. Blume, J. Bunders, L. Leydesdorff and R. Whitley (eds.), The Social Direction of the Public Sciences. Sociology of the Sciences Yearbook, Vol. XI, 1987, 89–115.

They have promulgated "ecology" as the science of the future, which, more than any other scientific discipline, could unravel the intricate relationships between all the organisms of the universe. These ideas, put forward during the course of the late 1960s and early 1970s, were in accord with those expressed in the then-emerging environmental movement. Here a common basis was forged between the holistic ideas of the new science, the new ecology, and the worldview of the environmental movement, which stressed a more harmonious and integrated relationship between man and nature, and expert and layman, than could be found in Western industrial societies in general.

Advocates of a new type of science, then, could be found among professionals and non-professionals alike. Among natural scientists the name Eugene Odum stands out as one example, while Theodore Roszak can be named as an example of an extra-scientific proponent (1). Significant for the development of this new type of science was the seemingly close cooperation between practising scientists and non-professional activists. In light of this cooperation, one might expect that the new knowledge interests advanced within the environmental movement during the 1960s and early 1970s would have been able to affect the development of established science. The central question of this article is whether and how such effects could have occurred.

The article is divided into several sections. Our first concern is with the notion of "knowledge interests" and in showing why this notion is particularly useful in exploring our case. We then turn to the issue of how the specific knowledge interests of the environmental movement have been articulated and developed since the 1960s. Here we focus on the relationship between practising scientists and environmental activists, and on the ways in which shifts in this relationship have influenced the production of scientific knowledge. We conclude with a discussion which assesses the potential impact of the environmental movement on the development of science.

2. The Knowledge Interests of the Environmental Movement

The concept of "knowledge interest" has a long history of usage in the sociology of knowledge (2). We find this concept particularly useful in analyzing developments in the environmental movement and its potential for influencing the production of scientific knowledge, because it provides a way of clearly identifying the cognitive basis of this "movement", i.e., a core set of ideas and strategies concerning the production and use of knowledge. Once these particular knowledge interests have been identified, it becomes possible to make judgements about the relative potency of the "movement", including in this case the theme of the present volume: the potential of "outside" social forces to affect the development of scientific knowledge. First we will identify the specific knowledge interests of the environmental movement, then outline the various phases of their historical development and change, including their effects upon established knowledge production.

In order to unravel the various features of the knowledge interests of the environmental movement we will distinguish three dimensions: cosmology, technology, and knowledge production. Here, we will describe in schematic fashion the thematic contents of each dimension and elaborate on the relations between them.

The term *cosmology* refers to the ordering of reality, understanding how and why reality is structured as it is. By cosmology we mean the more or less taken for granted frameworks through which human beings impose meaning on the world and through which their social world is constructed. We understand human action to take place through culturally defined worldviews, through systems of meaning and interpretation which must be produced and reproduced through the very practices for which they provide the frameworks of meaning. Although (by definition) "taken for granted" and pre-theoretical, cosmologies or at least aspects of them can be problematized, made the objects of conscious reflection, discarded and adopted.

Cosmologies blend together normative and cognitive judgements. One's view of reality or of the relation between nature and culture (for example, how one defines what "nature" or the "natural" means and comprises) is combined in complex ways with values about why it is good or better to define reality in a given way. To take an example from S. Cotgrove's study of the values and beliefs of the environmental movement in England: the pro-nuclear lobby in England not only believes that nuclear power will generate economic benefits, but also that this is desirable (3).

Part of the knowledge interests of the environmental movement with respect to cosmology involves making problematic the dominant "modernist" worldview it opposes. On this point the environmental movement and the sociology of knowledge share common ground, for each in its own way seeks to make conscious what is taken for granted and, in the Habermasian sense of "emancipatory cognitive interest", to liberate human knowledge from distorting, ill-considered assumptions (4). As is the case with most alternative cosmologies, the ecological worldview formed itself partly as a critique of that which it opposed. Its own (positive) alternative was thus formulated at first in negative terms: in a dismissal of the traditional or Enlightenment view of nature, which regarded nature as an object to be controlled and dominated. The alternative called for a more communicative attitude to nature, which was to be reflected in a new understanding of the conservation and management of nature and in a new approach to science. This latter expressed itself in terms of a more qualitative and holistic view of knowledge and knowledge production.

The roots of this alternative cosmological outlook can be traced through the history of an older conservationist movement which goes back as far as the late 18th century (5). However, in contrast to traditional conservationism, the current environmentalist movement incorporates a much broader and scientifically more sophisticated notion of the relationship between humanity and environment (6). Traditional conservationist views were often bound up with romantic notions of a "sacred" nature and with apocalyptic visions of the gradual degeneration of nature and culture, sometimes tempered by utilitarian ideas concerning their "proper" use and protection. While traces of such notions can be found in the movement today, modern environmentalism has developed in dialogue with modern scientific ideas about "ecology" and the balance of nature, rather than the older organic views which underpinned much of conservationist thought (7). This qualitative shift in thinking was facilitated by such post-war developments as the introduction of general systems theory and more advanced methods of investigating cause-and-effect relationships (8).

This alternative cosmology of the current environmentalist movement has implications for the other dimensions of the new knowledge interests: technology and the production of knowledge. If the cosmological dimension of the new knowledge interests involves a definition of what is "natural" and defines the limits of the mediations between man and nature, the technological interest defines the means of that intervention, the scope and method, the proper means through which human beings and nature interact; while the third dimension, knowledge production, addresses itself to how that intervention should be organized: the social relations of knowledge production and its distribution. The three dimensions of the knowledge interests of the environmental worldview are thus internally related, forming a more or less coherent, if not entirely systematized, whole.

The broader framework of the ecological cosmology defines the limits within which the interaction between man and nature takes place; and the idea of "balance" is a central one in the constitution of this cosmology. The notion of balance also helps to define the proper means of the intervention. Balance easily translates into the notion of "small-scale", which has formed one of the main themes of the environmental movement regarding *technology.* Another key notion in this connection is "appropriate", i.e., that the interventions between the human and the natural worlds must not disrupt the inherent balance in nature. In order to ensure this, appropriate means must be devised; since by definition all intervention is disruptive, care must be taken to restrict interventions, and perhaps "small-scale" is the best way of ensuring this.

Conceptualising the "appropriate" means of intervention involves at least two further considerations: simplicity and beauty. In addition to being "small-scale" and thus "proper" in reference to the natural balance, interventions should be simple, not complex, and aesthetically pleasing. Neither unnecessary complexity nor ugliness is considered appropriate to a natural setting according to the cosmological consideration of the ecological worldview. Technology, the means of intervening and of mediating the "natural needs" of human beings, must take such considerations into account.

The third dimension of the new knowledge interests of the environmental movement concerns the way *knowledge production* itself is organized and distributed. This dimension of knowledge interests calls attention to social and societal relations and answers the question: "How should the (necessary) production of knowledge be organized?" The answer given by the environmental movement can be summed up in the slogan "science for the people". Key considerations here are with "participation" and "decentralization", and with a dissolving of the division of labor between those who possess knowledge/information and those who do not. These values in part follow from the "appropriateness" considerations discussed above. They apply not only to technology and the means of intervention but also to the way the movement should organize its own knowledge production.

In what follows we will analyze how these dimensions of the new knowledge interests of the environmental movement have gradually evolved over the last two decades, and how shifts in these knowledge interests have led to a changing influence on the development of science. We would stress, however, that from the point of view of the environmental activist these knowledge interests are at most usually only vaguely articulated and understood; especially from the point of view of a social scientist observer, they are not to be seen as clearly defined lines or strict guiding rules for action which can be easily and unambiguously applied in all situations. They are rather vague orienting values which are open to articulation and interpretation and often serve as a source of conflict rather than a universally accepted framework for its resolution. We hypothesize, however, that such conflicts are most likely to emerge explicitly when the movement is forced by external circumstances to put forward clearly defined policies and proposals in response to a challenge by its opponents, either in the state or in private industry. In such situations, implicit differences in emphasis in the strategic use of knowledge become explicit. At the same time, such situations open the possibility for what is usually referred to as resource mobilization: choices which confront a social movement regarding the use of its material and ideological resources (9).

3. The Historical Shifts in Knowledge Interests

Three major phases can be distinguished in the articulation of the specific knowledge interests of the environmental movement. Although the exact timing of these phases varies from country to country, the rhythm of the shifts between phases is remarkably similar throughout the industrialized world (10). Partly this can be explained by a range of international events like the United Nations Conference on the Environment in Stockholm in 1972, or by particular events of international impact like the accident at the Three Mile Island nuclear plant in Harrisburg, Pennsylvania, USA in 1979. These events, external to the movement, triggered an activity that varied in degree and style, but was similar in kind. In any case, these events and others like them contributed to political discussions within national environmental movements that led to shifts in the way common knowledge interests were articulated. It is just those shifts that we will now try to identify and to link with the nature and organization of knowledge production in science, particularly in ecology and related areas. These linkages are particularly important and striking because scientists, both in and outside the movement, have been crucial in this process of shifting knowledge interests. In this connection, the changing relationship between practising scientists and activists within the environmental movement is of central importance.

3.1. *Phase One: The Rise of Environmental Consciousness, 1962–1968*

Most observers of the environmental movement identify this first phase as beginning in the early 1960s. The "initial conditions" of the gradual emergence of the popular environmentalism that marks this phase can be traced to the structural changes which occurred in industrial nations after World War Two, and which are usually characterized as the emergence of the "welfare state" and the service-centered economy (11). Such changes helped to broaden the potential base of support for what had previously been a rather limited conservation movement; they also tended to increase the receptivity of the "general public" to the environmentalist position as it began to be enunciated in books by Aldo Leopold and in particular by Rachel Carson (12).

This first phase can perhaps most adequately be characterized as one of public and self-education in which a new environmental awareness gradually took form in the minds of scientists. The vehicles for this were primarily works of popular science, written for the most part by biologists active in the field. This literature, of which Rachel Carson's *Silent Spring* (1962) is the most important example, was at first vague in its knowledge interests. A "counter-science based on a biology of whole systems and a respect for nature's balance", as R. C. Mitchell has expressed it, was only implicit in this work (13), perhaps consciously so. A year later, Barry Commoner wrote a now famous book against nuclear weapons which contained much the same type of argument: a critique based upon the alleged malfunctioning of "unnatural" technology which had, in one way or another, gone too far (14). In both works the idea of setting specific limits to human intervention in nature was expressed, and yet not stated explicitly in the form of a principle or as a new environmental ethic. Rather, this was the assumption, more or less taken for granted, that stood behind and motivated these warnings.

There was, of course, more to the rise of environmental consciousness than the writings of biologist popularizers. Also characteristic of this first phase was a growing awareness of environmental issues in the media, and the discovery of nature and the environment as a subject for public debate. Carson's book is significant in this respect also, for it signaled, inspired, and even fueled this debate. The discussion of her book, and of similar books written in various national contexts, as well as journalistic reports from biologists in the field, proved to be as important as the book itself. The environment, that arena of human action previously taken for granted, was being problematized and made into a topic of social concern and potential conflict. If there was no clear solution to the problems being raised and discussed at this stage, and as yet no new political response, there was at least a growing awareness of the dangers inherent in the dominant technological system, and the articulation of some of the negative aspects of this "system".

It was also in this phase that the criteria for environmentalist technics — what we have called technological knowledge interest — began to be formulated. Murray Bookchin was among the early advocates of an environmental technology. In his book *Our Synthetic Environment* (1962), written under the pen-name of Lewis Herber, Bookchin pointed to some of the criteria that would later be developed in a more political vein (15). The issue of scale was already central to this early work, and was placed in opposition to artificiality, to synthetic substances which were said to have replaced "natural" substances. For biologists like Carson and Commoner and for polemicists like Bookchin, chemicals in the environment were the prime enemy. The environment they articulated was not merely the untouched, pristine nature of the old conservation movement; it was also the suburban environment, the "nature" of the weekend fishing trip. This was a significant development, for in addition to problematizing an area of life previously taken for granted, it also expanded the idea of "nature" into areas which concerned a larger popular audience. For Bookchin, and soon for others active in the labor movement, the notion of environment and environmental protection also included the workplace and the home, where the use of chemicals — that miracle of modern life — were for the first time placed in a negative light.

Thus, as a result of the publication of alarming books by concerned scientists, a growing environmental consciousness can be noted, but it is too early to speak of a new social movement (16). During the 1950s and early 1960s environmentalism represented primarily the largely non-political concerns of professional scientists, and was understood as a problem aggravated by a lack of information on the part of the general public, which had allowed the situation to deteriorate to such an alarming point. The aim, then, was to educate the public at large in order to alert them to the dangers connected with the uncontrolled use of environmental resources. In addition, these books also sent a message to the larger scientific community about the necessity of more specific knowledge about the possibilities and limits of the natural environment. Such arguments and claims about the lack of sufficient knowledge made possible developments in scientific research connected to a new understanding of the human uses of nature.

Up until right after the second world war, environmental research had been a more or less neglected area in the natural sciences, which more and more turned their attention and resources to the new growth disciplines of chemistry and physics and to the connection between scientific knowledge and technological development. It was not until the early 1960s that a noticeable shift towards environmental research occurred, in both national and international contexts. In this connection, committees to design environmental research programs were organized in a number of countries (17). The United Nations also began to act, establishing environmental offices and programs in several of its major affiliates: the Food and Agriculture Organization (FAO), the United Nations Educational, Scientific and Cultural Organization (UNESCO), and the World Health Organization (WHO) (18). In addition to these more applied, problem-oriented international research programs, initiatives were also taken to develop programs concerned with basic environmental research. One such attempt was the International Biological Program (IBP) formally launched in 1964 at the initiative of professional biologists and financially supported by national governments (19).

The initiation of such programs came just at the time when drastic changes were occurring in the environmental sciences, particularly in ecology. Aided by the introduction of new measuring instruments (e.g., radio isotopes), ecology gradually evolved in the early 1960s from a descriptive to an experimental science (20). Inspired by (among others) the work of Lindeman and the brothers E. and H. Odum (21), studies of ecosystems were received with growing interest on both sides of the Atlantic, in Japan and Australia, and in several socialist countries. The concept of "ecosystem" was itself coined in 1935 by the plant ecologist Tansley to mark the complex interchange within a system between organisms and their physical environment (22). After World War Two, Tansley's concept achieved unexpected popularity and was adopted for use by many ecologists, especially in the period of growing environmentalist consciousness of the 1960s and 70s. The "ecosystem" approach was then seen not only as a promising way of developing ecological theory, but also as useful and immediately applicable in environmental management (23).

Although in this early phase the environmental cause had already begun to win acceptance and influence in national and international governing bodies as well as in the scientific community itself, its specific influence on the production of scientific knowledge became clearly manifest only later, in our next phase. In sum, in phase one, themes in the cosmological and technological dimension are articulated in very general terms, giving the appearance of unified knowledge interests. On the dimension of knowledge production, the specific knowledge interests of the budding environmental movement are only just beginning to be spelled out in more practical terms.

3.2. *Phase Two: The Transition from a Scientific to a Social Movement, 1969—1973*

The period 1969—1973 can undoubtedly be regarded as the heyday of environmental concern in most industrially developed countries. Indicators such as the degree of coverage in the media, journals and literature, and survey data of public opinion polls clearly show a peak of interest in environmental issues in society at large (24). New environmental organizations emerged simultaneously, often initiated by younger biologists, environmental scientists and landscape architects. These organizations consciously distanced themselves from older conservationist societies; their main activity consisted in disseminating the kind of environmentalist information/popular science that Rachel Carson and others had been offering in an individual way.

They were, however, more activist in orientation than the old conservation societies, often organizing media campaigns and even demonstrations against particular environmental problems; in the United States, Earth Day in 1970 led to the mock burials of automobiles on a number of American college campuses. In Norway, new environmentalist groups mounted a protest march to a planned hydroelectric power site. There was thus an echo of environmental movement-building, of the demonstrations against the Vietnam War; and in some countries the influence of the so-called New Left on ecology groups was already present in this phase. The new environmental protest groups emphasized participatory strategies to establish a social constituency for environmental concerns (25). Most of these efforts were devoted to changing basic societal values and norms. This "educative" strategy can be contrasted with later attempts to develop more directly political strategies. What was most characteristic of this second phase, however, was the explicit elucidation of an ecological worldview, for this was, more than anything else, the age of ecology. It lasted only briefly, and was as much a culmination of the first phase of environmental consciousness-raising as the initiation of a new kind of social movement. In a sense, it was the period of transition from a scientific movement to a social movement; the main contribution of this period was in making explicit the alternative knowledge interests of the new environmental protest groups.

These knowledge interests were still largely framed within the idiom of professional science; that is, the worldview was primarily formulated on the cosmological level, as an alternative representation of nature. This alternative definition of the natural was opposed to the cosmologies that were purported to stand behind the polluting and wasteful and large-scale technologies of the post-war industrial epoch. In this phase, ecologist Barry Commoner's book *The Closing Circle*, published in 1972, combined perhaps more than any other book a cosmological perspective emanating from ecosystems ecology with a small-scale technical interest and a participatory organizational interest. Commoner could make use of research results derived from one of the new environmental groups, the Citizen's Committee for Public Information, to establish a material basis for the new ecological consciousness. He could show that the cause of the environmental crisis, as it began to be called at this time, was in the particular type of technology that had come to dominate the post-war epoch; the synthetic chemicals and plastics that were the main polluters had also been the industrial products that showed the fastest growth in their rates of profit.

But Commoner was not alone in trying to formulate a worldview for the new environmental consciousness; in a sense, we can speak of this phase as a period of paradigm debate. Commoner's more "radical" explanatory scheme was opposed by Paul Ehrlich's population explosion model/ideas and by the Club of Rome's one-dimensional victimization of exponential growth curves (26). They all shared a reliance on ecosystems ecology and a holistic view of nature. Howard Odum took this "new ecology" view the furthest in his book *Environment, Power and Society* (1971), where he went so far as to prove the existence of God by means of energy-flow diagrams (27). Even the populist *Blueprint for Survival* (1972) from England tried to prove the need for a change to a small-scale, decentralized society by means of arguments taken from ecosystems ecology (28).

The period of transition from about 1969 to 1973 marked a temporary convergence of interest between practising scientists and environmental activists. Ecosystems ecology, in its various shapes and sizes, seemed to fit the needs of both; its proponents saw it as a framework that could be used to oppose the reductionism of modern science and technology, i.e., a cosmology that seemed to represent a holistic alternative. They also found here the basis for an alternative set of technical criteria, derived from cybernetics and systems theory, that could be imposed on nature as a new system of environmental engineering regulating the flows of energy and nutrients through any and all ecosystems by means of intricate feedback mechanisms (29). Finally, in calling for interdisciplinarity and team research, the ecosystems ecologists could converge with the new environmental activists in challenging the social relations of knowledge production, although this convergence, perhaps more than the others, would soon show itself to be little more than an illusion.

Even though much of this convergence in knowledge interests between practising scientists and environmental activists would dissolve in the ensuing years, the claims that were made for ecosystems ecology in the years 1968—73 provided a sense of identity for environmentalists (30). Looked at from the other direction, professional ecologists (backed up by the environmental movement) could also legitimate the importance of ecosystems research, thereby acquiring some extra financial and ideological support for developing this new research. The result of this dialectic between science and movement was a substantial increase in the volume of ecosystems research. This new research was not, as previously, restricted to a few research units, but now spread over a great number of academic and (semi-)governmental research institutions (31).

During this period, both state and scientific establishments were receptive not merely to implementing ecosystems research, but also to the growth of environmental science in general. Primarily initiated by scientists concerned with environmental issues, new educational and research programs were established, such as human ecology, environmental science, and landscape ecology.

It was in this second phase as well that the criteria for an alternative or appropriate ecological technology began to be formulated more explicitly. In the first phase, alternative technology had been little more than a slogan; but already by the late 1960s, groups and individuals in both the industrialized and the developing countries of the third world had begun to work with concrete experiments in building technologies based on "ecological" criteria. This alternative technology movement, particularly in the United States and Great Britain, was in many respects part of a broader "back to the land" movement, and represented an attempt to develop a personal kind of political activism.

The alternative technological interest of the environmental movement was defined in this second phase in works by E. F. Schumacher among others, as well as in the pages of journals such as the English *Undercurrents* and the American *Mother Earth News* (32). Many of the early efforts to create alternative technology centers — such as the New Alchemy Institute in the United States or Biotechnic Research and Development (BRAD) in Great Britain — were explicitly ecological; and the New Alchemy Institute, for example, remains a working experiment in ecological technology to this day. Specific research in alternative techniques — particularly in biological or biodynamic methods of agriculture and renewable energy — was in this phase still a result of the joint efforts of practising scientists and environmental activists, but it would soon be taken over by more established research institutions. This process was particularly apparent in the 1970s. Those which survived this takeover and remained "alternative" centers of research came in the same period to be more and more dependent upon the financial support of industry or private research foundations.

In this phase, however, one could still observe a close partnership between practising scientists and environmental activists. At this stage their joint efforts were primarily geared towards informing the general public through popularizing scientific findings; their common knowledge interests were here articulated on the very general level of values and ideas, which made possible the relatively easy formation of coalitions around a seemingly integrated set of knowledge interests. The actual scientific theories and ideas on which their activities were based, however, were for the most part those developed by the professional environmentalists, the "experts". The environmental movement concerned itself mainly with spreading knowledge rather than with producing it. Those professional environmentalists who acted as allies of the environmental movement participated in public debate and decision-making mainly as counter-experts. Despite this division of labor, the practising scientists and the environmental activists together formed a united front. Consequently, their mutual influence was relatively great, especially as the potential for divergence in knowledge interests had not yet emerged. This fragmentation in knowledge interests and the gradual split between professional and political environmentalism would, however, become much more apparent and acute in the next phase of this history.

3.3. *Phase Three: The Fragmentation of Knowledge Interests, 1974—1986*

The tensions that were building up internally in the environmental movement in the early 1970s appear, in retrospect, the growing pains of a new social movement. Most of the environmental groups, and alternative technology groups as well, had been started by public figures who had gained prominence in the first phase of environmental education, or by "concerned scientists" and/or science students for whom environmental activism was part of a professional vocational activity. But in the late 1960s and early 1970s there were a number of other social forces at work, which for the sake of convenience we can lump together as the student movement, that gradually forced a more explicit political standpoint on the environmental groups.

There were, of course, other external forces that were also pushing the environmental issue into a somewhat more central political position, and thus attracting to the environmental cause a large number of people motivated more by political ideology than by scientific ecology. In 1973, the oil-producing countries in OPEC dramatically raised the price of oil by 300 to 400 percent, thus making energy supply an issue of central importance in every country in the world. The United Nations Conference on the Environment in Stockholm in 1972 also had the effect of bringing environmental issues to broader public attention.

The environment by 1974 had thus become a political problem; and it became necessary for the environmental groups to take explicit political stands on a variety of issues that had previously seemed marginal at best to the concerns of the "movement". Internal political conflicts that were implicitly present in the loosely formed coalition of environmentalists in earlier phases were now stirred up by the specific demands of pragmatic politics and participation in the established political culture (as well as by attempts to redefine the limits of that political culture), by the necessity of finding acceptable means of accommodation with established institutions, and by the need to make knowledgeable judgements on both political and technical issues.

In some countries, this process of politicization led to splits in environmental organizations and to the formation of new organizations with more clearly defined ideological and political positions. In other countries, the political "debate" was carried out within the original organizations themselves, as was the case in Holland and Denmark, for example, where the effect was a change in orientation and, perhaps more importantly, a change in membership motivation. Environmental activists turned increasingly during this phase towards coercive and power strategies (the attempt to influence and coerce behavior) and away from the participation strategies (the attempt to change attitudes and induce voluntary compliance) that had been dominant in earlier phases (33). These developments had very significant and unintended consequences for the environmental movement.

As the movement gradually shifted strategies from participation to coercion and power, many scientists who were previously at the vanguard of the movement began to feel less committed. The effect was to place more politically motivated activists at the forefront of the movement and to create a schism between practising scientists and environmental activists. In addition to this shift in political strategy there are also a number of other reasons for the gradual withdrawal of professionals from the environmental movement.

In its initial phases, many professional scientists had reacted with enthusiasm to the cause of environmentalism and the solution of environmental problems. They threw themselves actively into the political arena and acted willingly as counter-experts. As D. Nelkin notes in her study of American ecologists: "After a brief burst of public activity, however, many of the scientists went back to the lab; 'pure scientists' once again, their traditional attitude towards politics as 'an alien element, essentially destructive of scientific endeavor', (was) reinforced" (34). Scientists began to feel torn between traditional scholarship and the demands placed upon their time and energy by participation in political debates (35). As a result, many found it quite easy, especially when their engagement now also demanded confrontation with the non-professionals in the movement, to return to the quietness of the laboratory and its more established traditions and practices.

A second important factor occurring at this time and affecting these developments was the creation of environmental protection agencies in various countries and the enactment of protective legislation. This was partly inspired by the United Nations conference, and partly by the new general awareness and concern with the environment. During the course of the 1970s, this new legislation was put into practice in most industrialized countries. The new agencies and the implementation of new legislation required new cadres of environmental experts; many of these experts were recruited from the "concerned" students and scientists affiliated with the environmental movement. Experts formerly or potentially associated with the movement were now being incorporated into state bureaucracies, and thus developing new sets of loyalties, or at least potentially doing so. The following remark of a Dutch fresh-water ecologist nicely illustrates this point: "When you work at a provincial agency of the Water Works, you are a person of the Water Works" (36). Many governmentally employed ecologists seem to feel that working for the government means distancing oneself from involvement in environmental groups and actions.

Thirdly, beginning in the 1970s but more clearly manifest in the 1980s, one can find a decrease in the size of research budgets on environmental problems at universities and other research institutions. Encouraged in part by an economic crisis, governments began to tighten budgets generally and to develop new science policy measures and regulations, including the stimulation of innovative research in micro-electronics and biotechnology, which were to have further consequences for the environmental movement. Particularly in academic institutions scientists were expected to pay increasing attention to their scientific output and be less distracted by "outside" concerns. This changing intellectual climate which made it more difficult to combine "career" and "politics", has, however, been tempered in some national settings; Holland is one example, through a policy aimed at crediting academic scientists for "socially relevant" as well as more internationally accredited research.

A last factor affecting the links between the professional scientist and the environmental movement has been the disillusion with the achievements of cybernetic ecosystems ecology itself (37). This new outlook in the biological sciences, which served as a binding element holding together all streams of environmentalists, gradually began to fall apart. Ecologists have become sceptical about the merits of the system-dynamic modelling approach which in the late 1960s had been the "trend" popular among ecologists and scientists in general. Within the scientific community at large this approach began to raise questions and doubts of its value as a useful research program (38). Thus, one of the bases of the coalition between the practising scientists and the environmental activists faded away.

In short, changes in institutional arrangements as well as in attitudes and developments at the cognitive level widened the splits in the environmental movement between practising scientists and environmental activists. Further, the shift in movement strategy from participatory to a more policy-oriented approach led to a divergence of underlying knowledge interests: pragmatic political practice often forced environmentalists to limit themselves to the fight against pollution and thus to restrict debate to very concrete issues. This had the effect of reinforcing a more technocratic orientation as opposed to the original "deep ecology movement" ideas referred to by Arne Naess (39). What this meant essentially was that ecological debate tended to become centered around "formal knowledge" rather than the wider claims of an alternative cosmology. One example of this more pragmatic way of thinking is the debate on nuclear energy.

Oil price increases helped to bring energy questions to the centre of the political stage. Energy issues, and in particular the issue of nuclear energy, grew in importance almost overnight. Of course, these issues had already been addressed within the environmental movement; and nuclear energy had already accumulated a critical literature (40). But it was the oil price rise that somehow brought these questions out of the closet as governments suddenly began planning to increase their commitment to nuclear power, or at least increase the pace of construction, and the many unsolved problems with nuclear energy began to come to public attention. By 1975 most countries had new organizations in opposition to nuclear energy, and the literature on the subject had increased enormously. Nuclear energy became the issue upon which the movement was built; it came to symbolize the enemy, but in so doing, it also sewed the seeds of fragmentation.

The fragility of the movement against nuclear energy was due to its specificity: the claims of an alternative knowledge interest that had inspired the movement in its formative years in the early 1970s became too concrete, too specific. The slogan "Nuclear Energy, No Thanks" was absolute enough to appeal to activists of all stripes; but it was too negative to form the basis of a social movement with a new worldview. The movement developed, we might say, in spite of itself, in spite of nuclear energy; but the result was a fragmentation of concerns, a fragmentation of organization, and a diffusion of the original knowledge claims.

The value of nuclear energy as an issue was that it forced the movement to make itself believable: it forced the movement to be practical. The arguments against nuclear energy, particularly when referenda were announced in a number of countries, had to be made comprehensible to the entire population; the ecological worldview had to be broadened and extended to relate to the experience of the entire society. Here the movement could utilize its legacy of ideas; at times it could emphasize the safety aspects, at times the artificial aspects, and at other times the economic aspects. Some groups opposed nuclear power because it symbolized centralized power, while for others it symbolized complexity, for others danger, for others masculinity, and for still others it represented bad economics, i.e., it was too expensive.

The result of this was that some environmental groups stressed the importance of developing alternative technologies, such as wind and solar energy, and attempted to put such alternatives to practical use. Others stressed instead the safety risks involved in nuclear energy and began to specialize as counter-experts, while still others sought to develop radical political alternatives more directly, such as "blowing up the system". These differences in emphasis concerning the knowledge interests in the environmental movement can be found in other areas as well, for instance in food production, industrial waste, etc. The divergence in political strategy which began in phase two became institutionalized in phase three. The "movement" now was split into various organizations with different tactics implying different knowledge interests. Therefore we can conclude that the specific demands of pragmatic politics not only caused a fragmentation in knowledge interests between practising scientists and environmental activists, but among the second group as well.

Before turning to our concluding discussion about the capability of the environmental movement to influence the production of scientific knowledge, a final word about the organization of knowledge production by the movement itself seems appropriate. The shift we noted, from an original set of knowledge interests articulated in very general abstract terms to concrete research projects connected to practical political issues, affected movement interests about the production and distribution of knowledge. While knowledge traditions on the cosmological level could be taken from various sources and were relatively open to general interpretation and debate, practical applied research had to be left to the professional scientist, to the expert rather than the layman. The movement thus became increasingly dependent upon the knowledge produced by such experts. The mobilization of expert knowledge for the environmental movement has varied from country to country. In the Netherlands, and now in Denmark and West Germany, the creation of "science shops", which seek to mediate between the knowledge requirements of social movements and established academic research institutions, help to ease this dependence on the commitment of individual professionals (41). Yet, as revealed by some of the other papers in this volume, problems remain. The call for a "science for the people" made in the early phases of the environmental movement has lost much of its appeal in the atmosphere of professionalism which dominates today. The problems this creates for the movement are discussed in the article by J. Bunders contained here (42). Yet the story is far from over, and the problem of mediating between the environmental movement and the production of knowledge may be solved in a number of innovative ways, of which the science shops are only one example.

4. Discussion

In this article we have developed the notion of knowledge interests in order to identify the elements which united the environmental movement and gave it a potential for influencing in significant ways the development of established science. We have argued that the unifying force provided by a common set of knowledge interests has become fragmented. Further, we located at least one source of this fragmentation in the practical imperatives of contemporary political action and another source in the gradual split between practising scientists and environmental activists.

What can we conclude from this in relation to our central question concerning the ability of the environmental movement to affect the development of science? Does the fragmentation we have identified destroy this capacity? We think not. Even in this fragmented condition, environmental organizations remain a significant force in the shaping of public policy and in influencing public consciousness. At the level of practical knowledge production, the environmental movement also remains an important source of inspiration for environmental scientists who have been responsible for the development of research programmes. (43).

In a less direct way, these scientists continue to have an impact on general environmental consciousness, one which has stimulated national governments and international organizations to support research programmes in the area. No government or political party in the industrialized countries would today deny the importance of such research. This is quite a step forward from the unlimited growth mentality of the 1950s and early 60s.

We can perhaps best present our conclusions about the influence of the environmental movement on the development of science by taking each aspect of knowledge interest separately.

Cosmology: The original environmentalist worldview centered around the establisment of a more communicative relation between humanity and nature, both in scientific research and in environmental management. This new relation was meant as an alternative to the dominant instrumentalist mentality of science, and resulted in a more holistic approach in the form of ecosystems ecology. Remnants of this new approach remain and are not likely soon to disappear from the present cosmology of science. However, its potential to transform totally the future development of science is no longer as obvious to its advocates as it once was. Although ecosystems ecology still presents a more complex notion of controlling nature than many other sciences, its technological attitude toward nature remains *control*, a position it shares, rather than opposes, with the dominant scientific cosmology (44). In the course of the 1970s and 1980s the common ground provided by the shared cosmology of ecosystems ecology began to disintegrate. Particularly in the scientific community itself, doubts began to emerge about the cybernetic nature of ecosystems theory, and therefore about the very fundamentals of the "new" ecology. This implies, to us at least, that the original and alternative ideas of cybernetic ecosystems ecology are slowly fading in the field of science itself. This development, however, may have unexpected results, in that its replacement, a more evolutionary approach, may acquire the same "alternative" flavor as ecosystems theory attained in the early 1970s (45).

Whether this new evolutionary worldview, as expressed by some professional ecologists, will lead to a reunification of knowledge interests between practising scientists and environmental activists at a cosmological level is an open question. Whether such a worldview will become a new force for influencing the direction of knowledge production also remains to be seen. The withdrawal of a great number of professional ecologists from the environmental movement, in combination with a more practical attitude towards environmental problems, means that the alliance between these groups is no longer self-evident. One of the few exceptions in regard to the pragmatic orientation of the environmental movement is F. Capra, who in *The Turning Point* (*1983*) promulgates a new holistic paradigm containing both science and spirit (46).

The idea of a more communicative approach to nature, which used to be one of the central themes of the new environmental movement, seems less clearly articulated than ever before. What exactly would be implied by a return to a notion of "sacred" nature, or a more sophisticated conservationist management linking science and technology in a new and better way, has not been settled. The current fragmentation seems more likely to increase dispute in this connection rather than settle it.

Technology: The alternative technology movement has had a visible impact upon the development of science; established science has taken over many of the ideas originally produced by this movement, and the movement has also stimulated the development of new technologies. These developments first occurred through the joint efforts of practising scientists and environmental activists connected with the environmental movement; today they have become a part of institutionalized and established research programs, which have themselves the flavor of being "alternative". Most industrial countries have developed programs which encourage and employ alternative technologies as one way of absorbing innovative alternatives into the dominant streams. Thus, one can conclude that the ideas of the alternative technology movement have had a clear impact on the science system. But, one can ask, what is left of its potential influence as a "real" alternative? Influencing the development of science by being incorporated into its main stream seems hardly the kind of alternative the environmental movement had in mind in its earlier days. Although internal differences of opinion in and about the alternative technology movement were already visible in the 1970s, an attitude critical of modern technology was crucial for the environmental movement as a whole.

In the early 1970s one could detect a general tendency in the environmental movement towards the outright rejection of "modern" technology as oppressive, overly centralized, and so on. Since the emergence of new technologies such as micro-electronics and biotechnology, attitudes towards technology among activists in the movement have been changing. Now one openly hears talk about the possible positive effects of such new technologies; that, for example, microelectronics may lead to reduced energy consumption, or that biotechnology may lead to the neutralization of some poisonous chemical compounds, and so on. Such changing attitudes may themselves result in a new kind of "alternative" conception of technology within the environmental movement; but at present more confusion than clarity prevails. This lack of clarity, we feel, weakens the movement's potential for influencing the development of science in a critical way.

Organization of Knowledge: The idea of a "science for the people", so central to the knowledge interests of the budding environmental movement, has today become institutionalized in science shops, free health clinics, and the like. However, in spite of such innovative developments, one can observe a tendency away from the democratization of science towards the demand for practical research, carried out, as we mentioned, by scientists in the movement's name. This shift from a demand for the democratization of science towards a demand for practical research is related to the shifting strategy of the environmental movement from participatory to power politics, and to the increasingly less sceptical view of experts to be found in society in general.

It is, however, especially on the level of the organization of knowledge that the potential influence of the environmental movement on the development of science is today most visible. While in earlier phases the main strength of the movement in influencing science was in its power to stimulate environmental research *programs*, in particular in ecosystems research, at the paradigmatic level, it now lies in the development of concrete research *projects*. Whether this shift in potential influence will have an impact upon the science system itself in a more profound way cannot be said. Much depends upon developments in particular countries, their science policies, and the receptivity of scientists towards socially relevant research. As we mentioned earlier, the Netherlands is one example of a country where national science policy has turned towards stimulating the production of socially relevant knowledge. How successful this will be is not yet known; whether other countries will follow suit, especially those with very competitive technological strategies, remains to be seen. Here again, a lot depends upon the strength, cohesion, and direction of national environmental movements.

In light of such developments, we can conclude that even if the integrated knowledge interests of the environmental movement have fragmented over the course of the past decade, thus appearing to have limited its potential to influence the development of science in fundamental ways, there are still signs of new "alternative" ideas. We have already noted several of these: the new evolutionary approach in ecology, the new conception of technology, or new ideas concerning the implementation of concrete research for the movement. Whether such developments can provide the basis for a new ecological paradigm, which could thus bring forward a new phase of unity, is an open question. Despite this uncertainty, we can still conclude that the environmental movement has significantly affected the development of science. Moreover, the possibility remains alive that it could at some future time reintegrate and regain its even greater potential towards that end.

Acknowledgements

We would like to thank Joske Bunders and Richard Whitley for their valuable comments on an earlier draft of this article; and we would also like to thank the Swedish Research Council for the Humanities and the Social Sciences (HSFR) for the funds they provided to make this article possible.

Notes and References

1. E. P. Odum, *Fundamentals of Ecology.* Philadelphia: W. B. Saunders, 1971, and "The emergence of ecology as a new integrative science", *Science* **195** 1977, 1289—1293; T. Roszak, *Where the Wasteland Ends.* London: Faber, 1972.

2. The concept of knowledge interests can be traced in various forms back to the origins of the sociology of knowledge. Two distinct lines of development regarding knowledge distortion and development were apparent right from the beginning in this subdiscipline of sociology. On the one side were those, most notably Karl Mannheim (*Ideology and Utopia*. New York: Harcourt, Brace and World, 1936), who concerned themselves with describing historically related influences upon the production of knowledge and made no claim about the achievement of objective truth through the application of this method of reasoning. On the other side, most notably Marxists, were those who hoped that the uncovering of, for example, structurally induced distortions to knowledge would provide grounds for emancipated and enlightened political action rooted in a true understanding of social conditions. Where Mannheim and his followers traced the distorting influence upon knowledge to "ideologies" and "utopias", i.e., those wider cosmologies which either support an established social order or seek to transcend it and in which various "interests" are latent, Marxist thought, including the critical theory of the Frankfurt School, traced distortions in knowledge to class interests. For examples and discussion of these matters, see M. Horkheimer, *The Eclipse of Reason*. New York: Seabury Press, 1974; J. Habermas, *Knowledge and Human Interests*. Boston: Beacon Press, 1971; R. Eyerman, *False Consciousness and Ideology in Marxist Thought*. Stockholm: Almqvist and Wiksell and Humanities Press, 1981; A. Jamison, *National Components of Scientific Knowledge*, Lund, Research Policy Institute, 1982, especially Chapter 2.
3. S. Cotgrove, *Catastrophe or Cornucopia, The Environment, Politics and the Future*. Chichester: J. Wiley and Sons, 1982, especially Chapter 3. For a more general discussion of ideology and political action see R. Eyerman, ibid.; for a discussion of worldviews and values connected to nature and science: M. Douglas, *Purity and Danger: An Analysis of Concepts of Pollution and Taboo*. London: RKP, 1966; and S. Toulmin, *The Return to Cosmology*. Berkeley: University of California Press, 1982.
4. The concept of emancipatory knowledge interest is developed in Habermas, *op. cit.*, 1971, Note 2; criticism of this notion can be found in T. McCarthy, *The Critical Theory of J. Habermas*, London: Hutchinson, 1978, pp. 75ff. Our approach to knowledge interests is connected both to Habermas and to the earlier sociology of knowledge. However, our concern is neither with the structurally induced distortions upon knowledge nor with supposed invariant aspects, but rather with the historical emergence of worldviews and their concomitant knowledge interests. We conceive of knowledge interests as connected to defining a way of life, which includes criticizing other forms. What interests people have in knowledge, in the sense of what they find interesting, is intimately connected (as Habermas claims) to their specific being in the world. However, as opposed to Habermas, we conceive of that specific being in the world as an emergent, rather than quasi transcendental phenomenon. Thus we conceive of knowledge interest in relation to the rise of a social movement and its attempts to "emancipate" itself and others from the dominant worldview which defines modern industrial society.
5. D. Worster, *Nature's Economy: The Roots of Ecology*. New York: Anchor Books, 1979; E. Schramm, *Ökologie-Lesebuch: Ausgewählte Texte zur Entwicklung*

Ökologischen Denkens, von Beginn der Neuzeit bis zum Club of Rome (1971). Frankfurt am Main: Fischer Taschenbuch, 1984.

6. D. E. Morrison et al., "The environmental movement: Some preliminary observations and predictions", in W. R. Burch et al. (eds.), *Social Behavior, Natural Resources and the Environment.* New York: Harper & Row, 1972, 259—279; L. K. Caldwell, "Environmental Policy," in F. Sargent II, *Human Ecology.* Amsterdam: North-Holland Publishing Company, 1974, 403—430.
7. Morrison et al., *op cit.*, 1972, Note 6, p. 262.
8. Caldwell, *op. cit.*, 1974, Note 6.
9. For a useful account of "resource mobilization" theory, see J. Craig Jenkins, "Resource mobilization theory and the study of social movements", *Annual Review of Sociology* **9**, 1983, 527—553; and "Socio-Political Movements", *Handbook of Political Behavior* **4**, 1981, 81—153; also M. Zald, "Issues in the theory of social movements", *Current Perspectives in Social Theory* **1**, 1980, 61—72.
10. While many of our examples are drawn from our research in Holland, Sweden, and Denmark, we do claim more general applicability for our discussion.
11. A. Touraine, *The Voice and the Eye.* Cambridge: Cambridge Univ. Press, 1981; R. Eyerman, "Consciousness and action" *Thesis Eleven*, Nos. 5 & 6, 1982, 279—288; and "Intellectuals and popular movements", *Praxis International* **3**, 1982, 185—198.
12. A. Leopold, *A Sand County Almanack.* New York: OUP, 1949; R. Carson, *Silent Spring.* Boston: Houghton Mifflin, 1962.
13. See the discussion in R. C. Mitchell, 'Since Silent Spring: Science, technology and the environmental movement in the United States', in H. Skoie (ed.), *Scientific Expertise and the Public.* Oslo: Institute for Studies in Research & Higher Education, 1979, 171—207.
14. B. Commoner, *Science and Survival.* New York: Viking Press, (1963), 1967, and *The Closing Circle: Confronting the Environmental Crisis.* London: Jonathan Cape, 1972.
15. L. Herber, *Our Synthetic Environment.* New York: Alfred A. Knopf, 1962.
16. What exactly constitutes a "social movement" is a matter of definition and controversy. We use the term here in a pragmatic and empirical sense, not as a form of political judgement. On this, see R. Eyerman, "Recent social movements", *Acta Sociologica*, forthcoming, and A. Touraine, *op. cit.*, 1981, Note 11; R. Eyerman, "Social movements and social theory," *Sociology* **18**, 1984, 71—82; R. Turner and L. Killian, *Collective Behavior.* New Jersey: Prentice-Hall, 1957.
17. A. Jamison, "On the politics of environmentalism in Scandinavia", Paper presented at the VII World Conference on Future Studies, *The Future of Politics*, Stockholm: World Futures Studies Federation, June 6—8, 1982.
18. Caldwell, *op. cit.*, 1974, Note 6.
19. E. B. Worthington, "The International Biological Programme (1964—1974)", in E. B. Worthington, *The Ecological Century: A Personal Appraisal.* Oxford: Clarendon Press, 1983, 160—177.
20. R. P. McIntosh, "Ecology since 1900", in B. Taylor and Th. White (eds.), *Issues and Ideas in America.* Oklahoma: University of Oklahoma Press, 1976, 353—372.

21. R. L. Lindeman, "The trophic-dynamic aspect of ecology", *Ecology* **23**, 1942, 399—418; E. P. Odum and H. T. Odum, *Fundamentals of Ecology*. Philadelphia: Saunders, 1953.
22. A. G. Tansley, "The use and abuse of vegetational concepts and terms', *Ecology* **16**, 1935, 284—307.
23. L. K. Caldwell, 'The ecosystem as a criterion for public land policy', in R. L. Smith (ed.), *The Ecology of Man: An Ecosystem Approach*. New York: Harper & Row, 1972, 410—420, and G. M. Woodwell, "A confusion of paradigms (Musings of a president-elect)", *Bull. Ecol. Soc. Am.* **57**, 1976, 8—10.
24. F. Sandbach, "The environmental movement", in F. Sandbach, *Environment, Ideology &Policy*. Oxford: Basil Blackwell, 1980, 1—41.
25. Jamison, *op. cit.*, 1982, Note 17.
26. P. Ehrlich and A. Ehrlich, *Population, Resources and Environment*. San Francisco: Freeman, 1970; D. L. Meadows, *The Limits to Growth*. New York: Universe Books, 1972.
27. H. Odum, *Environment, Power and Society*, New York: Wiley. 1971.
28. *The Ecologist: A Blueprint for Survival*. Middlesex: Penguin, 1972.
29. McIntosh, *op. cit.*, 1976, Note 20.
30. E. Goldsmith, "Whatever happened to ecology?", *The Ecologist* **15**, 1985, 90—91.
31. See for example: J. Cramer and R. Hagendijk, "Dutch fresh-water ecology: The links between national and international scientific research", *Minerva* **23**, 1985, 43—61.
32. E. F. Schumacher, *Small is Beautiful*. London: Blond & Briggs, 1973.
33. Morrison et al., *op. cit.*, 1972, Note 6, p. 264.
34. D. Nelkin, "Scientists and professional responsibility: The experience of American ecologists", *Social Studies of Science* **7**, 1977, p. 81.
35. Woodwell, *op. cit.*, 1976, Note 23, p. 8.
36. J. Cramer, "The behaviour of Dutch fresh-water ecologists in response to environmental concern", *Hydrobiological Bulletin* **2**, 1985, 207—216.
37. J. Engelberg and L. L. Boyarsky, "The noncybernetic nature of ecosystems", *The American Naturalist* **114**, 1979, 317—324; C. L. Kwa, "Representations of nature mediating between ecology and science policy: The case of the International Biological Program", in W. Callebaut et al. (eds.), *George Sarton Centennial*, Ghent: Communication & Cognition, 1984a, 233—236; and "De relatie tot de natuur in cybernetische en evolutionaire ecologie" (The relation towards nature in cybernetic and evolutionary ecology), *Kennis en Methode* **8**, 1984b, 25—40.
38. See for instance similar debates about this issue among economists: L. C. Thurow, "Econometrics: An icebreaker caught in the ice", in L. C. Thurow, *Dangerous Currents: The State of Economics*. New York: Vintage Books, 1984, 104—123.
39. A. Naess, *Ekologi, Samhälle och Livsstil*. Stockholm: L. T. Förlag, 1981; and "The shallow and the deep, long-range ecology movement", *Inquiry* **16**, 1973, 95—100.
40. S. Novick, *The Careless Atom*. Boston: Houghton Mifflin, 1969.
41. J. Cramer et al., "Science shops in the Netherlands," *Science for People* **45**, 1980, 8—10. See also L. Leydesdorff and P. van den Besselaar, "What we have learned from the Amsterdam science shop", in this volume.

42. J. Bunders, "The practical management of scientists' actions: Causes and consequences of cooperation between university biologists & non-scientific groups", in this volume.
43. For some concrete examples, see L. Leydesdorff, "The Amsterdam science shop and its effects on science', forthcoming in G. Eckerle (ed.), *Forschung, Wissensanwendung und Partizipation*. Baden-Baden: Nomos Verlag, 1986.
44. J. Cramer and W. van den Daele, "Is Ecology an 'Alternative' Natural Science?" *Synthese* **65**, 1985, 347–375.
45. Kwa, *op. cit.*, 1984 (a and b), Note 37.
46. F. Capra, *The Turning Point: Science, Society and the Rising Culture*. Toronto: Bantam Books, 1983.

THE SCIENTIST, THE FISHERMAN, AND THE OYSTER FARMER

GEORGES BENGUIGUI

Groupe de Sociologie du Travail, Université Paris VII, Tour Centrale/6e étage 2, place Jussieu, 75251 Paris Cedex 05, France

By focusing on the way scientific research is conducted, the new sociology of the sciences has raised questions not only about the social context of science in general but also about the influence society may exert on the content of scientific knowledge. Typically, this has led to the study of how social interests and structures affect the struggles, alliances, and negotiations between researchers. But for me it is precisely this last expression "between researchers" which creates a problem. If we reject the notion of a separate social context, i.e., the idea of a separation between the world of science and other worlds, then there is no theoretical reason why these other worlds should be excluded from the analysis. There is no reason to reduce the field to the study of relationships between researchers and their behaviour. Therefore in this paper I shall try to reintroduce actors other than researchers into the study of scientific work and institutions. I wish to show that under certain conditions, the non-scientific people who actually use scientific theories and results may actively intervene in scientific quarrels and negotiations.

I shall attempt to analyse this behaviour using the example of research into the new focus on aquaculture in France. In my opinion this case is very interesting, because, if it is true that various groups of scientific actors compete to impose their own definiton of "what is problematic and what is not", as M. Callon (1) would put it, it is also true that groups of fishermen and oyster farmers try to do exactly the same thing. Therefore I would like to stress from a theoretical point of view the fact that not only do the world of the sciences and other worlds remain permeable (which is obvious), but also that these relationships must not be seen in terms of mere flow. It is impossible,

S. Blume, J. Bunders, L. Leydesdorff and R. Whitley (eds.), The Social Direction of the Public Sciences. Sociology of the Sciences Yearbook, Vol. XI, 1987, 117—133.

for example to claim that the world of sciences produces knowledge (sometimes at the request of the other worlds) which the external markets then consume. It would be more correct to say that the worlds inter-act and define each other, that their borders fluctuate and are more or less transparent.

Before going any further, the so-called "new" aquaculture must be defined. In France there exists a traditional aquaculture which produces mainly oysters, mussels, and fresh-water trout. It is a well-developed economic activity. The so-called "new aquaculture" was intended to develop the rearing of new species; this is almost exclusively done in the sea, and concerns fish and molluscs as well as shell-fish. In this paper I will analyse the relations between the scientists working on the new aquaculture and the fishermen and oyster farmers affected by the new aquaculture. This is an example of oriented research whose target is to launch the rearing of new sea species in France. Like other authors, I purposely use the expression "oriented research" and not "applied research", because in the new aquaculture the object is not to apply knowledge already acquired, but to produce this scientific or technical knowledge. My hypothesis, in this situation of oriented research, is the following: if the field of investigation in which the scientists are to work is already occupied by non-scientists, the latter will probably be in a good position either to combat the scientists by restricting their freedom of action, or to influence and orient their work according to their interests. If the field of research is already occupied by another category of scientist, tension between the two teams is more likely than collaboration. A good example of this latter case is provided by R. Hohlfeld (2) in his discussion of oriented research on cancer in West Germany. In this paper, however, I will not discuss this latter type of case. I must add that when I speak of the research field, I am referring to two things: first the field itself, the physical territory; but also the sphere of research. I am therefore using the word "field" in both senses simultaneously.

In discussing marine aquaculture it is clear that the field in which the scientists work is not empty; it is already occupied by coastal fishermen and oyster farmers. The problem of the relations between scientists and these groups is also present automatically, since sooner or later research must be carried from the laboratories to the sea, to be continued through experimentation that can only be many-sided in view of the number of species and ecosystems involved. The first part of my presentation describes research in the new aquaculture and the behaviour of the scientists in response to the "professionals". The second part presents an analysis of the behaviour of these occupational groups themselves. In the last section, I will try to draw some conclusions from these analyses.

This paper examines the dominant research organisation and its scientific staff, because historically this organisation has played a key role in the new French aquaculture, and all the other actors have positioned themselves in relation to this organisation. The data used in this article come from numerous interviews carried out among sea professionals from the South-East and the West of France, with scientists from all the organisations involved in aquaculture, as well as from numerous written documents.

I. Research in Marine Aquaculture

In France, oyster and mussel farming started to develop on a large scale during the nineteenth century, and developed into traditional aquaculture. In the meantime, a few scientists were trying to breed other marine species. References to this subject sometimes include interesting commentaries with astonishingly modern connotations. But at the beginning of the twentieth century, this scientific work on the breeding of new species stopped and fell into oblivion. Some people have explained this as a result of the appearance of motorised fishing boats, which increased catches enormously and diminished interest in aquaculture. Nevertheless, it was not until the end of the 1960s that the Japanese idea of breeding marine species other than oysters and mussels was rediscovered by a few scholars and managers. At that time, two big companies — Les Salins du Midi and Générale Transatlantique — moved into aquaculture for different reasons, often taking Japan as a model. In the same period, General de Gaulle's government was launching large technological programmes (in space, computers, the Concorde, etc.). Within this ambitious framework, the government also decided in 1967 to create a new agency for the Oceans, called the "Centre National pour l'Exploitation des Oceans" (CNEXO). This agency rapidly established a very important working programme based on five main themes:

— the exploitation of living resources;

- the exploitation of minerals and fossils;
- the study and development of the continental shelf;
- the fight against pollution; and
- studying the action of the ocean on the climate and on meteorological conditions.

In terms of the first theme, the exploitation of living resources, a great emphasis has been put on the launching of the new aquaculture in France. People used to declare emphatically that the new aquaculture would put an end to starvation in the world. Many other prophetic sentences were pronounced concerning the promises of the sea, corresponding to the "utopias of newly emerging disciplines", as Hagstrom would say. In reality, the emphasis put on aquaculture was explicitly for economic reasons: the object of the exercise was to reduce the French trade deficit. For this reason, CNEXO chose to give priority to expensive species which were economically interesting rather than those which might have been more useful (in terms of production of proteins) for underdeveloped countries. The main point to remember here is that a new economic activity was to be created *ex nihilo*; it was a political act of will which claimed to determine a scientific and technical policy (3). The major scientific and technological choices — the choice of the species on which to work, the type of aquaculture to adopt, etc. — were made from the beginning and largely on a political basis. As the leading biologist of CNEXO put it, this was a case of "research directed by administrators." In fact, the people of CNEXO (which is a centralised and technicist state organisation directed for a large part by people coming from the Navy and the Commissariat à l'Energie Atomique), thought that in order to obtain a very productive aquaculture, it was necessary (and possible in a very short time) to choose an almost exclusively intensive form of aquaculture (4).

Among the consequences of this choice was the necessity of mastering the entire biological cycle of the species to be cultivated, which considerably multiplies the scientific and technical difficulties. The directors of CNEXO concluded that it was necessary to collaborate with the big companies. It was a great industrial adventure. One doesn't build nuclear power stations with small plumbers, and one doesn't build efficient aquaculture farms with coastal fishermen or small oyster farmers. Besides, in the eyes of the promoters of the new aquaculture, these groups, who very seldom own a big firm, have a hunter-gatherer mentality which is not compatible with the state of mind required to run an aquacultural farm successfully. At this time, CNEXO was reinforced in its decision to develop heavy aquaculture with big companies by the fact that two big firms had just become involved in new aquaculture. This strategy potentially freed CNEXO from all environmental constraints and from the people of the sea. In a sense, one could say that thanks to this strategy, all that the CNEXO needed were basins, taps, pipes, and pumps.

Yet two facts forced the CNEXO to change its policy, at least partially. The first is that the two companies entering the new aquaculture decided rather quickly to withdraw from this sector. They had encountered numerous difficulties, and the profitability of the operation was not sufficient. Since then, no big French company, with the single exception of the E.D.F. (5), has become involved in the construction of aquacultural farms in France. From an industrial point of view, the CNEXO was like an orphan.

The second fact is that if the birth of the CNEXO had the advantage of attracting a few scholars to engage in research on the new aquaculture, it also initiated a real competition between the CNEXO and other public organisations, including on the one hand the Institut Scientifique et Technique pour les Peches Maritimes (ISTPM), and on the other, the organisations belonging to the Ministry of Agriculture. The ISTPM is an old institution, created in 1905, and deals essentially with fishing and oyster farming, although nothing prevents it from dealing with the new aquaculture as well. Because of the nature of its missions (technical assistance and sanitary control essentially), the ISTPM is very well acquainted with the fishermen and the oyster farmers, and the latter are even represented on its Board (which is not the case at CNEXO)). But when the CNEXO was created, nothing was said concerning their respective fields of activity, and this silence inevitably became a source of conflict.

The organisations belonging to the Ministry of Agriculture include mainly the CEMAGREF (6), which is officially considered as technical, and the Institut National de la Recherche Agronomique (INRA), considered more scientific. Unlike many countries where the same ministry is in charge of salt waters and fresh waters, in France an administrative frontier exists between these two types of water: the Ministry of the Sea is in charge of salt water and therefore of marine fishing and aquaculture, and the ISTPM and CNEXO belong to it;

while the Ministry of Agriculture is responsible for fresh water and therefore for fresh-water aquaculture. However, fresh-water angling is under the responsibility of a third Ministry. Of course, this administrative division is sometimes absurd: where do brackish waters stand? And what about anadromous species like the salmon for example, which live alternately in fresh and salty water? Consequently, the CEMAGREF (taking advantage of these mixed cases) and the INRA (considering itself a scientific organisation which does not have to recognize the salty/fresh distinction) have extended the boundaries of their activity to sea water and marine aquaculture. The scientists of these organisations, coming from the agricultural world and from the management of Waters and Forests, have been able to present an integrated conception of aquaculture founded on the concepts of the ecosystem and extensive rearing (7). This has facilitated their relations with the fishermen and oyster farmers, which has been an essential factor enabling the professionals to exert pressure on scientists, and actually forced CNEXO to do research on rainbow trout.

In sum, the state decided to initiate a technological innovation, the new aquaculture. Immediately, different groups of actors tried to turn this situation to the advantage of their own new territory. This implies specific choices. For example, the choice of intensive aquaculture at the same time defines elements such as the scientific field of research, financial partners, professional skills, etc., all elements apart from those implied by extensive aquaculture. So it is not surprising to see the CNEXO in its first program (1968) choosing intensive aquaculture and at the same time rejecting extensive aquaculture in the following terms: "extensive aquaculture fattens, it does not breed."

At the second stage, the CNEXO remained without any industrial partners with whom it could have developed intensive aquaculture on a large scale, especially as large-scale success was long in coming. What is more, the CNEXO was left to compete with public organisations which had chosen different directions and which often had much better relations with the people "below" than did the CNEXO. Therefore the CNEXO reacted partly by transforming itself into an entrepreneur. Since it could not find any big firm interested in the business, one of its branches attempted to demonstrate the technical and economic possibilities of aquaculture. This branch produces about eighty tons of salmon per year, but up to now no one has followed its example.

Additionally, the CNEXO has started to work with fishermen and oyster farmers more than before. Contrary to its official doctrine, the CNEXO allowed a part of extensive aquaculture to develop in its bosom. In 1979 the CNEXO even organised a conference with the ISTPM on “extensive aquaculture and restocking.” In fact, this means that the CNEXO has been obliged to work in an already occupied field (the fishermen and the oyster farmers) more than during the first stage. The CNEXO now has to attract small fishermen and oyster farmers to aquaculture and the breeding of new species (even though they are not the only people that CNEXO is trying to reach). The CNEXO is trying to convince the coastal people that their interest lies in aquaculture. It has supplied them with subsidies and with technical and scientific information. In a sense, one can say that the CNEXO is trying to “buy” the people of the littoral; but it is more than that. In some cases, the CNEXO tries to “domesticate” them (i.e., to change hunter-gatherers into farmers), while at the same time it tries to domesticate new marine species. And contrary to what one might think, this is true for the oyster farmers as well. They are learning, for example, how to fight, every day and very meticulously, against the crabs which eat the young clams. Another more spectacular example concerns the scallops in Saint-Brieuc in Brittany, where natural production has collapsed in recent years: CNEXO scientists who were experimenting in restocking tried to convince the fishermen to respect a very strict discipline. They first urged fishermen to self-discipline, then they used maritime control, and at last they used airborne control! They had to “convince” the fishermen, just as they had to “convince” the scallops to reproduce themselves more efficiently (8). In addition, in many cases one has the feeling that these field experiments are not so much ordinary scientific experiments as an attempt to socialise the fishermen into aquaculture.

In the meantime, a number of researchers from the CNEXO tried to escape from the field occupied by non-scientists. They emphasised the superiority of their scientific qualifications and asked that teams of specialists in the development and communication of knowledge be put together to facilitate relations with the professionals of the sea, who were able to exert serious pressure on the scientists. In fact the researchers of CNEXO do not enjoy the same legal status as other researchers of the large French public research organisations, such as the CNRS for example. The CNEXO works like a private company, and its researchers are evaluated and promoted through a hierarchy which is not always made up of scientists. The CNEXO researchers’ strategy has thus been to constitute areas of autonomy by trying to "scientificise" their work as much as possible. Indeed, how is it possible to attack a researcher who has just submitted a doctoral dissertation or who publishes in a good scientific journal? A researcher told us: "Thank God we have our publications!" The success of their strategy was made easier by a number of failures caused, it seems, by a lack of attention to scientific problems shown by the management of CNEXO.

In short, in CNEXO there are three types of relations with the professionals of the sea:

(a) avoiding the people of the sea by choosing intensive aquaculture;

(b) "buying" and sometimes "domesticating" the people of the littoral; and

(c) escaping from the coastal people by withdrawing into science and the laboratories.

In fact these three behaviour patterns were not chosen solely because of problems in relations with coastal fishermen and oyster farmers. It is obvious, for example, that it was not the desire to be rid of the coastal people that was the primary reason for choosing intensive breeding. In the same way, it seems true that the CNEXO was aware beforehand of the need to develop fundamental research. Nevertheless, from the coastal people's point of view, the result is the same. They had the impression they were being left out. They were told that they had neither the money, the right mentality, nor the ability to submit themselves to the division of labour required by intensive aquaculture. Not only did the CNEXO remain silent about what seemed to be its primary concern, restocking, but its predictions of production were so high that it threatened to become a very dangerous competitor. As one coastal fisherman told us: "we were being traumatised by those new cultures!"

II. The Actions of the Fishermen and Oyster Farmers

Confronted by this situation, how did the people of the sea react? In practice we will see that several types of behaviour occurred, which sometimes blended together.

1. A restocking-oriented behaviour: restocking, which is a form of extensive aquaculture, is one of the fishermen's great preoccupations. This is at the opposite end of the CNEXO management's official doctrine, which even after the period of "dashed industrial hopes", as

one of the scientists expressed it, continued to bet on intensive aquaculture. The coastal fishermen explained in their "Mediterranean Charter" published in March 1982 that for them, the expression "marine cultures" must be extended to include the broader notion of administering marine resources through different measures such as restocking, artificial reefs, and reserves. In selecting these measures, the dominant concern has been the disappearance of certain species. The basic problem was to protect and renew the stocks. Two possible solutions were considered. First, there is stock management, which implies that (in the case of the lobster, for example) everybody (including the holiday-maker) is highly disciplined, which is not always feasible. In any case, stock management, with its catch quotas, calendars, non-fishing areas, and permits, by definition prevents previous levels of catches from being matched.

The second solution is to restock by returning post-larvae or juveniles obtained by different means (such as hatcheries or sprat collection) to the sea. This solution, which is an old nineteenth-century idea, seemed very appealing. The principle appeared easy and, what is more, it was garlanded with Japanese prestige, for it had been applied there successfully a number of times. Unfortunately for the fisherman, almost all the French scientists expressed serious reservations about it, even those working on the subject. During the conference on the evaluation of research in aquaculture in Brittany and Normandy organised by the Ministry of the Sea in March 1982, scientists spoke of the "ambiguity of the notion of restocking and of the risks of extensive restocking, etc." Conversely, all the scientists were favourable to stock-management practices based on population dynamics. The fishermen thus (with difficulty) started practising stock-management, and, in certain cases, restocking practically by themselves. On this subject, the story of the fishermen of the island of Houat in Brittany is worth telling. In Houat, shellfish represent about 80% of the catch. In the 1970s the fishermen published a short manifesto called "The Blue Belt of Brittany". They urged the need to protect and develop the living resources of the coast, and asserted that "the sea belongs to the fishermen". They then decided to develop lobster restocking. They sent two biologists on a research mission to Japan, built a hatchery with their own hands, and then poured the post-larvae back into the sea. The problem was that rejections began before any pre-study of the stocks had been undertaken and the post-larvae could not be marked. It was thus extremely difficult to evaluate seriously the results of this experiment, and besides, this evaluation must be done over the long term. Other similar experiments have been undertaken, but they were not coordinated.

It is easy to criticise these experiments, and quite rightly the scientists have done so. But two things have to be remembered:

(a) The real fascination of restocking for a large number of fishermen. As one Southern fisherman has said: "with aquaculture we have the possibility to remake the sea as it was before". He added that restocking would allow them to regain their liberty, to escape the quotas, etc. Another fisherman added: "aquaculture used as a means of restocking is worthwhile, but otherwise it is much too expensive". One day, in the Mediterranean, some fishermen asked a television crew to come and film 1000 fishes, left over from a small aquaculture experiment, being thrown back into the sea. This number was absolutely ridiculous, but contrary to what the scientists who told us the anecdote were thinking, it was not a stupid gesture: it was a symbolic one expressing this fascination and desire to replenish the sea.

(b) The fishermen's fascination with aquaculture also exerts pressure on the researchers. As one scientist, who is a lobster specialist, wrote: "French fishermen want the operations of throwing fries back to the sea to be continued and intensified without even waiting for the efficiency of this method to be demonstrated. This is a fact which one must reckon with. Under these conditions, it is necessary to carry on the research . . ." (9). Moreover, it seems that the same sort of pressure existed in Japan as well (10). This pressure exerted on the researchers is of course possible only because the fishermen occupy the field, have started restocking experiments, and are fighting for the survival of these experiments in spite of the severe criticism they have incurred.

2. The second type of behaviour exemplifies the fishermen's desire to remain autonomous; what the scientists did not want to do, they did on their own. For example, some fishermen experimented, practically on their own, with the breeding of mussels in long lines. Very often, the people of the sea carry on these experiments in order to keep control of their territory: "We are not opposed to aquaculture, but we do not appreciate the arrival of scientists on Maritime Public Property (11) who expel the professionals from their business and put just anybody in their place. We want to keep our traditional identity (. . .) The CNEXO wanted concessions, but we refused because we wanted aquaculture to

be ours as a complement to our fishing" (a Southern coastal fisherman). Thus, the professionals who carry out these experiments often ignore the big public research organisations, or they explicitly act against them in order to prevent them from occupying their field. In most cases, the experiments are carried out through associations which are more or less linked to local institutions such as local councils, and are financed by them. In certain cases, this occupation of the field by the people of the sea can make the scientists dependent on them, and even go so far as to disturb them in their work, at least for a while. For example, when the CNEXO launched a national experiment of clam production in different regions at the same time, it needed the participation of the oyster farmers. In practice, the CNEXO met with great difficulty in finding oyster farmers who were willing to take part in the experiment in the West. This is a paradox, because this is the primary region for the production of oysters. The main reason for this refusal was that this region had initiated its own experiment for producing clams.

The quest for autonomy and for the preservation of a "traditional identity" can even express itself through a more or less brutal rejection of aquaculture. Worried about the potential competition that aquaculture represents in their eyes, or impatient of its results (stock-management and restocking represent heavy constraints and are contrary to the fishermen's short-term interests, being very slow to produce results), some fishermen went so far as to sabotage aquacultural experiments in the Mediterranean and the Atlantic, and buildings belonging to CNEXO in Corsica (12). In a word, the fishermen simply refused to let themselves be "domesticated" and denied the right of the scientists to carry on their work. As a consequence, the system of control had to be reinforced, which increased the costs; and in certain cases, the researchers withdrew even more into their laboratories.

3. The third behaviour pattern exerted direct pressure on the scientists' choices. We have already seen, in the case of restocking, how the fishermen can exert indirect pressure on the scientists; we shall now see that the people of the sea can apply direct pressure as well. In a certain number of cases, the fishermen, considering that they had been pushed, or at least strongly advised by CNEXO to accept the new aquaculture, and more precisely to accept the production of certain species, felt that they had the right to ask CNEXO to orient its research around the difficulties they encountered. One Breton cooperative is a good example of this type of behaviour. After a period of instability,

this cooperative specialised in the production of trout reared in the sea (13). It has become the first firm in France for the production of this type of fish (about 100 to 120 tons per annum). Unfortunately, the trout have difficulty surviving the summer because of local conditions (too high a temperature and saline content). This prevents the fish from growing to a useful size and requires the producers to get rid of their stocks before the summer, which is of course a commercial drawback. Consequently, the people of the Co-op, who have their own biologist, exerted pressure on CNEXO scientists to develop research on this particular point: how to survive the summer. However, the scientists of CNEXO were convinced that the problem would be more easily solved if instead of producing rainbow trout, the people of the Co-op switched to Fario trout. Unfortunately, Fario trout are more difficult to breed, and the pisciculturists, who are the sea-breeders' usual suppliers, do not produce this species very much. Thus, the sea-breeders rejected this type of trout.

Additionally, the fishermen have exploited the competition between the public research organisations. They asked a team of the INRA, the main research organisation of the Ministry of Agriculture, to launch a new study in order to solve the problem of the summer passage among Rainbow trout. INRA researchers told us that they felt a bit embarrassed to be pitted against CNEXO, but this did not prevent them from accepting the contract. Not only did the people of the Co-op play one organisation off against the other (14); they also chose to favour one discipline over the other. In effect, the INRA team chosen by the Co-op specialises in fish genetics, and it was research in genetic selection that the Co-op wanted, while the researchers of CNEXO do most of their work in physiology. Trying to solve the problem of the summer, the Co-op chose genetics over physiology. The Co-op people thus have a genuine scientific strategy which they try to impose on the scientists. They do not simply pose a problem, but also propose ways of solving it.

Let us take another example, albeit a less elaborate one, of this will to put direct pressure on the scientists. In the South of France, in the Languedoc-Roussillon Region, we have seen that some fishermen have performed a variety of small breeding experiments in order to occupy the field. But that was not all. This object was to force the scientists into a real collaboration: "we don't reject the scientists, but we, the professionals, also have our word to say. Besides, there are some things that the scientists cannot do without the professionals." Another fisherman added that without the professionals, the scientists could not do anything: "we can stymie anyone. If the scientists deny the professionals their right to take part, nobody will do anything, we'll stump them with the water pumping and the Maritime Public Property." Indeed, no one in France has the right to pump sea water or to settle on Maritime Public Property without a permit which is delivered only with the agreement of the sea professionals. The sea professionals possess here an excellent means of exerting pressure on the scientists.

Generally, the coastal fishermen's attitude towards the new marine aquaculture is very clear (unlike the oyster farmers' position): new aquaculture, well and good, but it must be done for and by the fishermen. The fishermen's trade unions (the CGT has a majority in the South, and the CFDT in the West) are particularly clear on this issue. The CFDT fishermen published a very interesting booklet in February 1978 entitled: "Aquaculture (Breeding and Marine Restocking), a Political Choice." In this brochure, the CFDT said: "considering that aquaculture is a complement of fishing activities, it must be carried out by the fishermen, for our aim is to ensure employment on the coast." The CFDT declared itself in favour of fishermen's cooperatives and against capitalistic factories where the fishermen would be converted into unskilled workers. The CFDT said it was favourable to aquaculture, but within the framework of a coastal development policy. The same idea is found at the CEMAGREF and at the ISTPM (where, at least in theory, aquaculture is associated with coastal development, which is not the case at CNEXO). Similarly, in another more recent document (1982), the CFDT wrote: "aquaculture must be integrated with coastal fishing within the framework of a true development policy." Yet it must also be noted that for the CFDT, aquaculture is at the same time both intensive and extensive (specially where restocking is concerned); this is reflected even in the title of this booklet. As we have seen, the same theme appears in the Mediterranean Sea Charter published in 1982. It is truly a theme common to all fishermen.

The logical consequence of these attitudes is a clearly attested will to influence research organisations in their scientific choices. As the CFDT wrote in its brochure: "fundamental research, experimentation, and production can not be separated; it is essential to maintain a certain number of experiments scientifically supervised by public organisations near the aquacultural cooperatives." Furthermore, the CFDT opted for genetic selection, which was not popular among scientists at the time.

More precisely, the CFDT, in a letter addressed to the President of CNEXO in January 1983, asked the scientists to recognise the sea professional's ability to define the orientations of scientific research in the same manner as members of the CFDT recognise the scientists' competence. More crudely, a CFDT fisherman declared in an interview: "sea professionals are always the pretext and justification for the scientists' pretty reports. We know very well what problems they should be working on, but they won't listen to us." Nevertheless, some modest satisfaction has been granted to the sea professionals. In 1984, when the CNEXO and the ISTPM merged to create a new organisation, the IFREMER (15), a committee for living resources was established in its midst, with fifteen members (out of twenty-one) chosen from among people proposed by the sea professionals. Normally, this committee is consulted whenever the orientation and implementation of research programmes are discussed. Even before it was constituted, however, the CFDT protested against the existence of three different councils within the IFREMER (the Board, the Scientific Council, and the Living Resources Council). The CFDT feared in effect that the Living Resources Council would be used as a means of marginalising the sea professionals, and it called for the implementation of a single council.

This will to be linked, even indirectly, to research organisations, can be detected in another example. The different aquacultural organisations were led in many cases to hire biologists, with financial help from various sources, in order to have scientists on their side. But these hirings were precarious. As a result these associations demanded that the field biologists be integrated into public research organisations and then lent back to them. As for the scientists, they seem to have wanted the creation of a special body of advisers in aquaculture (as in agriculture) that would be separate from them, remaining outside the research organisations. The professionals' argument in favour of the integration of biologists into the research organisations was the following: the separation between basic research and applied research must be rejected. Besides being instrumental in research, these biologists also generate other research which is more fundamental, and there must not be one so-called research at the top and another at the bottom. Here again we find the debate between the professionals who want to keep control of research and the scientists who want to escape this control in the name of Science. Yet most of the problems raised by the fishermen can only be solved through fundamental research in genetics, pathology, etho-ecology, studying currents, etc. The conflict is not just a conflict between fundamental research and applied research. In reality, the object of the conflict turns around who is going to orient the research.

To sum up, one can say that the sea professionals have stated very clearly their preference for an open milieu and rather extensive aquaculture, which would also serve their own interests. Their action in this field has had a noticeable effect on research. Furthermore, some of them know how to exert direct pressure on the scientists' work. They know how to take advantage of their contradictions, how to establish alliances with some of them against the others. In a word, they actually have a research policy (to define priorities) and a scientific policy (to define work processes) which are more or less well integrated. They then try to impose them on the scientists, in different ways and with varied success.

Conclusions

In this paper I have tried to show that people who use scientific results, here the small-scale sea professionals, can contribute very concretely to the orientation and definition of scientific work: sea professionals can help scientists who want to work in extensive rearing, they can impose subjects of research and even propose ways of finding solutions (through the use of genetics, for example). I have also tried to show that this is possible only because the sea professionals were the initial inhabitants in the field in which the research is being done, and because in marine aquaculture, contact with the sea can not be avoided. This is very different from a situation where public opinion can eventually step in. In the case of cancer research for example, public opinion can ask for intensified research, but it can not specify what type of research or strategy is needed. This is not simply due to the fact that public opinion constitutes an amorphous group, for after all, structured lobbies can be created; the reason is rather that public opinion or lobbies do not occupy the field: a group of actors may want to favour a certain direction of research, for example epidemiology, but it has little means of imposing it on researchers.

This article has mainly focussed on the relations between the CNEXO and the sea professionals. In reality, the picture I have tried to draw involved many other people. To caricature it somewhat, if we put

CNEXO in the centre one could say that above CNEXO there is the State and the policy of its representatives (16), and below there are the sea professionals. Parallel to CNEXO there are the other research organisations. Each of these groups claims to know what the right research work in new aquaculture should be. Each group's reasoning is linked to its situation within the overall power relationship. This means that scientific work in its precise definition is the object of conflicts, negotiations and alliances, etc., which involve not only scientists. Fishermen and oyster farmers cannot be reduced to the simple status of the "social context"; they can be scientific actors, even if they are not labelled scientists. As one CFDT leader put it in 1977: "One cannot put Science on one side and the work force on the other. In the contact, scientists become a bit more like fishermen and fishermen a bit more like scientists." The world of the sciences and the other worlds indeed interact and are mutually constitutive.

Acknowledgements

This article stems from research on the effects of technical innovation on the sciences and professions carried out with D. Chave, P. Rivard, and P. Tripier. I thank my colleagues for their help, but of course they are not responsible for what I have written here. This project was financed by the S.T.S. Programme of the C.N.R.S.

Notes and References

1. M. Callon, "Struggles and negotiations to define what is problematic and what is not", in K. Knorr, R. Krohn, R. Whitley (ed.), *The Social Process of Scientific Investigation*. Dordrecht: Reidel, 1980.
2. R. Holfeld, "Two scientific establishments which shape the pattern of cancer research in Germany: Basic science and medicine", in N. Elias and R. Whitley (eds.), *Scientific Establishments and Hierarchies*. Dordrecht: Reidel, 1982.
3. The consequences of this State exercise of power are analysed in G. Benguigui and D. Chave, "L'État et les petits poissons", *Sociologie du Travail* **27**, 1985, 3.
4. Aquaculture is said to be more or less intensive or extensive according to the degree of human intervention in the biological cycle of the cultivated species.
5. But the EDF (Électricité de France) is interested in the new aquaculture in a very special context: to use the hot water coming from the nuclear power stations. The public relations aspect in favour of these nuclear stations is not to be neglected.
6. CEMAGREF: Centre d'Étude du Machinisme Agricole, du Génie Rural et des Eaux et Forêts. It includes a section on "littoral development and Aquaculture". The ISTPM's aquacultural section also includes the expression "coastal development" in its denomination.

7. It must be remembered that the aquaculture sections of the CEMAGREF and the ISTPM include the expression "littoral development".
8. M. Callon, "Some elements of a sociology of translation: domestication of the scallops and the fishermen in St Brieuc Bay", in J. Law (ed.), *Power, Action and Belief. A New Sociology of Knowledge*. Sociological Monograph Series, London: Routledge & Kegan Paul, 1986.
9. J. Audouin, "Aspect technique des écloseries de homards", in CNEXO, (ed.), *Aquaculture extensive et repeuplement*, Brest, 1981.
10. J. Querellou, "L'expérience japonaise de repeuplement", Mémoire CTGREF **10**, 1977.
11. In France, the coast and the coastal sea are called Maritime Public Property. This property is State-owned, and one needs an administrative permit to use it.
12. Perhaps, in the case of Corsica, the sabotage had a political and nationalist significance.
13. This breeding consists in taking classic fresh-water rainbow trout and putting them for a shorter or longer period into the sea.
14. Somewhat like when the fishermen take sides for the ISTPM against CNEXO.
15. IFREMER, the Institut Français de Recherche pour l'Exploitation des Mers.
16. I have not insisted on this point because I have discussed it elsewhere. Cf. G. Benguigui, D. Chave, *op. cit.*.

Translated from the French by A. Benguigui and G. Varro.

WHAT WE HAVE LEARNED FROM THE AMSTERDAM SCIENCE SHOP

LOET LEYDESDORFF
Department of Science Dynamics, University of Amsterdam, The Netherlands
and
PETER VAN DEN BESSELAAR*
Social Science Informatics, University of Amsterdam, The Netherlands

Introduction

It has been the programme of the sociology of science refashioned in the late 1960s "to open the black box of the production of science and technology" (1). New questions were raised, such as: How are scientific results brought about? How are knowledge claims honoured? How are the sciences organized intellectually and socially? And, in relation to the question of the "steering" of science: what exactly in the content of science can be externally influenced?

One reading of the results of many of the studies carried out in the past two decades can be summarized under the following headings:

— there are major differences between disciplines;
— there are important differences between the dynamics of the emergence of new scientific specialties on the one hand, and the developments of existing specialities on the other (i.e., we have to conceptualize "phases of development" in discussing these dynamics);
— science is socially stratified; whilst this social stratification leads to the formation of elites, it also plays a role in the social and intellectual integration of the sciences. This means that the sciences have specific internal power structures;
— relations with the social environments of science can be regarded as "negotiated", and these negotiations are accessible to analysis.

On the basis of this general perspective of "sciences as knowledge production systems" (2), the question *whether* non-scientists can influence science has to be rephrased. It is now rather a matter of examining the extent to which, the level at which, and the dimensions in which lay

S. Blume, J. Bunders, L. Leydesdorff and R. Whitley (eds.), The Social Direction of the Public Sciences. Sociology of the Sciences Yearbook, Vol. XI, 1987, 135—160.

people can influence the different sciences. What mechanisms are important? Which barriers can be overcome? How do (and how can) such influences become institutionalized, and at which stages of development of scientific specialties?

The complexity and multidimensionality of the interactions between science and society make it impossible to answer these questions once and for all. Although one of the more important conclusions of science studies over the last decade has been that scientists make trade-offs between "knowledge interests" and other (social and economic) interests, most of these studies have focussed on the science system itself, and have tended to treat external influence, or "external demand", as something in itself unproblematic, but the *function* of which needs to be explained (3). However, when we want not only to account for external influence on science, but to explain the *success or failure* of external demand, it becomes necessary to make problematic the nature and the composition of external influence or demand.

To that end we focus here on one such external group, the trade unions, and particularly on the question whether the trade unions can use the collaborations with scientists which are made possible by institutes such as the science shops for the programmatic development of a "labour oriented science and technology" (4) as opposed to the current development of S and T which is heavily linked with industrial interests.

This we will try to do essentially on the basis of two strategically chosen case studies of long-term collaborations of the Amsterdam Science Shop with the unions in completely different sectors of society (the chemical industry and banking). Because in these cases both different technologies and different markets are involved, we will be able to generalize our conclusions about how technology affects the positions of workers, and about the possibilities for trade unions to translate these consequences into what we will call "technological demand": the specifications of a technological research programme which may produce labour-oriented technologies.

Towards a Labour-Oriented S & T-programme

In the aftermath of the student revolts of the late 1960s Europe witnessed a revival of Marxist theorizing with special attention to science and technology. The changed relations between "capital" and "science" were a common focus among many different theories of that time. For example, some theorists declared that "science and technology" had become a productive force (5), revolutionizing the dialectics between the structure and superstructure of society (6). Others emphasized the ideological power of modern science and technology, implying the need for a critical theory (7). Much attention was also paid to the theory of changing class relations between workers and intellectuals, leading eventually to Mallet's thesis of "the new working class" (8) and Braverman's thesis about the degradation of work in the scientific-technological revolution (9).

In the 1970s these new ideas led to attempts to establish coalitions and collaborations between scientists and workers — whether unionized or not — in order to explore practically and empirically the niches of scientific and technological developments in the capitalist system from a labour point of view (10). In Holland, the lack of cognitive content in such cooperation was soon perceived as a problem, both by the scientific community and by union leaders. In 1973, union leaders dismissed an offer of cooperation from the newly founded Scientific Workers' League out of fear of interference in the union by intellectuals without well-defined roles (11). This problem could later be overcome through the proposal to create Science Shops: these university-based institutes would take the questions of their clients as external demands primarily for the university research system (12).

The University of Amsterdam accepted this concept of a mediating institute in 1977, and it succeeded in gaining union support for the initiative when the government of that time tried to turn down the small budget (Dfl. 30,000) requested for it (13). Ever since, the Dutch unions have been heavily involved in the further development of the Amsterdam Science Shop into a science policy instrument (14).

When in 1977 we established the Amsterdam Science Shop, we acted on the belief that our society was going through a period of change in power relations caused mainly by the rise of science and technology as sources of production, power and legitimation, and that therefore *access* to science and technology might be a crucial resource. The Science Shop was intended as a specific instrument to give access to science to groups underprivileged in this respect. Unlike earlier more humanistic experiments which focused on the distribution of knowledge, the notion of access by contrast refers rather to the *process* of the production of knowledge: to the knowledge production system (15). Reviewing the experiences of the Amsterdam Science Shop thus provides the possibility of analyzing external demand: in other words, of analyzing the role of identifiable external groups in the process of the production and the diffusion of scientific results.

In earlier articles we have dealt with the lessons which can be drawn from the Science Shop experiences for trade union participation in technology policies at the corporate level (16), in university research policy (17), and in national science and technology policies (18). In this article we focus on systematic attempts to explore the institutional and cognitive translations which are necessary to establish an enduring cooperation based on substantive mutual interests. We will examine instances in which unions, concerned by the social effects of new technologies, were led to address the Amsterdam Science Shop. We take these instances as input to a translation process, and we look at the possibilities which existed to feed these social effects back into the R & D-process.

Social Demands

The problem can be seen in relation to recent discussion of the nature of technological innovation and in particular the relations between "demand" and "supply"-side factors.

Mowery & Rosenberg, in a critique of the idea of "market demand" as the crucial determinant of technological innovation (19), showed how "market demand" had to be reinterpreted in other terms (such as "need specification") to be made accessible for empirical investigation, whilst such reinterpretation at the same time implied a translation of these "demand" factors into terms which can be dealt with from the "supply" side. Similarly Langrish et al. had concluded that "perhaps the highest-level generalization that it is safe to make about technological innovation is that it must involve synthesis of some kind of need with some kind of technical possibility" (20). Implicitly, Mowery & Rosenberg plead for the conceptualization of innovation as the outcome of creative combinations by actors who have access to information about (future) markets and technological developments (21).

From recent work in economics these emerges a picture of local concentrations of resource mobilisations leading to innovative activities in distinct submarkets, which under certain conditions can gradually gain the momentum of a "technological trajectory" (22). Dosi (23) stresses the room for various forms of *stricto sensu* non-economic interests of private actors and state interventions in these processes. However, the inducing mechanisms and focusing devices for such concentrated efforts over longer periods of time have been described as typically capitalistic in our type society, offering the prospect of long-term profit maximalization through labour-saving inventions (24). Braverman's thesis that technologies are as much social as technical, aiming explicitly at the control of labour, is well known in the sociology of work and industrial relations (25). Elsewhere (26) we have argued that such a thesis is over-politicized. Although at the operational level technologies are used to strengthen control over work organization and labour in general, it is essentially the directing influence of management over technological development, and not the very nature of science and technology, which gives technological development this character. From the study of relatively successful forms of technology interfaces in government technology policies (27) we can conclude that actors other than management can influence technological developments, but only when they build up a comparable commitment to goal realization by technological means. A major step in this process is the specification of "need" in terms which are accessible to programmatic research activities.

This then is the perspective from which we shall look at the experiences of the Science Shop, and its role in the "specification" and "translation" of the "needs" of the trade unions.

It is perhaps necessary to spell out some of the limitations of the argument in this paper.

We do not intend to explain technological development as such. We focus on the possibilities for external *actors* within the economic subsystem (management, unions) to influence technological development from their (normative) point of view. Nor do we intend to say anything about possible influence on science or technology in general, but only about science-based technologies. And because of the capital-intensive way in which this technology is being produced, we pay special attention to the knowledge-based multinational firm. Hence we will not discuss specific problems which emerge in collaborations between scientists and unions in the social sciences and economics (28), or in ecology (29).

We do however, want to clarify our conception of what "alternative" technology can be from a union's point of view. In the discussions on alternative science and technology, many different perspectives can be discerned. In their most radical form, alternative S & T are supposed to have different epistemological characteristics (30). (Among social scientists it is sometimes even claimed that the "social ontology" of the different theoretical schools determines the normative content of their research (31).) Another demarcation is the question of methods: alternative S & T should use other methods than mainstream S & T (for example, action research (32)).

To reiterate: we are interested in the question of the extent to which the unions can act as what is sometimes called a "leading edge consumer" specifying decision criteria, functional and eventually technical specifications for R & D programmes and projects. But as the reader will have appreciated, our concern goes beyond the analytical. Can we combine the trade unions' access to the relevant industrial relations, on the one side, and the high standard of R & D facilities at the Dutch universities, on the other, to create perspectives for alternative programmes which would at least shape the contours of technological alternatives?

An example may help to make clear the nature of our concern, beyond that of pure analysis. In many chemical industries, workers are engaged in shift-work; this means that they regularly have to work in the evening, at night and during the weekend. Social scientists have shown the negative consequences of this type of work-organization for health and for social life. But at this moment, in many cases it is not possible to stop using shift-work because of the technical characteristics of chemical production-processes.

Unions have — broadly speaking — two ways of handling this problem. They can either try to negotiate — accepting the existing technologies — about the negative consequences of shift work, and hence demand compensation (higher wages) or labour-time reduction for the workers involved (five instead of four shifts). Or, at another level, they could demand the formulation of a research-programme that should lead to new (chemical) process-technology that can be started and stopped more easily than the technologies of today. Only this latter demand could generate an input to the development of an R & D program in chemical technology and lead to the notion of a technological alternative which can again be dealt with in economic terms.

Left on their own, unions will choose the former strategy. To establish interest in the latter strategy at the level of S & T policies is a task in itself, which requires a mutual build-up of interests between unionists and scientists.

Trade Unions and 'Science-Based' Technology

As the Science Shop emerged from the academic year 1977—1978, we had to conclude that the demands put forward by the unions were (i) nearly always technical and not scientific, (ii) occasional and not general, and (iii) required service and not research. A scientific or technological problem "behind the questions" was never obvious.

From our science studies background we were aware that this was not accidental but systematic, so we decided to take a programmatic step. In March 1978, we launched a project on "Natural Sciences and Trade Unions"; from then on, we looked systematically for situations in which the unions were confronted with the effects of science and technology, and in which they could be expected to have to take these effects into account in their policies.

This more active attitude towards the problem of translation quickly led to success; through the local union we established contact with a shop stewards' committee at the large Amsterdam site of a major diversified Dutch chemical corporation, AKZO. The shop stewards' committee of that plant requested help on issues which *they* believed to be "technological", because they were facing a loss of about 50 jobs per year. Between May 1978 and February 1982, we collaborated on some 13 projects.

In the context of this paper we cannot go into details about the different projects, which were aimed primarily at discovering whether, and if so, how the unions were indeed confronted with the effects of S & T (33); our primary objective here is to establish analytically what conditions have to be fulfilled in order to enhance the integration of unionists' demands with R & D.

Strategic Management of Technological Development

A model of relatively successful management of the interface between S & T and external demand is available in "science-based" industries. To some degree, at the level of long-term planning, management is able to account for the obsolescence of the product portfolio, and consequently to anticipate new technological developments. The more it does so, the more it can take advantage of the leading edges it has developed in its own R & D laboratories.

Therefore, in contrast to a former stage of development when the main function of the industrial R & D laboratory may have been to keep in touch with the pool of knowledge and science, and to make it possible for the company to profit from it, the control of parts of these markets has now become a central objective of management. This has important implications for the organization of the relevant sciences, and in particular for the division of labour between industrial and public science. The relations between these two segments of science have themselves become the object of (industrial) control. Increasingly, industries, and not the institutions of science, control the flow of scientific and technical information (34).

In many knowledge-intensive sectors in which international corporations are in oligopolistic competition with each other, a crucial condition of success is the management of the interface between R & D and marketing (35). To that purpose, the company has to build up a control structure of its R & D facilities, which must allow it to establish a "leading edge" in at least a few of these science-based technologies. This implies that the organization can accomplish a superior degree of integration of the relevant streams of information, among which the relevant research fronts are most preeminent. To do so, the company's own R & D has to be strong, and it must be also strongly linked to the most reputable R & D facilities in the relevant areas.

It is not enough for a company like Philips to have its own research laboratories, and it is significant that these laboratories have been where the most advanced solid state physics has been done in the Netherlands in recent decades. At the same time, the company has to watch its prime competitors (such as Bell Laboratories), to stimulate the Dutch goverment to organize solid state physics in a number of other locations, to provide universities with professors who can conduct research by international standards, and to be a centre of intellectual activities which allow the company to claim its share in the reputational control system of solid state physics.

Although the position of Philips in the Netherlands may be an extreme case, the same patterns and the same attempt to achieve this sort of integration can be found in all major industrial corporations. S & T are no longer incidental to the production process; they have become a central concern. The emergence of internal R & D policies within the corporations has stimulated an awareness that a company cannot sustain competition if it is not able to organize the public sciences actively. This is not to say that these sciences do not have their own intellectual standards, their own internal stratification and selection procedures. In solid state physics, we have ourselves found that people in Dutch university laboratories were heavily engaged in what they believed to be "pure science", without industrial or direct social relevance. At the same time their colleagues at Philips were working on the same subjects with the same qualifications for strategic reasons (36). The existence of an academic community with its specific ideologies should not blind us to the extent to which the academic community has been integrated into the modern industrial system.

Barriers to Workers' Influence

The ability to control highly differentiated and highly diversified structures is essential to the success of a science-based multinational corporation. Such a system has to be supported by institutional structures which embody the principle of *divide et impera*: if industrial relations are of minor importance at the strategic level, institutional provisions have to prevent workers' influence at that level. To that purpose, it has become common practice among multinational corporations to run their local companies through national daughter companies whose formal organization does not correspond to their functional integration. For example, AKZO is run according to the scheme in Figure 1 (p. 144).

In this scheme the limited company AKZO Chemistry Netherlands Ltd. is *presented* as part of the AKZO *Netherlands* Ltd., although in reality it is a functional part of — and has about the same board of directors as — AKZO *Chemistry* Ltd., which is the international division for chemical specialities.

Once a segregation has been brought about between the national level and the international level, integration between R & D and marketing is concentrated in the international branches. Knowledge-intensity and internationalization belong together.

At the national level, operational planning and execution are the main issues. Management's explicit task at this level is to deal with local and national authorities and to bargain with the unions.

This model of organization severely limits the types of insights which can be brought to bear by the participants in alternative circuits such as those organized by science shops, etc. The labour force is pushed back

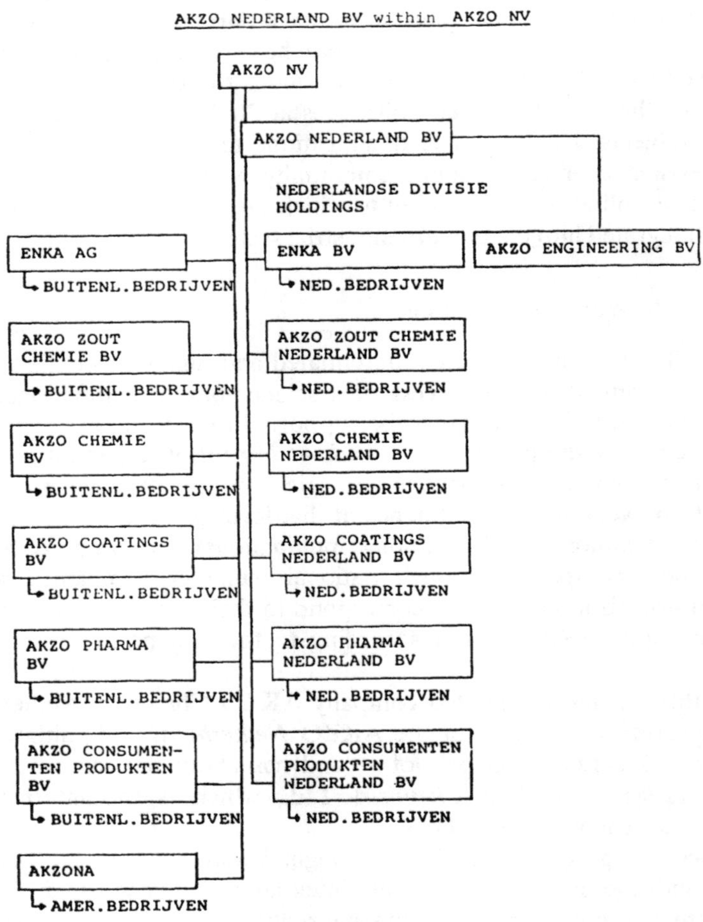

Fig. 1.

into a dependent role with respect to strategic information by the erection of new institutional structures. The unions have no access to the processes of strategic decision-making in which technological options are matched with future company needs. The problems they are confronted with are at the executive level, negotiated with management, and therefore problems of industrial relations. *As a rule the union will not have the possibility to develop its own interface with science and technology at company level "over the heads of their industrial partners".*

After this conclusion was drawn from the AKZO case, we wanted to know whether it was generalizable to other industrial sectors. To that end we conducted a survey of 37 alternative corporate plans ("workers' plans") by unions in Dutch industry. These "alternative corporate plans" follow the Lucas Aerospace model (37), and hence intend to spell out an alternative strategy for the firm with respect both to the market and to technology, in opposition to management's strategies. Assessing these plans, we found that workers' plans are always elaborated in reaction to a threat (of reorganization, closure, etc.) which is in fact generated by an earlier strategic decision of a corporation to disinvest in that activity (38). The decisions responsible for the threat of unemployment were made at higher levels of the company, and were thus beyond the control of the local unions. Hence, it was impossible for workers to propose feasible technological alternatives, even when they were relatively successful in establishing relations with external researchers and engineers.

However, all our cases were chosen from industrial sectors where what was at issue was innovation in basic technologies. It has been argued that the new information technologies which aim at systems development are more "flexible" and hence more accessible to workers' influence (39). Moreover, many industrial products are sold on world markets with heavy competition, whilst in other sectors, such as services, local factors might well be more important. It would therefore follow that other actors could be more influential, too.

These considerations led us to seize the opportunity offered in 1981 when the National Service Workers' Union asked the Amsterdam Science Shop for advice about the plans for a National Payment Circuit among the commercial banks and the Post Office. In this case we had office automation as a new technology in another empirical domain and in a market without strong international competition, and with the traditionally strong position of state-owned public services. Would our conclusions from the industrial case still hold under these conditions?

In 1981, after several discussions with the union representatives, we formulated a common project on "Trade Unions and Electronic Funds Transfer Systems" (EFTS).

Electronic Funds Transfer Systems

Although the emergence of electronic funds transfer systems is a process taking place in all western countries, its specific form and the speed of its diffusion differs from one country to another according to differences in the existing payment systems. It is beyond the scope of this paper to discuss the different EFTS developments (40). In general, the introduction of information technology in funds transfer is a development with important consequences for society at large. It implies the rise of a new (financial) infrastructure with consequences for such different things as the monetary control system, the structure of the banking sector, and the internal management information and control system of each bank.

A concomitant feature of all these developments is the list of *social* problems which accompany the "electronification of the payment system". Among these the following problems are to be mentioned: (i) the change in market relations, (ii) privacy, (iii) the accessibility and pricing of funds transfer services, (iv) the quality, reliability and safety of the system, (v) legal and juridical problems (who is responsible in case of failure?), (vi) the impact on the organization of banking institutions, (vii) aspects of the costs of EFTS, (viii) employment, (ix) the quality of labour (41).

Among the service sectors banking is particularly relevant for our subject because the automation of retail banking eliminates a large part of the manual control in funds transfer systems. Actually, banking is a labour-intensive sector which will gradually be transformed into a capital-intensive sector by the introduction of the new technology. The immense employment effects of this technological change make it likely that personnel policy will in this case have a strategic character: the speed of introduction of the new technology will be partly determined by the ability of the banks to get rid of their personnel (42).

Major automation projects for retail banking have been set up since the early 1970s. In the Netherlands, a leading project was initiated in 1972 by Philips, in cooperation with its home-banker AMRO (Amsterdam-Rotterdam Bank), to develop a datacommunications network.

Retail banking is predominantly an internal market. In Western Europe, and particularly in Holland, this market is to a large extent controlled by state-owned postal services. As early as 1975, the Dutch government founded a Steering Group to direct the design of an information-processing network between the different banks and the Post Office (43). Such a system was expected to integrate the different giro-systems between banks, on the one hand, and the Post Office, on the other, and to become at the same time an infrastructural provision for the exploitation of new electronic funds transfer services like Automatic Teller Machines (ATMs), Point of Sale Terminals (POS), etc.

It would take the Steering Group till 1981 to publish its first Green Paper for an integrated National Payment Circuit (NBC) (44). The conflicting interests between the banks and the Post Office could finally be brought to a trade-off when the Post Office made concessions in exchange for the condition that the NBC would use the public information processing network which was to be exploited by another division of the Post Office.

Technology Assessment of EFTS

The first studies dealing with the social effects of information technology became available only in the late 1970s, mostly from the social sciences. They emphasized the *different* nature of this new technology: in addition to and closely linked with the hard- and software, "orgware" had to be taken into account (45). This concept stands for the organizational knowledge which has to be brought to bear to let the new technology work. Because software can be built into the architecture of the hardware, and orgware seems to be highly integrated with software in systems design, it appears that one encounters here a direct operational interface between science-technology and its organizational and social effects.

The unions in this area, concentrated in the National Service Union and the Christian Service Union, both became aware of the impending impact of these technological developments only in the late 1970s. In order to develop "alternatives", both of these unions actively sought alliances with political parties, the Post Office, the unions of civil servants *and* the universities. One of the union leaders, himself a graduate of Amsterdam University, took up these issues and soon became a member of the Daily Board of the (larger) National Service Union. He stressed particularly the importance of a systematic search for alternatives such as "user-oriented systems design", technology agreements, etc. (46). From 1978 onwards he systematically addressed such questions to the Amsterdam Science Shop.

However, it took some time to agree upon a suitable topic. In our research we were primarily interested in the question of whether the social problems, i.e. the unemployment effects which could be foreseen, could have had some bearing in one phase or another on the design of the system. Of course, for the union which had been refused participation in the Steering Group, the quantitative and qualitative effects on employment and work were the most important aspects in terms of which the plans had to be assessed. However, next to this direct purpose, a thorough technology assessment of the NBC plans would be useful to enable the unions to influence public decision-making on the NBC (Parliament had still to deal with it!), and they hoped that such an effort could also lead to the formulation of alternatives.

In a certain sense, we had to gain access to the domain for our analysis by doing a technology assessment of the NBC for the trade-unions. The most important results of this study can be summarized as follows. First, we were able to provide a very detailed assessment of the employment consequences of the NBC plans. Combining information about the envisaged savings with information about the investments needed for the NBC, we could show convincingly (using the figures of the Steering Group itself) that they had played down the loss of jobs with their estimate of 700 jobs. As we received more precise information about the actual workflow in the banks and the proposed function of the NBC, we were able to predict with detailed arguments a loss of about 2500 jobs (47). Besides, we pointed out that this would only be a "modest" beginning: if the new electronic services like POS and ATM were to be implemented in The Netherlands and accepted by the public (which is a crucial variable here), job losses in other sectors could be several times higher (48). From our analysis, we were also able to specify in detail which categories of workers would suffer most from the NBC.

Social Effects of What?

Although the union was very successful in using our reports in the press and in Parliament, the central question for us — whether it really was *technological development* which caused these effects — had to be answered negatively.

In our investigation of whether other NBC designs could be more labour-friendly, we had to conclude that the employment effects of the NBC were affected by its technological aspects only to a small extent. Actually, the integration of the different giro-circuits is not primarily a technological but an organizational affair, which is facilitated by the emergence of the new technology. One can think of ways of integrating the two giro-circuits administratively, without information technology as a medium. When we actually calculated the employment effects of such an "organizational NBC", our detailed computations showed that in that scenario the reorganizations would cause the loss of almost as many jobs as the "technological NBC". Moreover, the same categories of workers would suffer from this "organizational NBC"!

Therefore, in these cases of office automation, it is not the implementation of new technology as such that causes unemployment, but its organizational form.

This also explained with hindsight why neither in the design for the NBC nor in the secret — but nevertheless available — minutes of the Steering Group could any arguments be found in favor of the thesis that the management of any of the participating institutes was implementing the NBC for the purpose of establishing better control over the workforce. In no instance was the decision-making on the NBC essentially influenced by considerations of the close connections between the new technology and its consequences for labour. Nor was decision-making influenced by the manpower problems which the Pactel report had claimed were of strategic importance particularly in this sector (49).

If it was not the technology itself, nor the world market — as we explained above, this was a domestic market — what then did guide the dynamics of this development?

A more detailed analysis of the specific market relations and competitive positions of the main actors made clear that their strategic considerations with respect to their market positions had nevertheless been crucial for the choices they had to make — ten alternatives were discussed — and for the compromises they were willing to accept. The introduction of EFTS made it possible to work out another arrangement between the banks and the Post Office. The Post Office could accept the integration of the two circuits — very much to its disadvantage — in exchange for some concessions from the banks regarding the structure of the sector. The major technical point in this — there were political points, too — was the realization of the NBC through the public information processing network, which was to be exploited by another division of the Post Office.

However, the social problems involved were, as in the industrial case, left over for the "operational" levels of each participant, to be worked out with their employees and the unions.

Of course, the next question to be raised is whether the organizational form of the technological development is not itself a part of the technology in systems design?

The answer is "yes and no": the high-tech side of the organizational problem is oriented towards the market, while the interface with the workforce gives rise to more trivial (not technological) problems, such as the organization of the remaining tasks.

Let us illustrate this with an example:

The actual *network* which has to carry the data can be drawn either as a "star" with central functions, or just as an infrastructure through which every participating institute communicates directly with everyone else. This choice has major implications for the participating institutions, but they are not of primary importance for the workforce because it all ends in a terminal, which in either case can have almost the same functions.

Because the banks already had a central institution to clear their mutual transactions and to prevent money from leaving their circuit (the Bankgiro-centrale), paradoxically the Post Office had an interest in decentralization which would break open these central functions. The Post Office feared that if the NBC were to be a "star-shaped network" and as a consequence deal with many central functions, then the Post would continue to suffer from the arrangements between banks. Therefore, this organizational aspect of the NBC, which comes down to a choice between technological alternatives, is of crucial importance for respective market positions.

However, it is not easy to imagine the consequences for the workforce from the choice of either of these possibilities. The essential choices at their end of the line involve issues such as whether to integrate in one function (and terminal) both the service to the public and the cashier function (50). To very large extent, such choices have nothing to do with the technology involved.

Conclusions from the EFTS Case Study

Of course, technological developments lead to an increase in the productivity of labour; but these effects are mediated through organizational changes which can, as in this case, themselves be the major source of the social consequences.

The choice between technological options can be vital for the entrepreneur, since the consequences of these choices affect the position of the company in the market; at the same time, however, the social consequences for the workforce are not considered relevant for this type of decisions, and the choices which can be made are actually rather indifferent with respect to workers' interests.

In this case the influence of the unions in decision-making was also blocked by the institutional arrangements between the banks and the Post Office. But even if they had not been blocked, it would have been very hard to think of alternatives because the technological options are coupled to something the unions are not directly interested in (market positions) rather than capital-labour relations (jobs).

Not only can firms deal better with the management of technologies when they differentiate between strategic planning and operational planning, but also the other way round: science-based technologies tend to be more accessible for those actors who are able to internalize the different (technological and organizational) dimensions of the problems in their organization.

Although we originally had strong reasons to expect differences between the case of technological developments in the chemical industry and the implementation of the new technologies in banking, we found our experiences in the former case very fruitful for the explanation of the developments in EFTS. As in the former study of AKZO, we tried to investigate the way in which technology affects the positions of workers, and the possibilities for trade unions to translate these consequences into what can be called "technological demand": a technological research programme which produces labour-friendly technology. The conclusions in this case are negative for both questions. First, the effects of the NBC on employment are caused not by the technology, but by the organizational choices which are made in the process of automation (51). It is not capital-labour relations, but market positions that are primarily related to *technological* choices. Secondly, as in the advanced industrial sectors, the unions in the banking sector are incapable of formulating research programmes in the way management can — translating one's goals in terms of R & D (as in the choice of an electronic NBC and of a certain type of network) — because their internal structure is shaped with respect to a local environment, in which they are institutionally reinforced.

In knowledge-intensive sectors, the possibilities for organizations to translate their goals into R & D, to produce (technological) knowledge and to implement it (influence its diffusion) depend on the position of the organization in the inter-organizational network and on the structure of the organization involved. In the case of the unions, we see that their position lies too much outside the relevant structures of decision-making for them to be able to articulate a functional interest in the technological decisions as they are actually being made. However, such a commitment is a precondition for success in influencing technological development (52). Of course, one can dream of a world in which Labour would be able to build up this level of differentiation within its organizations; but in the cooperation with unions in our present situation it is of the utmost importance to be aware of these severe limitations on the "demand" which can be brought forward by the unions (53).

Unions on High-Tech Markets

Although not central in our sociology of science perspective, the question whether unions as an alternative may have the possibility to exercise influence on technological developments by way of changing the market, either directly or through pressure on the government (e.g., legislation on health and safety), is relevant to the normative issues raised here.

Under the conditions of very typical neo-corporatist arrangements (54), the Scandinavian, and notably the Swedish unions, seem to be able to influence government policies in such a way that specific segments of markets are created. The best-known example of gearing this power to technological options is the UTOPIA project, in which unions and researchers collaborated to specify and to design new technology for the graphics industry (55).

In the project, experiments were undertaken with different forms of man-machine interfaces to develop requirements for graphic technology from a worker's (and product-quality!) point of view. Cooperation with the (state-owned) firm LIBER, which is producing graphic equipment, led to the development of equipment that satisfied the requirements of the UTOPIA team to a large extent. Hence one may conclude in this case that the unions were able to direct the development of an important element of graphics technology, the man-machine interface.

However, market forces necessarily come to play a role in the development phase: LIBER (and the unions involved) hoped to achieve a strong position on the Scandinavian market for graphic tools through the local graphic unions, which could force employers in their sector to buy these tools.

The firm decided to follow also a further strategy to expand from this very special segment of the market to the main market for graphic tools, which is predominantly the US market. (The whole Scandinavian market is less than 2% of the American market.) To that end, it had to adapt to the requirements of that larger market, which among other things meant lowering standards for quality. Because it could not integrate these two type of specifications, LIBER followed a "double" strategy for the two markets. Eventually the firm was not able financially to meet the two sets of standards at the same time, and failed in both markets. The equipment of the UTOPIA project did not pass its experimental phase.

In our opinion, this case can show that under very special conditions, local resource mobilization by unions can lead to technological change relating to specific aspects of (information) technology at the man-machine interface. The front-end character of this aspect, somewhat comparable with "health and safety" issues in the chemical industry, makes it accessible to union intervention, provided the unions can count on state support. Therefore, we can expect once in a while a counter-example to our general claim about the inability of the unions to integrate market perspectives and technological options within their organization — and we actually did find in our survey of the "workers' plans" some instances of successful collaboration between unions and scientists in the early phases of the elaboration of innovative ideas (56). However, in such cases unions and nation-states in the western world are extremely badly equipped to manage the process of technological innovation up to the phases of market introduction.

Conclusions

The idea of a possible integration of unionists' demands with university R & D was based on the conceptualization of science and technology as a knowledge production system: a socially contingent work organization which might be managed with different objectives. It was thus implied that management uses technological developments in order to bring about effects on employment, and that employees are in principle able to counteract these effects by organizing their own interface with S & T. However, we have argued that these assumptions are no longer valid for advanced "knowledge-based" sectors:

(i) although technological developments indeed have effects on employment, only in exceptional cases are these effects significant at the strategic level of managerial decision-making at which S & T are systematically incorporated;
(ii) employees are not in a position to organize alternative R & D policies which could counterbalance the economic integration of R & D and markets in complex and knowledge-intensive corporations.

The special role of employees, which can be legitimized by their presence in the enterprises, will have to be put in perspective: the experiences of workers are today *not* (or no longer? (57)) a point of access for a better understanding of the mechanisms which drive technological developments. (Of course, this is even less so in the case of issues raised in negotiations between unions and management, which are by their nature of a global character.)

What can be expected from cooperation between unions and scientists, given this state of affairs?

In our opinion, our conclusions have serious implications for the joint projects of unions and scientists in public institutes such as those advocated by science shops. Research on the social *effects* of technological developments has to be distinguished from *technological research* that is directed toward reaching social goals.

In so far as social scientists try to cooperate with unions in "technology assessment" (A in Figure 2), there exist possibilities for implementing a union's point of view in research questions. However, in that case the development of science and technology is effectively assumed, and the main purpose of the study is to explain — or even to predict — the social effects of given developments (such as, for example, office automation). These effects can be studied at the level of specific enterprises, at the level of a branch, or at the level of society at large, each requiring its own form of social scientific analysis.

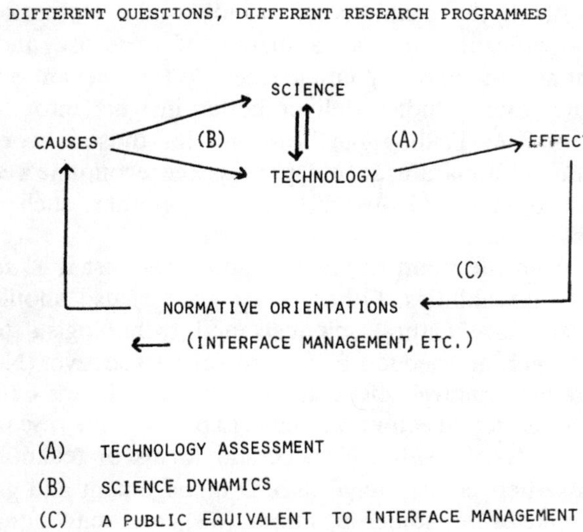

Fig. 2.

The primary aim of such studies is the better assessment or even the prediction of these social effects. Technology and natural sciences are relevant as sources of information. Researchers in these latter disciplines are needed for such cooperation primarily as experts on the relevant future developments. From the point of view of these researchers, this task has more to do with knowledge transfer than with real research.

The cooperation with external groups is a natural complement to such studies. As scientists are used as experts, a need emerges for counter-expertise, which can be provided by science shops and similar institutes.

A second group of studies (B) (of which the present account is one) belong to the sociology of science and technology. From this perspective, S & T and their practitioners are not mere sources of expertise, but the objects of study. In these studies we do not focus on the social implications *of* new technologies but on social influences *on* the development of S & T. To what extent and in what ways is the development of S & T to be understood by means of the social contexts of S & T? We would like to call this programme the "science dynamics" programme within the whole area of S & T studies.

In this context, the opinions of scientists and technologists have a different significance: not as a matter of expertise and counter-expertise, but as a method of gaining access to the relevant domains. To the extent that these studies deliver better insights into the steering mechanisms of S & T, they can be useful for those who exert some power on R & D apparatuses and possess the economic resources to stimulate developments in one direction or another, such as governments and boards of science-based enterprises.

However, from the point of view of public interest it is *desirable* to broaden these possibilities. Other social groups also should be in a position to stimulate natural scientists and technologists to develop technologies which are needed from a social perspective. (Note that in such a programme natural scientists are not only objects or sources of knowledge but the actual actors who have to perform the research!)

From the study of relatively successful forms of technology interfaces like those between strategic R & D management and government technology policies in some instances (58), we may conclude that the integration of insights from "technology assessment" studies and "science dynamics" studies into a normative perspective demands a special effort. Specific conditions are required which may vary with differences in various dimensions, such as the character of the organizations involved, the structure of the market one operates in, the time scale of the planning process, and the relevant disciplines. As we have argued, to achieve this in the knowledge-intensive sectors, the intervention of a strong and organizationally sophisticated actor is required.

The problem, in our opinion, is that it is doubtful whether parties other than those which already dispose of their own substantial R & D facilities, or can exert power on the market for new technologies, can generate the precise mixture of cognition and organizational power which seems necessary to act upon the S & T system. It is also doubtful, for the same reason, whether over longer periods such external partners could become functional participants in interorganizational arrangements exerting influence on technology, even when such an arrangement is backed up by the technology policy of the state. Without their own experience with the production and diffusion of scientific knowledge, these groups cannot stay in touch with the substance of the process except on a normative level; their contribution to the decision-making becomes formal, or degenerates to wishful thinking.

The disappointments over the role of experiments such as the Science Shop, the Colloque national, etc., in developing alternative S &

T policies can in our opinion be explained largely in terms of a lack of clarity about the analytical discrepancies among the intellectual and organizational questions involved in the envisaged integration (59).

Notes and References

* No order of seniority implied.

1. Cf. R. D. Whitley, "Black boxism and the sociology of science: a discussion of the major developments in the field", *Sociological Review Monographs* **18**, 1972, 16—92. See also: H. M. Collins, "The sociology of scientific knowledge: studies of contemporary science", *Ann. Rev. Sociol.* **9**, 1983, 265—285.
2. R. D. Whitley, *The Intellectual and Social Organization of the Sciences*. Oxford: Oxford University Press, 1984.
3. E.g.: M. Callon, "Struggles and negotiations to define what is problematic and what is not", in K. Knorr, R. Krohn, and R. Whitley (eds.), *The Social Process of Scientific Investigation*, Sociology of the Sciences Yearbooks, VI. Dordrecht: Reidel, 1980; K. Knorr, *The Manufacture of Knowledge: An Essay on the Constructivist and Contextual Nature of Science*. New York: Pergamon Press, 1981. See also: S. Woolgar, "Interests and explanation in the social study of science", *Social Studies of Science* **11**, 1981, 365—394.
4. See e.g.: M. Cooley, *Architect or Bee*. Slough: Langley Technical Service, 1980; LO, *Forskning för arbete och demokrati*, Stockholm: Tidens Förlag, 1982; P. Löw-Beer, *Industrie und Glück*, Berlin: Klaus Wagenbach, 1981. See also: L. Leydesdorff and P. Van den Besselaar, "Squeezed between capital and technology. On the participation of labour in the knowledge society", *Acta Sociologica* (forthcoming).
5. E.g.: E. Altvater, "Produktivkraft Wissenschaft?", in E. Altvater and F. Huiskens (eds.), *Materialien zur Politischen Oekonomie des Ausbildungssektors*. Erlangen: Politladen, 1971.
6. H. Marcuse, *One-dimensional Man*. Boston: Beacon Press, 1964, pp. 22f. See also: R. Richta et al., *Politische Oekonomie des 20. Jahrhunderts*, Prague/Frankfurt a.M.: Makol, 1968.
7. J. Habermas, *Technik und Wissenschaft als 'Ideologie'*, Frankfurt a.M.: Suhrkamp, 1968; J. Habermas, *Erkenntnis und Interesse*, Frankfurt a.M.: Suhrkamp, 1968; L. Althusser, *Pour Marx*, Paris: Maspero, 1965.
8. S. Mallet, *La nouvelle classe ouvrière*, Paris, 1963.
9. H. Braverman, *Labor and Monopoly Capital. The Degradation of Work in the Twentieth Century*. New York/London: Monthly Review Press, 1974.
10. See for empirical work e.g.: D. Gallie, *In Search of the New Working Class*. Cambridge: Cambridge University Press, 1978.
11. In 1973, at the Congress of the Scientific Workers League B.W.A., Arie Groeneveldt, the Chairman of the Industrial Workers' League — the largest Dutch union of that time — explicitly turned down the offer of external expertise, with the sole exception of expertise on health hazards from chemicals. *BWA-Ledenbrief* **5**, 1973/1, 9f.
12. BWA, "Instituten voor Maatschappelijk Gericht Onderzoek", *Wetenschap & Samenleving*, 1977/1, 125.

13. The government had supported the idea of alternative research facilities for public interest groups in a 1976 Green Paper on so-called *Sector Councils* for Science Policy. In these councils, the users of scientific results, government officials and researchers would advise on research priorities. However, in the changing economic climate of those days, the official policies were more and more reluctant to follow the "left of centre" university policies elaborating these ideas.
14. L. Leydesdorff, "Trade unions and university research-policy", *Higher Education and Research in the Netherlands* **24**, 1980, nr. 3/4. 54—58; L. Leydesdorff, A. Teulings, P. Ulenbelt, "Trade union participation in university research policies", *International Journal of Institutional Management in Higher Education* **8**, 1984/2, 135—146.
15. T. Ades, "Holland's science shops for 'made-to-measure' research", *Nature* **281**, 18 October 1979; L. Leydesdorff et al., *Philips en de Wetenschap*, Amsterdam: SUA, 1980. See also: L. Leydesdorff and H. van Erkelens, "Some social-psychological aspects of becoming a physicist", *Scientometrics* **3**, 1981, 27—46.
16. L. Leydesdorff and S. Zeldenrust, "Technological change and trade unions", *Research Policy* **13**, 1984, 153—164; Leydesdorff, Van den Besselaar, *op. cit.*, 1986. Note 4.
17. Leydesdorff et al., *op. cit.*, 1984. Note 14.
18. L. Leydesdorff, *Werknemers en het Technologisch Vernieuwingsbeleid*, Amersfoort: De Horstink, 1984.
19. D. Mowery, N. Rosenberg, "The influence of market demand upon innovation. A critical review of some recent empirical studies", *Research Policy* **8**, 1979, 102—153.
20. J. Langrish, M. Gibbons, W. G. Evans and F. R. Jevons, *Wealth from Knowledge*. New York: Halsted/John Wiley, 1972, p. 57.
21. Mowery et al., *op. cit.*, 1979. Note 19, 147—153.
22. Among others: R. R. Nelson and S. G. Winter, "In search of a useful theory of innovation", *Research Policy* **6**, 1977, 36—76; G. Dosi, "Technological paradigms and technological trajectories", *Research Policy* **11**, 1982, 147—162; M. Teubal, "On user needs and need determination: Aspects of the theory of technological innovation", in M. J. Baker (ed.), *Industrial Innovation. Technology, Policy, Diffusion*, London, etc.: Macmillan Press, 1979, 226—289. See for the dynamics of technological trajectories also: D. Sahal, "Technological guideposts and innovation avenues", *Research Policy* **14**, 1985, 61—82.
23. Dosi, *op. cit.*, 1982. Note 22, 160.
24. N. Rosenberg, "The direction of technological change: Inducement mechanisms and focusing devices", *Economic Development and Cultural Change*. Chicago: University of Chicago Press, 1969.
25. Braverman, *op. cit.*, 1974. Note 15. See also: "Technology, the labor process and the working class", *Monthly Review* **28**, 1976; D. F. Noble, "Social choice in machine design: The case of automatically controlled machine tools, and a challenge for labor", *Politics and Society* **8**, 1978, 313—347.
26. Leydesdorff and Van den Besselaar, *op. cit.*, 1986. Note 4.
27. R. R. Nelson (ed.), *Government and Technical Progress*. New York etc.: Pergamon Press, 1982.
28. Cf. K. Fridjonsdottr, "Social change, trade unions and sociology of work", elsewhere in this volume.

29. See also: R. Eyerman, J. Cramer and A. Jamison, "The knowledge interests of the environmental movement and the potential for influencing the development of science", elsewhere in this volume.
30. G. Böhme, *Alternative der Wissenschaft.* Frankfurt a.M.: Suhrkamp, 1980.
31. A. Giddens, *New Rules of Sociological Method.* London: Hutchinson, 1976, 15ff.; R. Bhaskar, *The Possibility of Naturalism. A Philosophical Critique of the Contemporary Human Sciences.* Sussex: Harvester Press, 1979, 31ff.
32. H. Nowotny and H. Rose (eds.), *Counter-movements in the Sciences: The Sociology of the Alternatives to Big Science.* Sociology of the Sciences Yearbooks, 3, 1979.
33. Leydesdorff et al., *op. cit.*, 1984. Note 16.
34. D. Dickson, *The New Politics of Science.* New York: Pantheon, 1984; R. W. Schmitt, *Continuity and Change in the U.S. Research System*, Washington D.C.: School of Public Policy, George Washington University, 1985. Occasional Papers No. 1.
35. R. Rothwell, and W. Zegveld, *Reindustrialization and Technology*. London: Longman, 1985, 74—80.
36. See for an elaboration of the Philips-example: Leydesdorff et al., *op cit.*, 1980. Note 15.
37. Cooley, *op. cit.*, 1980. Note 4.
38. We have to make an exception for one case in which we are not sure what caused the deterioration of working conditions which in turn gave rise to that workers' plan. See for further details: Leydesdorff and Van den Besselaar, *op. cit.*, 1986. Note 4.
39. E. Mumford and D. Henshall, *A Participative Approach to Computer Systems Design*. London: Associated Business Press, 1979; U. Briefs, C. Ciborra and L. Schneider (eds.), *System Design For, With and By the Users.* Amsterdam: North Holland, 1983.
40. A. Bequai, *The Cashless Society. EFTS at the Crossroads.* New York: John Wiley, 1981.
41. R. Kling, "Value conflicts and social choice in electronic funds transfer system developments", *Comm. ACM* **21**, 1978, 8; K. King and K. Kreamer, "EFTS as a subject of study in technology, society and public policy", *Telecommunications Policy* **2**, 1978, 3.
42. "Employment legislation, trade union pressure and the banks' own recruitment policies will place constraints on the ability of the banks to change the number and type of staff they employ. (. . .) Banks who solve this problem will establish a competitive edged over their rivals. The whole area of manpower planning will present a major challenge to European banks in the 1980's." Pactel, *Automation in European Banking. 1979—1990,* Management Summary, 1980, 6.
43. In Holland, the Post Office has also its own R & D facility (the Dr. Neher Laboratories) which performs R & D at very high standards.
44. Stuurgroep Integratie Giroverkeer, *Onderzoek Voorontwerp Nationaal Betalingscircuit met gebruikmaking van het openbare datanet DN-1*, Amsterdam: De Nederlandse Bank, 1980.
45. G. M. Dobrov, "Systems assessment of new technology in decisionmaking in government and industry", *IIASA Working paper.* Laxenburg, Austria, 1977, 77—8.

46. W. van Gelder, *Automatisering de Baas*. Woerden: Dienstenbond FNV, 1983.
47. A. Ruiter, *De werkgelegenheidskonsekwenties van het NBC*. Woerden: Dienstenbond FNV, 1983. The figures are only indicative of the differences. Because with every discussion in the Parlimentary Committee new and higher figures became available, both the Steering Group and we had to adjust the estimates in each report. See for further details: P. Van den Besselaar, "Trade Unions and EFTS" (in preparation).
48. E. J. Kirchner, N. Hewlett and F. Sobirey, *Report on the Social Implications of Introducing New Technology in the Banking Sector*. Luxembourg: Official Publications of the European Communities, 1984.
49. Pactel, *op. cit.*, 1980. Note 42.
50. This example is also mentioned in H. Levie and R. Moore (eds.), *The Control of Frontiers. Workers and New Technology; Disclosure and Use of Company Information*. Oxford: Ruskin College, 1984. See for an elaboration: A. van Asch, *Case studie Nederlandse Middenstandsbank: automatisering, werknemersbelangen en bedrijfsinformatie*. Amsterdam: FNV, 1985, 93.
51. In Leydesdorff and Van den Besselaar (*op. cit.*, 1986. Note 4) we distinguished between two meanings of 'technological determinism: (i) technological development is a determined process, and (ii) technologies determine their social consequences themselves. Our point here is, that for labour technologies cannot be influenced in the first sense, but that there still is room left to influence the social consequences of new technologies.
52. Nelson, *op. cit.*, 1982. Note 27.
53. See for our more political conclusions: Leydesdorff, Van den Besselaar, *op. cit.*, 1986. Note 4.
54. Cf. Fridjonsdottr, *op. cit.*, 1987. Note 32. See also: P. C. Schmitter and G. Lehmbruch (eds.), *Trends toward Corporatist Intermediation*. London: Sage, 1979.
55. S. Böker, P. Ehn, S. Romberger and D. Sjören (eds.), *Graffiti. The UTOPIA Project*. Stockholm/Aarhus: Swedish Center for Working Life, etc., 1984. See also: P. Ehn, M. Kyng, Y. Sundblad et al., "The UTOPIA Project" in Briefs et al. (eds.), *op. cit.*, 1983. Note 39.
56. As has been said, we may have to make an exception for those specialties which focus on man-machine interactions in a very strict sense, such as "systems design", "quality of VDUs" and "health and safety" issues, because these issues can be dealt with without affecting strategic decisions about technologies.
57. "It would be possible to write quite a history of inventions, made since 1830, for the sole purpose of supplying capital with weapons against the revolts of the working class". K. Marx, *Capital I*. Moscow, 1961, p. 436. See also: Rosenberg, *op. cit.*, 1969. Note 24; Noble, *op. cit.*, 1978. Note 25.
58. Nelson, *op. cit.*, 1982. Note 27.
59. See also: L. Leydesdorff, "The development of frames of references", *Scientometrics* **9**, 1986, 103—125.

PART III

COLLABORATIONS IN NATIONAL CONTEXTS

THE ORIENTATION OF THE PUBLIC SCIENCES IN A POST-COLONIAL SOCIETY: THE EXPERIENCE OF INDIA

RADHIKA RAMASUBBAN and BHANWAR SINGH

Centre for Social and Technological Change, Bombay, India

Introduction

The forces which condition the development of scientific and technological activity in the late (or newly) developing countries embody several contradictory elements. Although modern science and technology were introduced into these countries in a limited way during the colonial period and their importance highlighted during the freedom struggles, it was only after the attainment of political independence that they were accorded the role of major knowledge producing institutions, i.e., where scientific knowledge is highly regarded and its production strongly supported. The State, in these countries, has sought deliberately to steer the scientific system under its own sponsorship, direction and management in order to turn it into a tool of economic development.

Unlike the industrialised countries where the relationship between the scientific system and industry has resulted from an organically evolved process since the days of the industrial revolution (1), in the former group of countries — the latecomers — the forging of a partnership between scientific establishments, industry, agriculture and other arms of the economy and polity became a major task to be accomplished through consciously defined policies and programmes. These conscious policies, again unlike the case of the forerunners, cannot presuppose the existence of strong university centres functioning as the anchors of a thriving scientific system wherein a relatively autonomous scientific community has evolved its own forms of organisation, controls, orientations and momentum for knowledge production and its validation. Rather, the transformation of tradi-

S. Blume, J. Bunders, L. Leydesdorff and R. Whitley (eds.), The Social Direction of the Public Sciences. Sociology of the Sciences Yearbook, Vol. XI, 1987, 163—191.

tionally agricultural and artisan societies into predominantly industrial societies through the development of a monolithic modern manufacturing sector, under the umbrella of State-sponsored knowledge producing institutions, became the overarching ideology during the 1950s and 1960s.

This economic transformation, however, has faced intractable problems, thereby shifting the main focus of social and economic analysis from "development models (2)" to the understanding of under-development. So, too, has the scientific system floundered, and the main challenge for social studies of science here lies in highlighting the complexities in the working of the scientific system and in identifying the processes which stifle scientific and technical activity in these societies.

Out of the disillusionment with the performance of the scientific system, particularly its failure to generate meaningful innovations to suit grassroots needs, there has emerged, since the onset of the 1970s, a wide spectrum of initiatives attempting to forge coalitions between members of the middle-class intelligentsia and the disadvantaged sections of the population, particularly in the countryside, with the former articulating the knowledge interests (3) of these coalitions. These interests range from a move to take science to the people through the propagation of rationalism and campaigns against superstition, to promoting rural development through the spread of awareness of improved techniques, to the concern to temper the negative effects of science and technology on people's lives. Only a very small number of these groups, however, have attempted to translate their disillusionment with the official science and technology strategies into an experiment with an alternative pattern of scientific organisation and development.

The scientific and technical activity under these coalitions is not organised around any single, uniform principle; it is rather governed by the interplay of a variety of factors. The replacement of the competitive principle by the harmony principle is one. This contrasts the ever-increasing destruction of the ecological balance through the exploitation of natural resources for industrialisation with the redressal of ecological balance through the controlled and collective use of natural resources by all the inheritors of the planet. The extension principle is another. Here, the notion of science feeding on science is discarded, and addressing scientific work directly to combating hunger, thirst and disease is seen as more important than scientific contribution per se.

That even original and significant scientific contributions can only come out of a scientific ethos grounded in the social reality is yet another principle at work. Finally, the development of non-hierarchical, mutually supportive working cultures as a more conducive climate for the pursuit of professional interests, compared with the hierarchic and administrative controls characteristic of State-sponsored scientific institutions, has functioned as a guiding principle in the organisation of this alternative pattern of scientific activity. Our main challenge lies in discerning the underlying strength of these processes through marshalling evidence on their crystallisation into distinct organisational forms, stimuli and responses.

Examination of the Indian experience is, in many ways, illustrative of the manner in which all these factors affect the direction of the public sciences in the post-colonial societies of the post-war period. We first present a background note on the Indian scientific system as it has evolved from the days of colonial rule. The following section outlines the institutional setup and the quantitative growth of scientific and technical institutions and personnel. This is followed by a discussion of the interaction between the social and the scientific systems, which discloses the process of tying the scientific system to the reigning development paradigm. We see how the latter has stifled and distorted the former, and how the management of the scientific system along bureaucratic lines resulted in research activity being plagued with problems of hierarchy, lack of coordination, and the distribution of personal patronage. The final section is devoted to the emergent grassroot coalitions, outside government or industrial control or direction, their stimuli and responses. In order to display their complexity we have attempted to construct a typology of these processes.

The Indian Scientific System: A Background Note

Modern science as an organised activity in India began during the period of British rule. Although organised primarily to meet the strategic needs of the empire — army, trade, and the welfare of European inhabitants — it served as an important base for the growth of scientific activity in India. The period in which the British Crown took over India from the East India Company (1857) provided the impetus for systematic scientific investigations to aid the expansion and economic penetration into the interior, in order to exploit the agricultural and mineral raw material resources of the country for export to Britain to feed the growing industry there. This required the establishment of a few research institutions and agencies, but the scientific personnel who manned them were entirely British. The imparting of scientific and technical skills among Indians was first necessitated by the expansion of railways and other public works, which required their involvement at the lower levels, and for this a few engineering and technical schools were set up (4).

It was only around the turn of the century, coinciding with the final phase of British rule in India and the decline, in global terms, of Britain as a major industrial power leading to intrusions into Britain's monopoly in India by other rising industrial nations, that scientific activity in India under the aegis of the colonial rule received its final spur. During this phase research efforts in agriculture were expanded in order to increase the exports of agricultural raw materials and food grains. The intense competition between the European colonial powers in seeking cures for major tropical diseases impeding the further colonisation and exploitation of the tropics led to the establishment of a nucleus of laboratories for medical research (5).

Pressures from the growing nationalist movement, which felt that the lack of facilities for scientific education and research was inhibiting the industrialisation of the country, were also responsible for laying the foundation of scientific education by the colonial government. The country also witnessed the beginnings of indigenous efforts to foster scientific activity, reaching its peak in the brief spurt of nationalist science arising out of the Swadeshi movement of Bengal. The potential revolutionary implications of science did not fail to move a section of the Indian intelligentsia and found expression in three inter-related streams of development: the establishment of forums for intellectual communication among the small embryonic Indian scientific community such as the Indian Science Congress (1914), the demand from Indians for more facilities for science education, and the stress on the need for harnessing science to promote industrial development and to combat disease, hunger and other forms of physical deprivation among the mass of the Indian people. The advancement of science and technology was an important plank in mobilising popular support for the nationalist movement in India which sought freedom from colonial rule.

Therefore, when in 1947 India achieved independence, it launched an ambitious programme for scientific and technological "self-sufficiency". The institutions which had been set up by the British were upgraded, diversified and strengthened. New institutions and organisations were set up including apex bodies to provide the lead in almost all major areas of scientific activity. Many of these institutions were endowed with special privileges and prestige through their statutes, financial allocations, and ease of communication with the political leadership. A large infrastructure for scientific research has come into existence. There are today more than a hundred and five universities in the country (including seventeen agricultural universities) which impart higher learning in science and provide facilities for research and fifteen centres of advanced study in science subjects, including five national institutes of technology. There are eight big research agencies: the Department of Science and Technology; Department of Electronics; the Department of Atomic Energy; the Department of Space Research; Council of Scientific and Industrial Research; Indian Council of Agricultural Research; Indian Council of Medical Research; Defence Research and Development Organization; plus thirty national laboratories and a network of more than two hundred other laboratories and research institutions organised under various government departments such as food, fertiliser, education, petroleum and mines, etc. The total number of scientific and technically trained personnel now exceeds one million (6).

However, the veneration for science of the early years (the 1950s and '60s) when these institutions were regarded as "temples" of modern India has gradually given way to disenchantment with this national resource. Exercises undertaken to evaluate the functioning of these various institutions reveal that scientific activity in the country is both stifled and distorted. The expression of this state of affairs came from the highest level of political authority, the Prime Minister of India, at a recent press conference. When asked whether the liberalised policy of his government towards the import of foreign technology and technical know-how was not going to have an adverse effect on the indigenous growth of scientific and technical know-how, the Prime Minister gave a categorical reply that the output of all this activity in India hitherto had been "rubbishy" research. This implies that the Government of India does not, under the given conditions, recognise the existence of any trade-off between the strengthening of its own scientific and technical system and the reliance on foreign technology. But the Prime Minister has also been known to make eulogised references to the progress of science and technology in India during the last few decades: a common ambivalence which arises because of the ostensibly praiseworthy achievements of the Indian scientific and technical system in "prestige" areas such as the independent construction of fast-breeding atomic reactors, the launching of satellites, the creation of a base in Antarctica, and the manufacture of relatively sophisticated military equipment. Perhaps in those areas which the ruling class regards as critical to its prestige, strength and international standing-effective and efficient scientific activity is achievable. But for the purpose of this analysis we limit ourselves to the organisation of the scientific system in relation to industry and agriculture. The environment of medical research in India has been discussed by one of us elsewhere (7).

The Infrastructure of Science

Organised scientific activity in India is divided between two major streams: the universities and technical institutions; and in addition, various research laboratories and institutes which work under the overall direction of the apex bodies such as the Council of Scientific and Industrial Research (CSIR), Indian Council of Agricultural Research (ICAR) and Indian Council of Medical Research (ICMR).

As during colonial rule, the main obsession of the universities continues to be with producing large numbers of graduates, giving rise to the misconception that India has the world's third largest complement of scientific and technical personnel and that therefore Indian science is doing well. To the extent that technical personnel are required for production activity in the various sectors, this need is met by these institutions. Teaching rather than research is their raison d'être (8). There are, however, some individual scientists and scientific teams within this system with considerable reputations for research and whose work conforms to the normative imperatives of an autonomous sub-system shielding itself from external influences (9).

The bulk of scientific research in India, however, is carried out in the laboratories and institutes working under the overall direction and supervision of the various apex bodies. Scientific activity here is primarily government sponsored, and is conceived mainly as a tool for social objectives. Since science and technology were seen as the major instruments in achieving the national socio-economic goals, the first task undertaken by the Government of India was to expand the organisational structure for science and technology by setting up these laboratories, institutes and councils. The establishment of this base, accompanied by a progressive increase in expenditure on scientific research, has been the most significant aspect of scientific activity in India during the last three decades (10). The expenditure on scientific and technical research and related activities since independence has multiplied by a factor of sixteen. As a share of the GNP it has increased from 0.2 percent to 0.85 percent. Out of the total expenditure of Rs. 12,375 million in 1982/83, only ten percent of the expenditure was incurred by the private sector. While the expenditure on scientific and technical research and related activities by the private sector in the last two decades rose from Rs. 1.5 million to Rs. 423.5 million, the expenditure by the Central government on these activities increased from Rs. 218 million to Rs. 12,136 million. Two-thirds of this expenditure was incurred by the apex bodies. Inherent in this logic of scientific activity and its organisation is a system of frequent appraisal or performance evaluation. We have the evidence of various committees and commissions on the performance of this system (11). As a result of these evaluations the organisation of scientific research has been undergoing continuous changes. The approach underlying these exercises has been essentially of an administrative nature, with attention given mainly to issues pertaining to administrative hierarchy, length of tenure, selection and promotion procedures, coordination, etc. (12).

Our argument is that the organisation of scientific research along bureaucratic lines has resulted in a stifled and distorted scientific activity in the country. Widespread frustration and dissatisfaction among scientists have been reported for reasons such as the pirating of the work of junior scientists by their bosses, the denial of promotions and research facilities to those who deserve them, the subordination of working scientists to scientist-administrators, and the granting of facilities and rewards for inappropriate reasons. The prevalence of a bureaucratic culture geared to political expediency has given rise to instances of false claims made before verification or the over-advertisement of favourable results setting aside the principle of organised scepticism. Much of the work consists of the inferior reproduction of known innovations, i.e., of reinventing processes or repetition of work done elsewhere.

Interaction between the Social and Scientific System

Notwithstanding the impact of the administrative approach on the functioning of the scientific system, scientific activity in India is undermined by the contradictions inherent in the process of economic development itself. As a result of relying on western models of economic growth, the existence of mass poverty in India was seen as a reflection of the over-crowding of the labour force into low wage and low productivity activities and sectors. The growth of the modern sector — manufacturing, processing, transport and service — therefore became the main plank of development planning in India. It was hoped that once the modern sector was bolstered through the acceleration of the rate of investment there would be a quantum shift of the labour force into high wage and high productivity jobs which would lead to amelioration of mass poverty through improved incomes and the inculcation of modern values and norms.

The first five-year plan (1951—56) was essentially a documentary dressing up of a set of projects most of which were already under way. The intellectual content of development planning in India was acquired only with the formulation of the second five-year plan (1956—61). The pattern of investment in this plan reflected a distinct preference for capital goods industries and was guided by the two-sector growth model developed by Mahalanobis (14). Under a set of particular assumptions, the Mahalanobis model demonstrated that a growth pattern which maximised the production of investment goods and minimised consumption in the early stages would ensure a high rate of growth in output and consumption in the long run. The plan had travelled not even half way when it was afflicted with serious resource bottlenecks and had to retreat to a list of "core" projects. Tax revenues could not be raised to desired levels, the public sector profits gave no evidence of relief, and the recourse to heavy doses of deficit financing pushed the general level of prices sky-high. The increase in demand for imported raw materials and capital equipment under the conditions of static export earnings drained the accumulated stock of sterling balances, resulting in a serious foreign exchange crisis. The third plan (1961—66) marked a significant intellectual improvement as it attempted to apply multi-sectoral models to examine balances in demand and supply. Larger investment outlays were allocated to agriculture, but the emphasis on heavy industry continued. By the end of the third plan the Indian economy started witnessing a downward trend in overall growth (15). The 1965 war with Pakistan, accompanied by crop failures, necessitated heavy imports of food. The balance of payments crisis was accentuated. At the end of the third plan there was a suspension of planning for three years, and there followed a period of industrial stagnation.

The inherent limitations of this approach to economic and social planning were fully exposed when after a period it became difficult to maintain a steady rate of investment — and in some spheres, even utilising the already created capacity became a problem. The ticklish problems of foreign aid, international commodity and technology markets and the structure of domestic demand became too severe a constraint to allow this strategy to run its full course (17). The modern sector jobs have not even kept pace with the net additions to the labour force (18).

The large infrastructure for scientific research was created with a view to aid the process described above. It is not surprising that special emphasis was given to building up institutions for research relating to industry, which had been neglected in the colonial period. It was also stressed that Indian scientific development should be of an order where it would soon meet international standards and be able to compete on an equal footing with the advanced countries of the West.

The import substitution strategy undermined the effectiveness of the scientific system at two levels: (a) competition from metropolitan science and technology, and (b) the inherent limitations in the import substitution strategy (some of which are outlined in the above paragraph), which do not allow it to run its full course in a developing economy.

In spite of the impressive infrastructure for scientific activity and the adoption of policy resolutions seeking a cross-fertilisation between research and industry, there exists a glaring hiatus between the research apparatus and its users.

As compared to the expenditure incurred by the CSIR, the earnings from the processes released to industry have been insignificant (2 to 3 percent) (19). This makes a sharp contrast with the situation in advanced countries like Sweden and the U.S. (20). According to CSIR officials this is due to the fact that it cannot obtain favourable terms regarding royalties, premia, etc., due to competition from imported know-how, which not only comes in the form of package deals, but also demands higher royalties because of the performance guarantees it offers (21). Under these conditions the CSIR has to go more than half way and, in order to have its processes accepted by industry, release them at a throw-away price: often free of charge. But the value of the products turned out by the processes released by the CSIR, as compared to the total industrial output, has also been negligible (0.3 percent); even the expenses incurred by the activities of the CSIR is four times the value of the output generated through its processes (22). What is more important is that none of the evaluation exercises have come out with any definite evidence of research programmes leading to a self-generating scientific effort and an independent technological base for Indian industry (23).

An explanation for this state of affairs in scientific research in the field of industry can be sought in the ensuing uncertainties in its role as an effective agent in the import substitution strategy. Uncertainties arose from wider social and economic factors. Apart from the ineffectiveness arising out of the weaknesses inherent in its organisational structure, which has evolved along the principles of public administration, industry's reliance on foreign technical know-how took away the sense of direction from the scientific system. Right from the beginning of the planning process conditions were created for importing advanced technology for the industrialisation programme (24). The absence of any definite policy and mechanisms to bring about the cross-fertilisation between industry and scientific research, together with the increasing dependence of both private and public sector industry upon foreign financial and technical collaboration due to the government's liberal policy, have resulted in the neglect of the CSIR laboratories by possible users. In the face of the denial of its predesigned social role, the scientific system has been left floundering. As a result, the choice of research projects in the national laboratories has for the most part been on an ad hoc basis. Reducing these laboratories and institutions to the position of a periphery is also manifested in the much greater expenditure on their administrative activities than on research activities (25).

The failure of the import substitution strategy to bring about the transition to a monolithic modern sector is in itself a contributing factor in the marginalisation of the scientific system, particularly when the latter has been designed mainly to meet the specific needs of such a course of transformation. As a result, the scientific system has a narrow base to operate on and the competition from imported technical know-how makes this even narrower. However scientific activity relating to industry has, during the last few years, witnessed a new lease of life. This is due to: (a) the demonstration effect of a science and technology push, at the global level, in the field of electronics; (b) input shifts — fossil fuels to bio-gas and bio-mass fuels — caused by physical shortages and price escalations; and (c) the responsiveness of expatriate Indian scientists in the U.S., Canada and Western Europe to bring in capital and technical know-how for setting up science and technology parks/villages in India. It would be premature to pass any judgement on these trends.

Agriculture

Scientific research in agriculture offers an interesting comparison with the research in industry on two counts:

(a) Though concerned mainly with exportable cash crops, the research apparatus in agriculture at the end of colonial rule was of greater magnitude than that in industry.

(b) While scientific research in the post-independence period in the field of industry has faced severe competition from foreign technical know-how, the agricultural research system has evolved mainly under the patronage of the latter. Since both systems operate within the same social framework, however, the net result has not been vastly dissimilar.

For the first decade and a half after independence agricultural research occupied a low priority. Institutional reforms concerning the ownership of land and community development programmes were considered of greater value in raising agricultural production than research-based technological changes (26). During the late 1950s two joint Indo-American study teams highlighted the need to strengthen agricultural research and education (27). The Government of India, while accepting their recommendations in principle, did little to follow them up. It was the accentuation of the balance of payments crisis and the decline in overall growth, along with the heavy import of food in the mid-1960s, which necessitated the redressal of the imbalance; and the development of agriculture had to be accorded its due place while fixing priorities for the planning process. The fourth five-year plan document (1969—74) made an explicit statement of this shift. The case for a more research oriented agricultural development policy was again revived and importance shifted from institutional to technological solutions to the food problem (28). The breakthrough in wheat technology — the miracle seed for short duration dwarf varieties — reported by the Rockefeller Foundation in Mexico found an immediate response in India. A package of practices that would radically improve productivity under tropical and sub-tropical conditions was now available for the wheat crop. The only disconcerting feature was that since this technological solution was fertiliser and irrigation intensive and would require greater managerial skills, it would be confined only to the better-off sections of the farming community in a selected number of agro-climatic zones. The foreign funding agencies which were actively involved in the agriculture extension services under the community development programme in India, of which the Ford Foundation was the most important, strongly advocated the need for "aggressive research" oriented towards revolutionising the production techniques of the competent few rather than research which focused on all categories of farming households, which would necessarily be simple and non-revolutionary in impact. An example of this latter research would be the emphasis, hitherto, on crop rotations, seed preparations, the rational use of composts and manure, and improved farm implements. Funds from extension services under the community development programme were being reallocated to research activities. The foreign agencies also shifted their support to agricultural education and research institutions. Six new agricultural universities were set up in collaboration with six American universities along the lines of the US Federal Land Grant Universities. The Indian Council of Agricultural Research (ICAR) was reorganised. Scientific research for cereals was strengthened and special emphasis was given to fostering a few "centres of excellence" (29). The situation of meagre financial allocations to agricultural research vis-à-vis industrial research was remedied. The position of agricultural scientists in terms of pay scales, career prospects and working conditions was improved. Physical facilities by way of the supply of modern tools — precision instruments and other laboratory equipment — and library facilities were upgraded.

This sense of buoyancy in scientific research relating to agriculture, however, gets considerably watered down when we look at the performance. Apart from the wheat crop which got a tremendous boost following the Mexican experiments, there is very little else to point towards the radical transformation of Indian agriculture (30). In wheat, too, the situation has reached a plateau. Further gains in productivity have become increasingly rare. Agricultural research has not shown any appreciable success in breeding varieties of seeds to cover other crops, such as coarse grains, oilseeds, vegetables, spices and condiments, or other environments — areas with heavy rainfall and arid zones.

The sudden catapulting into importance of the agricultural research structure in the mid-1960s has, on the contrary, resulted in hastily formulated changes in the organisational setup and unnecessary expansion. Research activity is plagued with problems including the ambiguous delineation of responsibilities and jurisdictions of the different institutions, the lack of effective coordination mechanisms, hurriedly formulated research schemes, the dominance of a political and bureaucratic culture, and the management of research for distributing personal patronage, which kills initiative and drive on the part of competent scientists (31).

The predominant feature of agricultural research since the mid-1960s is that it has been mainly involved in adaptive research through the various All India Coordinated Projects, working up largely on the know-how already developed in the international research institutions. There have been some positive features of this process itself. For example, in the case of wheat, *bajri* (a coarse grain) and maize (the principal crops covered by the HYV seeds) there has been a marked acceleration in the growth of output. The rise in output levels of some of these crops, particularly wheat, has resulted in substantial food stock reserves, and it is argued that Indian agriculture can now withstand the erratic behaviour of the monsoons (32). Further, there is some evidence which shows that improved agricultural performance is associated with reduction in poverty (33).

Considering the extent and severity of poverty in rural India, however, the main challenge of the agricultural research system lies in developing technologies both for the farm and the non-farm sector to suit the needs of different agro-climatic zones and environments. The Indian research structure has been ineffective where it has had to innovate for indigenous problems and locations (34). A recently completed study (35) by one of the present authors demonstrates that the response of new technologies in respect of poverty amelioration depends substantially on structural conditions, not just at the level of states or the agro-climatic zones of India, but also at the village level in line with institutional and social history. The poverty-technology link is determined by differences among types of villages, and the uniformities of behaviour and response within a village of a given type.

The scientific and technical activity spearheaded by grassroot processes assumes particular significance in this context.

Evolution of the Grassroot Processes

While the mainstream scientific research in agriculture kept itself firmly oriented towards seed-fertilizer-irrigation technologies because of the enhanced prestige it afforded through links with the American agricultural research system, and was, in turn, enmeshed in a bureaucratic muddle (36), the grassroot processes were in some measure sparked off by the shift in political rhetoric to "garibi hatao" or "banish poverty".

The economic crisis of the mid-1960s, as reflected in the balance of payments crisis, the existence of idle capacity, and mass starvation due to the successive failure of the monsoons and the stoppage of PL 480 food aid during the war with Pakistan, necessitated a shift by the ruling class in economic thinking and power base. A split occurred in the ruling party. The section which retained power nationalised the banks with a view to offsetting the humiliation suffered on account of the devaluation of the rupee, purportedly under pressures from the World Bank and the United States government. It was argued that the banks would now come to the aid of the poor and the agricultural sector which they had hitherto fought shy of. Direct assault on poverty became the dominant political slogan. Apart from emphasising the development of agriculture through research in agronomy, the role of appropriate technologies in farm as well as non-farm sectors such as rural arts, crafts and non-conventional fuels was highlighted both in official documents and political rhetoric.

Organised scientific and technical activity was too preoccupied with seed-fertilizer-water technology and it was the grassroot processes which responded more positively to the challenges of innovating for specific local conditions. The fallout of the democratic rhetoric was the infusion of a pro-poor rural bias into the science and technology policy debate (37). This perspective was spearheaded by a few prominent scientists from leading institutions in the formal sector who saw in this the possibility of orienting the public sciences in an alternative direction, i.e., a strategy of scientific and technological development which, by directly improving the lives of over half the population, would make for a more egalitarian society. They perceived the categorical imperative of such a strategy to be the integration between the social aspects and scientific activity, and between production techniques and the physical environment.

Basing themselves on the notion of appropriateness, they argue that the compulsions of the Indian situation call for the orientation of the public sciences towards the development of low-cost, ecological, and elegant technologies in the service of the hungry, thirsty, sick and unsheltered. The emergence of these processes in India also coincides with the build-up of alternative science at a global level, with decentralisation, appropriateness, and participation as its key elements (38). The problems to which these groups have addressed themselves are not peculiar only to a given region or location. It is in attempting to match local resources with local needs, and in making their choices of research problems from within the possibilities and constraints set by the local milieu (rather than in pursuing research directions suggested by a perusal of scientific journals, usually western in origin), that they derive their micro-level or bottom-up character.

Although the role of science in social transformation for the benefit of the poverty groups is the central theme for all those participating in these processes, the approaches differ. According to one point of view, the present sciences are undesirable since their effects are "elitist, exploitative and explosive". But these effects arise from the nature of the "structures" which sustain these scientific truths, and therefore critiques of modern science relying on the same structures cannot bring about convincing changes. Alternative truths require alternative structures. Some of the elements of these structures are : (a) that there are no distinctions between science and other knowledge systems; and (b) that alternative truths are for exploitation by man but not for "harmful exploitation of person by person or non-compassionate use of earth" (39). Most of the population being poor and outside the commercial mainstream, science must begin with the study and application of those resources which are obtainable outside this mainstream. These resources today are industrial and agro-based wastes, and renewable energy (40).

Another perspective takes the present sciences as the starting point but seeks to shape them into a tool for bottom-up development through appropriate technologies (41). A third perspective subscribes to faith in science and the scientific method and believes in arming the poverty groups with this knowledge. Equipped with the scientific method and the ability to analyse the forces making for social inequalities, the poor can effect social transformation in their own favour (42).

There are nearly three thousand groups (43) working outside the formal mainstream which have come to be known as the voluntary sector. The majority are social action groups concerned mainly with the implementation of various welfare schemes in the rural and urban areas, or with taking preventive action against further deterioration of the physical environment: deforestation, the displacement of tribals due to commissioning of large-scale irrigation, power, or other industrial projects, industrial pollution and health hazards. There are a few loosely cohering groups of educated youth, largely in urban centres, engaged in debate and discussion about the science and society dynamics. Some of these occasionally conduct demonstrations attacking superstitious practices and popularising scientific ideas.

There is only a very small number (44) of groups engaged in the systematic production of knowledge and whose coalition with the local population has acquired a relatively greater significance. We have investigated the working of three such institutions in the country. One is a unit of a leading formal scientific organisation, but is attempting to evolve alternative criteria of scientific and social relevance. The second one is a voluntary research organisation located in rural surroundings on the periphery of a metropolitan town and supported by an industrial house from a region reputed for its philanthropy and support in the field of education. This organisation has severed itself from the mainstream of organised science to pursue a vision of an alternative path of scientific growth. The third is mainly in the nature of a regional association of scientists, science teachers, and social activists interested in raising the level of mass participation to protest against the negative effects of technological schemes launched in the garb of industrial/economic development of the region.

The scientific activity of the first two organisations pertains to basic research, applied research, and the designing of simple, low-cost, and integrated technologies using local resources for local needs. Their work is on renewable sources of energy; the recycling of bio-wastes; low-cost and energy-saving building materials; water harvesting; photosynthetic bacteria (for the treatment of industrial waste, for use as natural fertiliser, and for enhancing bio-gas generation); use of algae as a protein supplement through pot culture as well as mass culture; and in the field of agro-processing, e.g., making cellulose fibre from groundnut shells and edible cellulose as well as sodium silicate from rice husk. The various devices and applications which are being studied/developed by them include bio-gas plants, windmills, solar driers, stills, cookers, and ponds, wood gasifiers, shelter belts as wind control measures, bio-dynamic gardening, village stoves, and a single pan wood stove as a substitute for kerosene stoves in urban areas. The work of the third group consists mainly in the field of protecting the environment and existing skills and activities from the onslaught of developmental schemes such as large hydroelectric plants and industrial establishments. They have mobilised protests against "unscientific" and "wasteful" electricity and irrigation policies of the government, industrial pollution, irrational drug formulations, harmful drugs, and neglect of occupational health and safety.

In order to unfold the complexity of the stimuli and responses in the production and validation of knowledge in these organisational forms, we have worked out a typology based on interviews with a cross-section of individual scientists in these three institutions. This typology of actors' perceptions and motivations cuts across group boundaries and affords vivid insights into the functioning of these processes. (Some of them had reservations about being quoted or compared with other groups; therefore we are maintaining the anonymity of the individual scientists as well as of the institutions.)

The first category represents the approach which places these processes in opposition to the formal or established forms of scientific activity. The forces which alienated these scientists from the formal system came both from their perception of the formal system and from a dawning social perspective. They worked in esoteric areas of specialisation motivated mainly by the final reward of a research paper's acceptance for publication in a "standard, reputed" (read international) journal. The first few euphoric years of notional participation in an international scientific community, by virtue of subscribing to an internationally accepted paradigm, gave way to a realisation that communication — the multi-channelled interaction that sustains day-to-day scientific activity — was missing from the Indian scientific scene, even in its higher reaches. No one read these published papers. Even if they did, it did not matter, since the ideas were borrowed, and in an area actively researched and highly developed in the West, they could neither make a fundamental impact there nor be of any real relevance at home where different conditions obtained. Nor was there the satisfaction of working in an area of basic research for the sheer excitement if afforded, because such a ferment of ideas did not exist here, unlike in the advanced countries where even in esoteric areas there is frequent and intense interaction. The scientists found themselves isolated from the vibrancy of the international community. Nor did they feel that they were part of a vibrant local scientific community. In a system where ideas were derivative and the culture of interaction was missing, the race was to ape the West. Such an insecure community does not feel the need to be mutually supportive, and in fact there is a pride in ignoring each other's work.

Their feelings of dissatisfaction and irrelevance found a fruitful avenue in the grassroot processes whose potential caught their imagination. They became sensitised for the first time to the inequities in Indian society and to the total failure of formal Indian science to address itself to the task of redressing the problems arising out of these inequities. According to these scientists, they first underwent a process of unlearning the "modernisation model" that the educational system so effectively sells. For some of these scientists, the possibility of retaining links with the formal scientific organisation while yet doing high-quality research relevant to the needs of the masses provided the final spur for abandoning the imitative path. What motivated them was the desire to demonstrate that if grassroots problems were chosen as the starting point to inform research activity, it was possible to make an impact both upon the growth of scientific ideas and in terms of developing technologies relevant to the conditions in the field. In their new fields of activity there is a conscious attempt to redress the problem of isolation that afflicts the formal system. The working style is of a collective nature, drawing upon one another's resources and overcoming narrow specialisations through a multidisciplinary group orientation which also enhances the scientific and social perspectives of individual members. There are regular internal seminars and an eagerness to communicate with a wider, often "non-scientific" audience, using channels they wouldn't have dreamt of using earlier, e.g., a chemist publishing in a social science journal, a scientist/engineer working on building materials writing for a journal of architecture, or the same professionals participating in workshops conducted by a rural development agency or a seminar on peasant consciousness organised by a social science research institute. These is also a conscious attempt to interact regularly with village populations in the extension centres adopted by these groups, from whom they draw their research problems and to whom they go back from time to time for validation of their findings. Pamphlets in the regional languages explaining specific innovations and how they can be produced/assembled have also been evidence of such attempts.

The scientists coming to the grassroot processes, however, face considerable career insecurities, and many of them stay there only temporarily. The opposition of the formal scientific system to those who fail to observe its norms is brought out by the case of a scientist who, after a period of secondment to such a group, applied for a higher position in his parent organisation only to be disqualified for having failed to add to his list of publications during the period in which he was working on rural technologies. Scientists in this genre also find it difficult to inspire research (doctoral) students to work on the problems they have identified in the course of their work, as this would amount to jeopardising career prospects which are sought mainly within the formal organisational setup. These scientists feel the need for a thoroughgoing reform to guide the scientific stream in the direction of what they see as relevant research. According to them, the changes must be made at the school level itself — stressing learning by doing and breaking down the distinction between scholar and craftsman, a latter-day "renaissance" process (45).

For those who have opted out of the formal scientific system, neither the stimulus toward nor the validation of new knowledge is to be sought within the formal system. They see that the formal system, subsisting on the periphery of the international scientific system, is willing to recognise only its own kind and to seek approval from it would be pointless. Not for them is imitative knowledge construction or the compulsion to publish, where the guiding principle is "past achievements rather than conflicts over the relevance and significance of research results for future work and so the direction of research" (46). Many of these scientists have the strength of past reputations to back them, but what sustains them in their present paradigm is "fine tuning" to their social context. By so doing, they are able to eliminate the first and second-level constraints that operate upon scientists in the formal system, i.e., the restricted problem choice due to borrowed criteria of scientificity which leads to the exclusion of many potential research problems, together with the dependence on foreign journals as sources of ideas and as publication outlets (which leads also to dependence on sophisticated instruments, usually imported, for solving problems posed by western research programmes). What guides the choice of problems by scientists in the grassroot processes is "practice", through which a congruence is sought between what is conceived and what is possible, i.e., as one of them put it, "know-what and know-when is more important than know-how". A great many of the laboratory instruments are also devised locally, using local craftsmen with their long tradition of innovation with available resources. One scientist we talked to said that he had gone through a stint of village living when he had deliberately isolated himself from books, telephones, recorded music, and mechanical transportation to "teach his hands to think" and to nurture an ability to communicate meaningfully with the local craftsmen and mechanics. In situ research is an article of faith among this category of scientists. As a result, they have come out with fundamental innovations with high acceptability which have nevertheless been denied the label of scientific respectability since they neither grew out of past known research nor used accepted methods and techniques, e.g., using algae as a food supplement for human consumption as against leaf protein, which is known.

The second type of approach to the bottom-up processes is represented by those scientists who regard their work at the grassroot level as just another variant of scientific activity and who discount the possibility of any real conflict with the formal scientific system. For them, the main significance of grassroot activity lies in shaking this system out of its inertia and in affording an alternative venue of work free from administrative controls and considerations of hierarchy, and an alternative notion of rewards and work satisfaction. Some of these scientists who joined local universities after obtaining advanced degrees from western universities were soon disillusioned with the "all pervasive frustration among colleagues" and the "pettiness of university politics". They found the grassroot organisations to be much more hopeful and dynamic places, and the exposure to new perspectives and disciplinary orientations was refreshing and satisfying. Though the situation with regard to career prospects in terms of salaries, perquisites, and upward mobility did not hold much promise, and at times financial stringencies restricted the flow of facilities, many of them expressed the feeling that they never felt dominated by the negative aspects. Free from the sluggish environment and the hierarchical and administrative controls of most other conventional organisational structures, some of these scientists have been able to pursue significant lines of research in both basic and applied areas and develop a multidisciplinary perspective. That participation in unorthodox research teams is beneficial to the individual research worker is also evidenced by the pattern of out-migration from the grassroot organisations. For the bright but not necessarily extraordinary products of the Indian university system, participation in such teams — which provided strong focusing figures, individual freedom, the initiative to take decisions at every stage of the research process . . . which, indeed, provided motivation and facilitated individual and collective rewards — brought them both early recognition and offers from bigger formal institutions, and an edge over their peers from the formal system in the competition for jobs. To give a few examples, one scientist who participated in the early research on algae in one of the groups was made an offer by a US multinational planning a major programme on algae, an offer he found difficult to refuse. A biotechnologist joined a private firm in the country on the strength of his work experience in a grassroot organisation. Another scientist joined a reputed group of newspapers in the country as its chief science correspondent thanks to the multidisciplinary perspective he acquired during his work at the grassroots. For some of the scientists there does not seem to have been a fundamental shift in the area of specialisation as a result of joining these organisations. If they were active before, they continue to be active now, continue to be invited to symposia, meetings, and conferences, to receive reprints from all over the world and drafts of papers for comments, because they are specialists in the field. "You are valued because you work in a particular area, not because you work for this or that type of organisation". They do, however, admit to coming across odd situations on account of their working outside the formal organisational set up. For example, one scientist who applied for funds to an agency funding scientific research for work on a particular project was turned down on the grounds that the organisation hosting this study was a "rural development" organisation not falling into the category of research organisations. The grant would have been made available if the scientist concerned had been doing the same research in a university.

The third type of approach goes even a step further. Some of the scientists involved in these processes see in themselves the potential of producing relevant and far-reaching intellectual innovations and of thus performing, eventually, the role of pace-setters for the organised/formal scientific system itself. For these scientists, the formal norms of science must be adhered to even while working at the grassroot level. "Even a rural technology must fulfil the criteria of scientific and engineering skill and efficiency" was the comment from an engineer/scientist who specialises in heat transfer in rockets and is trying to apply those principles to designing a simple stove and developing a wood gasifier. This group of scientists believes that it is by adhering to the "scientific norms" in their work that they will be able to attack the "sneering" attitude of those working in the formal scientific system with regard to any talk of social relevance. Once it is established that scientific activity informed by social relevance and conducted outside the formal system can also be of high scientific value, more and more scientists will feel drawn to it.

The fourth type of approach to the bottom-up processes is in the nature of building up a mass movement for the social control of science. It is only with a redressal of the inequalities in the possession of scientific knowledge and control over the direction of its growth that science can become an instrument of social advancement. The inequality in the control over science and technology leads to inadequate or wrong applications of science, resulting in the adverse impact of "developmental" activities upon both the well-being of the mass of the people and of the environment. Irrational drug formulations, harmful and useless drugs made available to the people, is also an outcome of such inequality. Agitation for a safer environment and for a better health policy, therefore, have been the main thrusts of this movement hitherto. One of the hallmarks of the movement is its base among a wide range of professionals — scientists, engineers, and doctors — who provide the scientific and technical back-up for putting pressure upon government and industry for a more people-oriented development strategy. The other hallmark is the strengthening of the mass base by putting "science" in the hands of the common people, i.e., the ability to critically analyse their physical and social environment and to appropriate courses of action to transform this environment. This is done through the imaginative use of the spoken and written word, art and folk forms, public debates and lectures, scientific publications — books and pamphlets — in the regional language and innovatively designed science magazines again in the regional language, catering to the needs of school students and teachers. While the movement does not perceive the possibility of a thorough-going shift, under the present social system, in the orientation of establishment science and technology which it characterises as "elitist, foreign-oriented and irrelevant" (47), it does concede that individuals and groups of scientists motivated to engage in active research on problems relating to the poor do have an important role to play in the mass movement through their demonstration that innovations need not come from outside and that feasible technologies can be generated through interaction between scientists and the people for raising the living standards of the poor. The movement stresses this interaction of the scientist with people — at his extension centre as well as with mass movements, voluntary groups and political parties — if he is not to lapse into another kind of isolation and if he is to ensure that the social struggle takes place in favour of the technical solutions offered. But the scientist must needs be subordinate to the activist.

The sources of funds supporting the work of the three grassroot organisations we have studied range from membership contributions and donations from philanthropic foundations to project funds from government funding agencies such as the Department of Science and Technology and State Councils of Science and Technology and nationalised banks, and to block grants from international organisations/agencies such as the United Nations University and the International Federation of Institutes of Advanced Study.

Over the last decade and a half, a number of grassroot voluntary groups have arisen working in the area of rural development. The leading figures of many of them are engineers, scientists, and doctors who have dropped out of the formal system. While the activities of these groups are mainly in the area of nonformal education, particularly science education, health, and rural development, they provide an audience for the work by the scientists in the grassroot research organisations. To facilitate this kind of communication the research groups send out fairly regular newsletters in the form of technical notes communicating work in progress. They also have regular monograph series, and besides the production of manuals and pamphlets (often in the regional language) on specific designs, they also publish papers in national and international journals and participate in regional, national, and international seminars and conferences. Nor have invitations been lacking to active members of these groups for memberships in government advisory and policy making committees.

The conventional rewards, therefore, have not eluded these groups.

High visibility awards have also come to some of these research groups, particularly in their first phase of activity, instituted by government and by autonomous/philanthropic bodies both within the country and outside, and their work has also acquired visibility through the media.

One significant factor which has contributed to the high productivity and initial successes of these groups would appear to be their adherence to the norms and spirit of "revolutionary science" in the Kuhnian sense, an openness of enquiry unconstrained by preexisting notions. This, combined with open, democratic non-hierarchical organisational structures and the stress on team work and mutually supportive interaction, gives these groups their characteristic vigour. Strong focusing figures who see the vision of a new path for Indian science have provided the motivation and the original ideas.

Conclusion

The Indian experience is illustrative in many ways of the social direction of the public sciences in post-colonial societies. Realisation of the full potential of scientific and technical activity for the growth of the national economy and for combating mass poverty has been a common theme, with varying emphases, of "freedom movements". Scientific and technical activity is assigned an important role (backed by financial allocations, statutory provisions, and social prestige) in nation-building activity. Concomitantly, scientific activity is controlled directly or indirectly by the government through its supervisory or funding powers and the bureaucratic norms which dominate the reward structure. Partly due to the internal functioning of the organised scientific activity and mainly because of the contradictions in the social and economic processes inherent in a post-colonial society evidenced, for instance, in the lop-sided priorities and imitative growth path charted for Indian science, scientific and technical activity is rendered ineffective and distorted.

The floundering of the scientific system among the latecomers can be attributed to the failure of the development paradigm itself. Due to the various intractable problems described earlier, the development strategy has failed to transform the traditional economic scene of the overcrowding of labour in low wage and low productivity activities and sectors into a modern monolithic economy through a quantum shift of labour into high wage and high productivity jobs which inculcate modern values and norms. Instead, it has resulted in a strong dualistic structure where a narrow-based modern sector coexists with a vast traditional sector. The absence of the required dynamism in the modern sector took away, in a sense, the life-support system of the scientific and technical infrastructure which was created mainly to meet the needs of the former. The competition from imported technology further undermined the role of the Indian scientific and technical system even within this narrow base on which the modern sector was operating. In the field of agricultural research there was no infringement by western technical know-how. In fact it was inspired and supported by the latter, and scientific research in agriculture has been limited only to the adaptation of known processes under Indian conditions. Inspired by new-found opportunities derived from its relation to the research system of the American land grant universities, organised scientific activity in agriculture moved inexorably in the direction of replicating advanced country-oriented research, and the sole purpose of the working scientist became one of contributing to journals. However, lacking the stimulation of communication and interaction present in the scientific milieus of advanced countries even these efforts acquire an extremely mediocre character. Insulated by their limited successes in evolving seed-fertilizer and irrigation packages in wheat, *jowar*, and *bajri* crops, and enmeshed in an organisational culture whose notions of administrative hierarchy and controls has vitiated the research environment (as reflected in the rampant favouritism, piracy, and unhealthy competition such as the premature announcing of results in the mass media), organised research in agriculture has failed to respond to the stimuli offered by the shift in political rhetoric towards banishing poverty. The fallout of this rhetoric was witnessed mainly in the emergence of the bottom-up processes.

These processes differ from organised science in the country mainly in the sense of "shared social purpose" (48) that distinguishes the grassroot initiatives. This sense has led them to draw a wide range of actors into a network of communication around their research programmes, in an attempt to forge an altogether new coalition between the scientist and the poverty groups. There are links with other non-scientific voluntary groups, the hard-core scientists are also meticulous about peer reviews, and the groups receive funds from various departments of the Central and State governments. They derive considerable prestige from persuading the government(s) to incorporate in their science policy documents the set of priorities which fit into their perception of relevance. They argue against maintaining vast bureaucratic scientific empires and against the pathetic struggle to stay in internationally leading areas, and for a decentralised research thrust drawing together concerned scientists in existing institutes, laboratories and universities to respond to local needs and problems.

The predominance of these factors in governing the internal organisation of the grassroot processes also goes far to explain the coming together of individual scientists, with markedly differing motivations and perceptions on questions of epistemology and the social relations of science, to work together as a group, and even to move fluidly between the organised system and the grassroot structures. At the present juncture it is not clear if, how, and when these differences in individual perceptions and motivations will become sharpened and resolved. What is clear, however, is the lack of any legitimate reason for interpreting them as a strong coalition between the scientists and the poverty groups. The adherence to an extension approach in their scientific and technical activity, rather than a participative process between the scientists and the poor, is the main factor divesting them of this strength. Their failure to effect widespread dissemination of the new innovations and thereby to continuously refertilise the research process may be ascribed to the fact that the real goals, needs, interests, and aspirations of the poor in their relation to technology have yet to be grasped fully. This would appear to be the key to whether intended coalitions such as these will work themselves out into what Blume has called "common social projects" (49) or whether they will remain intelligentsia-inspired and a gap-filling activity in relation to the formal system.

Whether there will arise any trends within the bottom-up processes to redress the imbalance and forge a workable coalition between the scientists and the poverty groups and thereby to sustain the sense of shared social purpose that underlies these initiatives remains to be seen.

Notes and References

1. Stuart S. Blume, "University-industry relations: A speculative inquiry into their meaning" in the same volume discusses the extent and complexity of these interactions.

2. For an authoritative and up-to-date account of the trends in development economics we cite Gerald M. Meier, *Emerging from Poverty: The Economics That Really Matters*. New York: Oxford University Press, 1984.
3. The theme of knowledge interests has been extensively discussed by another paper in this volume in the context of the environmental movement. See R. Eyerman, A. Jamison and J. Cramer, "The knowledge interests of the environmental movement and the potential for influencing the development of science".
4. These and other trends discussed in this section have been documented in Radhika Ramasubban, "The development of science and technology in India", unpublished Ph.D. Dissertation, Bombay University (1977).
5. For documentation and elaboration on this aspect, see Radhika Ramasubban, *Public Health and Medical Research in India: Their Origins Under the Impact of British Colonial Policy*. Stockholm: Sarec, 1982.
6. Government of India, Department of Science and Technology, *R and D Statistics*, 1982—83, New Delhi.
7. Radhika Ramasubban, "The environment of research in less developed countries: The case of human reproduction research in India", CSTC Communication Paper, 1984.
8. This is an oft-repeated criticism by educational and science policy planners which has acquired particular urgency in the 1980s. Among the most eloquent critics are Prof. B. M. Udgaonkar in his J. P. Naik Memorial Lecture, Bombay 1983, and Prof. CNR Rao in his Haldar Memorial Lectures, Bombay 1986. For an earlier critique see B. R. Seshachar, "Science in universities", *Mainstream*, January 1973.
9. The Indian Institute of Science, Bangalore, can be cited as an example.
10. Government of India, *op. cit.* (6).
11. The significant enquiry commission reports generated by the government are:
    CSIR, *Report of the Committee of Enquiry I*, 1970.
    CSIR, *Report of the Committee of Enquiry II*, 1971.
    Government of India, Administrative Reforms Commission, *Report of the Study Team on Scientific Departments*, 1970.
    Government of India, Ministry of Agriculture, *Report of the ICAR Enquiry Committee*, 1973.
    There also exist evaluations by Parliamentary Committees e.g., the Public Accounts Committee, Committee on Public Undertakings, and the Estimates Committee, which contain useful information about the functioning of various public bodies.
12. A detailed discussion of the changes is contained in Radhika Ramasubban, "What ails the CSIR", *Mainstream*, September 1977.
13. Government of India, Planning Commission, *Draft Seventh Plan*, New Delhi, 1985.
14. Jagdish N. Bhagwati and Sukhamoy Chakravarty, *Contributions to Indian Economic Analysis: A Survey*, Bombay: Lalvani, 1971, contains an elaboration of this model.
15. T. N. Srinivasan and N. S. S. Narayana, "Economic performance since the third plan and its implications for policy", *Economic and Political Weekly*, Annual Number 1977.

16. S. Paul, "Growth and utilisation of industrial capacity", *Economic and Political Weekly*, December 7, 1974.
17. For an authoritative account of these aspects of the Indian economy, we cite:
R. H. Cassen, *India: Population, Economy and Society*, London: Macmillan, 1978.
P. K. Chaudhuri, *The Indian Economy: Poverty and Development*, London:. Crossby Lockwood Staples, 1979.
An up-to-date review of literature on the poverty - echnology relationship is contained in:
Bhanwar Singh, *Agrarian Structure, Technological Change and Poverty*, New Delhi: Agricole, 1985.
18. For details, see Centre for Monitoring the Indian Economy, *Basic Statistics Relating to the Indian Economy*, 1984.
19. Ramasubban, *op. cit.*, 1977 (4).
20. Blume, *op. cit.*, 1985 (1).
21. K. K. Subrahmanian, *Import of Capital and Technology*, Bombay: People's Publishing House, 1972.
22. Ramasubban, *op. cit.*, 1977 (4).
23. *Ibid.*, for documentation.
24. Subrahmanian, *op. cit.*, 1972 (21).
25. Ramasubban, *op. cit.*, 1977 (4) documents the details.
26. Ashok Rudra, "Organisation of agriculture for rural development: The Indian case" in Dharam Ghai et al. (eds.), *Agrarian Systems and Rural Development*, London: Macmillan, 1979.
27. Indian Council of Agricultural Research, Report of the Joint Indo-American Team on Agricultural Research and Education, I: 1955: II: 1960.
28. Rudra, *op. cit.*, 1979 (26).
29. *Ibid.*
30. Singh, *op. cit.*, 1985 (17).
31. See Ramasubban, *op. cit.*, 1977 (4) for details.
32. Singh, *op. cit.*, 1985 (17).
33. M. S. Ahluwalia, "Rural poverty in India: 1956—57 to 1973—74", World Bank Staff Working Paper No. 279, 1978; and J. S. Sarma and Shyamal Ray, "Behaviour of foodgrain production and consumption in India 1960—1977", World Bank Staff Working Paper No. 339, 1979.
34. Singh, *op. cit.*, 1985 (17).
35. *Ibid.*
36. Seven suicides took place in the Indian Council of Agricultural Research (ICAR) network within the first few years of modernisation of the research structure. Dissatisfaction with the working atmosphere and discontent with the channels for upward mobility were central themes of the upheavals they caused. See also (11).
37. The first ever nationwide debate in which scientists participated with government policy makers was held in 1973. The forum for this debate was the approach paper for a proposed science and technology plan. National Committee on Science and Technology, *Approach to Science and Technology Plan*, New Delhi: 1973.
38. Eyerman, Jamison and Cramer, *op. cit.*, 1985 (3) discuss this in some detail.

39. "Towards a Calangute statement", Lokayan Science and Technology Meeting, Hyderabad, September 1982 (Mimeo).
40. C. V. Seshadri, "Recycling wastes and renewable sources of energy", Madras: 1980 (Mimeo).
41. A. K. N. Reddy, "Choice of alternative technologies: Vital task in science and technology planning", *Economic and Political Weekly*, June 23, 1973.
42. Kerala Sastra Sahitya Parishad, *Science and Social Activism*, Trivandrum: KSSP, 1984.
43. This number is based on the list compiled by the Consortium on Appropriate Rural Technology (CART), New Delhi.
44. Their number would not exceed a dozen.
45. Edgar Zilsel, "The sociological roots of science", *American Journal of Sociology*, July 1941, speaks of a similar lowering of barriers between scholar and craftsman as the spur to the birth of modern science in Renaissance Europe.
46. Richard Whitley, *The Intellectual and Social Organisation of the Sciences*, Oxford: Clarendon Press, 1984, p. 9.
47. Kerala Sastra Sahitya Parishad, *op. cit.*, 1984 (42).
48. Blume, *op. cit.*, 1985 (1).
49. *Ibid.*

WORKERS' FACULTIES AND THE DEVELOPMENT OF SCIENCE CADRES IN THE FIRST DECADE OF SOVIET POWER

HARLEY D. BALZER
Georgetown University

Introduction

Western governments' science policies do not generally encompass the total national educational policy. The extent to which universities will allow themselves to be guided by governmental concerns in formulating their admissions policies varies greatly in the West. Government policies on the one hand depend on the relations between universities and the state, and on the other hand on the relations between individuals and the state. Even when serious national needs are identified, government policies tend to take the form of encouraging societal and market forces in education, rather than directing activity from above. In socialist societies, where the state monopoly over education precludes alternatives and education is regarded as an instrument of social policy, the relationship of the education system to science policy becomes more complex.

Rather than merely trying to influence scientists, we find under socialism a case where the government has sought to create its own scientific cadres. This carries our analysis beyond science policy and the relationship of scientists to the state, requiring us to explore the interrelationship between science and the education system, including its connections with broader areas of social policy.

A national program of fostering scientific manpower also brings science into close relationships with two groups of non-scientists: the political authorities who administer the programs in accord with their perceptions of cadre needs, but also in response to shifting political requirements; and the mass of human "raw material" from among whom new scientific and technical cardres must be recruited. Each of

S. Blume, J. Bunders, L. Leydesdorff and R. Whitley (eds.), The Social Direction of the Public Sciences. Sociology of the Sciences Yearbook, Vol. XI, 1987, 193—211.

these groups has its own goals and priorities, almost inevitably differing from the program scientists themselves might prefer. And each group of non-scientists has an impact not only on formulation of the policies, but even more on the actual results.

I have argued elsewhere that education is an area of government policy, and of science policy, that is particularly resistant to direct political manipulation (1). It may under some circumstances be a relatively simple matter to close institutions or exclude individuals and even entire groups from an education system. But it is much harder to encourage or to coerce students to study diligently that which political authorities deem necessary for development. Encouraging science and technology requires positive inducements, and inevitably some power accrues to the students and to universities.

The attractiveness of educating new specialists despite the formidable difficulties involved stems in part from government unease when faced with the power of those possessing expert knowledge (2). In pre-Revolutionary Russia the government frankly admitted that it could not dismiss politically questionable professors at its polytechnical institutes because there were not adequate cadres of specialists to replace them (3). The Soviet government recognized the problem and made it a high priority. But it was not the only priority. And again, educational policy was not as easily enforced as other areas of Bolshevik social engineering.

The early Soviet experience has implications for the broader themes raised in this volume. Despite government efforts to use education and access to education for social engineering purposes in the hope of changing the social base of the science system, the success of the education system ultimately depends on individuals' receptivity. Yet it is impossible to deny that government policies have an impact. The more important questions pertain to long-term versus short-term impact, the effects of competing/conflicting government goals, and the ability of individuals and groups to deflect policies to suit their own particular situations.

The Soviet case suggests that while the "science system" may be altered by drastic government policies in the short term, its resilience in the longer term is quite impressive. New specialists frequently absorbed the professional culture of their disciplines. In fact, scientific personnel with solid proletarian and Communist credentials are in a stronger position to lobby for their interests, since they can not be dismissed as opponents of the regime. Science policy, especially in education, can not be separated from broader aspects of social and economic policy, often creating contradictions and conflicting pressures in policy formulation.

Early Soviet Science Policy

Following the Bolshevik Revolution, the new Soviet government introduced a broad series of measures designed to create a new, rationally planned science policy. While owing much to the European "rationalization" movement and wartime economic mobilization, the new policies superceded European states' efforts at developing national science policies. The main lines of the Soviet program included measures to preserve the existing "cultural fund," a policy of planning scientific development with special attention to its relationship to economic applications; a program of budgetary support for science; and a cadre policy intended to win the loyalty of bourgeois specialists while simultaneously preparing a new generation of communist scientific and technical personnel (4).

To some extent, the situation in each of these areas could be improved simply by allocating funds and proclaiming political priority. Cultural and scientific institutions could be preserved if the Party insisted and the state paid. Economically important scientific and technical research could be stimulated if resource allocations were directed to priority fields. (Although introducing the results of such research into production proved more difficult.) The efforts of the new Soviet state in these areas were impressive, even if the results failed to correspond to hopes and expenditures.

But the most difficult problem in instituting a Bolshevik science policy was the lack of Bolshevik scientists. No matter how many times the government admonished scientists to think in a new way or to respond to new initiatives, the active cooperation of scientific-technical experts was not to be achieved by coercion. Lenin recognized quite early that repressive measures had their practical and psychological value, but were not really suited to the task of encouraging scientific research. Threats might prevent a scientist or engineer from sabotaging a project, but to encourage scientists to spend long hours at economically valuable creative activity, some more positive measures were required.

There were two main options available to the Bolshevik regime. They could use the bourgeois specialists, either by granting them privileges and offering inducements, or, perhaps, by eventually winning them over to the socialist view. Or the regime could train new cadres of Bolshevik scientists. Immediate requirements forced them to follow both lines simultaneously. But since it could never be certain that a bourgeois specialist had become a committed Bolshevik, it was preferable to take committed Bolsheviks and try to turn them into scientists and engineers (5). Such a policy had the additional benefit of providing education and social mobility to supporters of the Party. However, this subsidiary goal raised the danger that scientific goals might be subjugated to be requirements of political and social policy. In the event, this happened quite often. A brief account of the history of the workers' faculties (*rabochie fakul'tety* or "rabfaks") will help us to deal with these questions on a more concrete level.

Rabfaks and Soviet Educational Policy

The rabfaks were a major element of the program to provide educational opportunities to previously disadvantaged groups and thereby recruit a new intelligentsia from among workers and peasants. A variety of rabfaks developed on the basis of initiatives "from below," and, after a period of trial and error, became a functioning part of the education system. Their purpose was to provide a preparatory education enabling workers and peasants — especially those who were Party or Komsomol members — to enter techical institutes and other higher schools (6).

The rabfaks represented one of the major ways by which the fledgling Soviet apparatus sought to create a proletarian intelligentsia. Yet for most Soviet leaders, particularly Lenin, the rabfaks were a minor initiative at a time when the requirements of economic reconstruction and state-building demanded utilization of all available resources. The creation and persistence of this form of preparation for higher education owed at least as much to the initiative of interested workers and peasants as to the efforts of the Soviet leadership. The rabfaks were both a government creation and a response to persistent demands for access to education on the part of the social group on whose behalf the revolution had supposedly been undertaken.

The rabfaks also became part of a feedback system influencing Soviet political life. Rabfaks and their students were major players in the effort to establish firm communist control over educational institutions. The rabfaks became enshrined in the regime's new mythology, attaining a symbolic importance out of proportion to their quantitative or qualitative role, and persisted through the 1930s despite being superseded by larger and more effective programs of proletarian recruitment into higher education. Rabfaks have been a component of the process of "socialist construction" in other Marxist states, giving them an importance beyond Soviet history and the theoretical concerns addressed here.

One of the first steps of the Bolshevik regime in education was to abolish all barriers to admission to higher education (7). The result was a brief, massive influx of individuals aspiring to advanced training. It quickly became apparent that most of the prospective students lacked even the rudiments of general education necessary for successful study in a higher school. The Peoples Comissariat of Education (Narkompros) responded by organizing courses for workers in basic education subjects, as a way to prepare them for higher education. There were perhaps a dozen such courses in Moscow and a few in Leningrad in 1918. But the idea was almost certainly doomed from the start. Massive organizational problems made it impossible to provide adequate instruction. Even if these problems could have been overcome, it was extremely doubtful that workers, many of whom were barely literate, could be provided with education equivalent to seven years of secondary school in a few short months (while still working full time!).

Yet lack of preparatory education was not a barrier to aspirations for higher education. Following their bitterly fought victory in the revolution and civil war, individuals from the proletariat and peasantry demanded immediate access to the rewards society might bestow. That a veteran of years of Red Army travail should be forced to spend six or seven years in a secondary school for children was not acceptable. People wanted higher education — the mark of status and culture in Central European society — and would settle for nothing less (8). Secondary schools and various other alternatives such as technicums (specialized secondary schools) did not conform to their image of an appropriate reward.

We see here the origin of an excess demand for higher education that has presented an almost continuous problem for Soviet education planners (9). The insistence on higher education in the early years after the revolution resulted in many teacher training institutes and agronomy schools being converted into higher schools. The number of higher educational institutions increased mostly as a result of upgrading existing facilities. Yet this did nothing to alleviate the problem of inadequate preparation of entering students, particularly those from the working class and peasantry. A variety of experimental preparatory classes succeeded chiefly in proving that there were no rapid shortcuts to acquiring basic general education.

Narkompros authorities, seeking a model for rapid remedial education for adults aspiring to higher schools, adopted a program of education for older students first introduced at the Moscow Commercial Institute. The story of this first rabfak illustrates the groping nature of Soviet educational policy at the time and the dual purpose served by the new preparatory programs.

The initial impetus for enrolling workers at the Moscow Commercial Institute was political, not pedagogical. In the autumn of 1918 elections were held for the school's student government. The Bolsheviks, in a minority among the students, sought to bolster their ranks by inviting workers at neighboring factories to utilize the open admissions policy and register as students, permitting them to vote in the election. Nearly 1000 electrical and metal workers responded to student agitation and presented themselves at the Institute. The result was a Bolshevik victory in the elections. A number of the workers, however, began to take their status as students more seriously, deciding to remain at the Institute to receive a higher education. (Use of the word "receive" rather than "obtain" is intentional: many of the workers initially regarded education as something they would be given, rather than as something they must acquire.)

In this situation, the workers' lack of general educational preparation quickly became apparent. Since dropping the new registrants from the rolls would have been too blatant an admission of political maneuvering, and would have offended workers who genuinely desired an education, some mechanism had to be found for providing the background knowledge normally acquired in secondary school. Special courses were set up, first for three months, then six, and finally, when experience revealed the true magnitude of the task, for two to three years (10).

This first rabfak opened at the end of January 1919, at a time when Soviet officials were searching for ways to provide preparatory education for workers. By September, Narkompros issued a decree "On organizing rabfaks at universities" adopting the system for use at the nation's elite higher educational institutions (*vysshie uchebnie zavedeniia* or VUZy).

By calling the new programs "workers' departments" (*rabochie fakul'tety*), Soviet authorities unwittingly created the basis for much subsequent confusion. Russian VUZy are divided into administrative departments (*fakul'tety*). The use of the appellation "department" implied that the new programs were on an equal footing with the existing departments at VUZy, and rabfak students were quick to assert their equality with other higher school matriculants. The result was a great deal of uncertainty, especially since Narkompros itself had not yet resolved just what the rabfaks were to be: crash programs for training specialists, preparatory courses for VUZ admission, or some other type of education.

The question of the rabfaks' role was resolved only in 1920, when Lenin signed a decree "On Workers' Faculties" specifying that their purpose was to attract workers and peasants into higher education. They were to be a part of the VUZy, but to occupy a special place preparing workers, peasants and Party personnel for further study. Effectively, this made the rabfaks rivals of the secondary schools. We have noted that Lenin's own attitude toward the rabfaks was less than completely enthusiastic. He regarded the tasks of winning over bourgeois specialists and building a normal educational system as more important. But it appears he was willing to tolerate the rabfaks as an additional educational program that helped satisfy demands for social mobility. Lenin may well have hoped that the symbolic importance of the rabfaks would be a major factor, while their quantitative contribution would not interfere too much with more pressing initiatives.

But the rabfaks were destined to become a more powerful force in Soviet education and politics. In 1920 a separate administration was created for the rabfaks, providing them with an institutional lobby that proved difficult to dismantle. The real source of support for the rabfaks, however, was among workers, peasants and others who desired access to higher education.

In the following years the number of rabfaks increased, often due to the demands and activity of potential students. Following the civil war, many demobilized Red Army veterans and Bolsheviks who had fulfilled important roles in the military and political struggle sought responsible positions in civilian life. Education was the path to these jobs, and they felt the regime owed them opportunities and support. To satisfy its supporters, the Bolshevik government sanctioned creation of increasing numbers of preparatory education programs. We see here a definite bottom-up process, in which the Soviet regime acceded to demands by its supporters to provide educational opportunities.

Yet, as the origin of the first rabfak indicates, these institutions simultaneously served an important top-down political purpose. There were few supporters of the Bolsheviks among the professoriate, and the few Bolshevik students had long since been mobilized for pressing military and political tasks. If the VUZy were not to remain hostile institutions, they had to be "taken" by Bolshevik personnel. Preparing workers, peasants and Communists for study in VUZy appeared a promising way to assert control over these important institutions. In the interim, the presence of pro-communist cadres in preparatory divisions at the higher schools was a way to leaven the student body with regime supporters. And, eventually, it was hoped the educated Bolshevik cadres would be in a position to "take" the scientific and technical institutions themselves (11).

Conditions at the rabfaks were difficult, and they were forced to compete with established programs for space, faculty, dormitories, stipends, books and supplies. Neither Narkompros nor the individual VUZy regarded the rabfaks as a high priority. Rabfak students were discriminated against by VUZ administrators, and well into the 1920s student assemblies at individual schools debated the question of whether rabfak students could truly be considered members of the VUZ student body.

However, the rabfaks did enjoy important political support from the Komsomol and from trade unions, encouraging their expansion. The number increased from 14 in 1919 to 45 in 1921 and 62 in 1925. They continued to be established most often in response to the insistence of workers and peasants who demanded their constitutional right to education.

While the number of rabfaks grew throughout the 1920s, actual enrollments fluctuated in response to changes in educational and social policy. Figures on school enrollments are questionable at best, and those on social origin of students are so unreliable as to have little value for a historical portrait of social processes (12). However, the "official" figures were the basis for regime policy decisions, and therefore have a historical importance regardless of their accuracy.

An example of the problems in assembling statistics can be seen in the very first years of Soviet rule. According to official tabulations, at the end of 1918 there were 157,819 students enrolled in VUZy. Yet as of January 1, 1919 only 55,412 were actually studying in higher schools. Much of the drop-off may be explained by the civil war, but it also reflects the inadequate educational preparation of many new matriculants (13).

Following the civil war, returning veterans and Party members organized many of the new rabfaks. Yet they were unable to guarantee that students admitted to these programs were in fact of proletarian or peasant origin. The result was a pattern of expansion and purge that persisted through most of the 1920s, sometimes complementing and sometimes contradicting overall education policy. We see here that educational policy and cadre development were intricately bound up with broader political struggles (14).

The situation was further complicated by local and regional differences. As in so many aspects of social policy in the 1920s and 1930s, central policy decrees were subject to varied interpretation by local officials. At some educational institutions officials had very different intentions from the central authorities, and their actions demonstrated these differences. Such local variations were abetted by conflicts within Narkompros itself over the best policy line, and by attacks on Narkompros' control of many types of VUZy by other institutional actors, particularly the economic planning administration (Vesenkha).

For our purposes, the fluctuating political fate of the rabfaks is less significant than their contributions to formation of the Soviet scientific-technical intelligentsia. Here, the quantitative role of rabfak students and graduates becomes important. Despite political complications and policy fluctuations, the absolute number of rabfak students increased through the 1920s. More important, the number of rabfak graduates entering VUZy rose steadily, as can be seen in Table I:

TABLE I

	Rabfak Graduates
1920	215
1921	2000
1922	3576
1923	5315
1924	6810
1925	7450
1926	7410
1927	5580
1928	5124
1929	6405

Source: N. M. Katuntseva, *Rol' rabochikh fakul'tetov v formirovanii intelligentsii SSSR*, Moscow: Nauka, 1966, pp. 48; 171.

Yet this does not tell us who the successful rabfak graduates actually were, much less how they fared once they attained places in higher education. Constant attention to the social composition of the student body and frequent purges of both rabfak and VUZ students suggest that it was extremely difficult to discern who was truly a proletarian or peasant. It was easier to identify Party and Komsomol members, and their representation among rabfak students was high. But virtually anyone could work for a year or two and present him/herself as a worker eligible for admission to a rabfak.

Scattered records of the commissions charged with determining social origin of students illustrate the complexity of the task. Was a child of a worker who had joined the Party and become a minor official to be considered the offspring of a proletarian or a white collar worker? Did the daughter of a petty nobleman who had spent five years working as a seamstress deserve to be considered a member of the working class? And what about an individual's attitude? Should the child of a bourgeois family who claimed to have repudiated his or her class background and joined the revolution be punished for parental transgressions? Stalin later rejected (at least officially) the idea that children answer for the behavior of their parents, but in implementing policies at the local level family background and old personal scores probably counted a great deal.

Purges occurred with an almost unimaginable frequency. In Leningrad, there were five purges of the VUZ student body during 1921–1922 (15). This reflects both official determination to impose proper class composition in the schools and the difficulty of determining individuals' social origin or guaranteeing compliance. Given the chaos and fluidity of the situation, it was relatively easy for a student excluded from one institution to simply register at another higher school. It would not be at all surprising to discover that the same individuals were removed from a number of VUZy during the repeated efforts to "cleanse" the student body.

In addition to the general purges of the VUZ student body, separate purges were carried out in rabfaks to insure that only proper worker, peasant and Communist cadres availed themselves of the privileged access to education. Each time the enrollments at rabfaks increased, it subsequently was assumed that individuals who were not really eligible had managed to gain admission. Each time, steps were taken to remove the "false" workers and peasants (16).

The rabfaks markedly increased the proportion of workers and peasants entering higher education. McClelland calls them "The only effective and lasting reform of this period" (17). However, whether they actually produced a significant number of new specialists in the 1920s is open to serious question. Official statistics on rabfak graduates' subsequent success in VUZy were not maintained. While such statistics might be embarrassing, it is more likely that lack of a major institutional incentive for collecting such data and the difficulty of monitoring individuals in diverse institutions over an extended period of time explain the gap. But we may infer that a large percentage of the rabfak graduates did not complete the full program of study at a VUZ.

The graduation rate at Soviet VUZy in general in the 1920s was probably in the neighborhood of 10% of the students per year. Before the revolution it had been normal for students to take six or seven years to complete the five-year VUZ course, and in the early years of the Soviet period conditions for most students were less favorable. There were frequent complaints about poor success rates and threats to oust those not completing their studies in the allotted time. Yet the authorities could not expel all the students.

If the success rate for all students was low, it is hardly likely that the rabfak graduates performed significantly better. Despite their age and motivation, they were not as well prepared for higher education as graduates of a complete secondary school, and in many cases their rabfak courses were barely adequate. It is likely that most of the rabfak graduates either left school or were still students when the Cultural Revolution began in 1928. One of the best sources gives a figure of perhaps 10,000 Communist graduates, including rabfak graduates and others, from ALL education institutions in the period 1918—1928 (18).

The quality of the education available at rabfaks was dubious at best. As experimental institutions, the rabfaks were beneficiaries/victims of the rampant educational experimentation that characterized the 1920s. After the effort to provide a full secondary education in three to six months proved unrealistic, educational authorities constantly sought other short-cuts. Rabfaks implemented the brigade method of instruction, with small teams responsible for joint work in laboratories. There were also provisions for "collective responsibility" in examinations, where the fate of a group depended on the capacity of its best student. These initiatives almost certainly did more harm than good to the basic goal of providing a solid general education background, though they did provide statistical "successes."

In material terms, some rabfaks enjoyed a relatively privileged position. Despite initial resistance from school authorities, rabfak students were entitled to receive dorm rooms, stipends, food, medical care, and even uniforms, as well as free access to cultural institutions and free railroad transportation. Rabfak students were exempt from tuition and were guaranteed admission to higher education at a time when secondary school students were required to pay increasingly higher fees and forced to compete for a declining share of openings in higher schools. In some years in the 1920s it was almost impossible for secondary school graduates to gain entry to elite VUZy in the face of competition from rabfak students and other specially chosen candidates (19). The benefits granted to rabfak students compounded the problem of screening applicants, since some individuals sought to enter the rabfaks as a way to be housed and fed (20).

But material benefits could not compensate for rabfak students' lack of basic education, or for the demands political work imposed on the time of Komsomol and Communist Party members. In an era when the

Party was desperately short of skilled cadres, local authorities were reluctant to permit their best people to spend years studying in another city. Often the individuals seconded to rabfaks and VUZy were not the most talented, most eager or most qualified, but simply the most expendable. And active communists at educational institutions were likely to be heavily involved in political work, leaving minimal time for study. All these elements contributed to the small number of rabfak veterans graduating from VUZy in the first decade of Soviet rule.

To some extent, judging the rabfaks on the basis of just a decade-long experience is unfair. When the course of study at Russian VUZy was five years, the three-year rabfak meant that a student would require at least eight years to complete the full program of preparatory and higher education, with nine or ten years a more realistic figure. Cadres entering the rabfaks in 1920 would have been graduating from higher schools no earlier than 1928 or 1929. Even when VUZ programs were reduced to four or three years, a rabfak student faced a minimum of six years of full-time study, with a longer program not uncommon. In some years rabfak graduates comprised half the students at VUZy. But again, we have no hard evidence regarding how many of them completed the VUZ course. Some sense of the tenor of the time can be gleaned from the following description of the Mining Academy:

> Even having been sent to study, we sought somehow to do something of real value for the country as quickly as possible, and worked wherever we could. Vania Tevosian studied and at the same time directed the Organization Department of the Zamoskovoretskii raion of the Party. The student Zaveniagin, who subsequently for seventeen years was Secretary of the District Party Committee in the Donbass, for all practical purposes occupied the post of Prorector of the Academy. He was called the right hand man of Ivan Mikhailovich Gubkin . . . Many of us arrived at the Mining Academy direct from the front of the civil war (21).

It was only in the mid-1920s that the Soviet regime was able to stabilize its policy and begin a serious effort at developing science cadres (22). In the following few years, much was done to establish a solid base for scientific-technical institutions and cadres. But political issues continued to intervene. This was also true in the case of the rabfaks. The rabfaks might have been in a position to prove their value in the years after 1928, when larger numbers of former rabfak students would have been graduating from Soviet VUZy. But in the era of forced industrialization even the relatively speeded-up education provided by rabfaks was not rapid enough to meet the demand for specialists. In July 1928 a Central Committee Plenum decreed mobilization of one thousand Party members for enrollment in higher education, along with the assigning of 3000 additional matriculants to the rabfaks. The overwhelming majority of these newly mobilized students entered engineering schools. In subsequent months the number of "Thousanders" swelled. In addition to an elite of some 20,000 Communist Party Thousanders, there were tens of thousands of other party cadres and workers brought into higher education during the First Five Year Plan. A Western historian who has studied the subject in depth suggests perhaps 110,000 communists and 40,000 other former workers were "promoted" into higher education in this period (23).

To accommodate the mass influx of students, the VUZ system was greatly expanded and the course was shortened to three years from the previous five. Shock methods analogous to those employed in the economy were instituted in education. If the Five Year Plan could be completed in four years, than certainly the new three-year VUZ course could be completed in two years. Pedology and other experimental programs reappeared in the schools, and the entire education system entered a period of chaos that made the tribulations of the civil war seem mild.

Beside the massive numbers of communist Thousanders and the special arrangements for facilitating their completion of higher education, the rabfaks faded into the background. They continued to extist, and even increased in size, but their importance as a conduit for elite higher education had ended. The three-year rabfak course was too lengthy for the requirements of socialist construction, and individuals with careers to make could not afford the time. The increased number of places in rabfaks were occupied by younger students who had not managed to complete their secondary education (24).

The Soviet regime never formally abolished the rabfaks. They persisted through the Cultural Revolution, despite being dwarfed in significance by other modes of *vydvizhenie* (promotion) into higher education. Rabfaks ceased to exist in 1940, by which time the Soviet secondary school was considered to be well-enough established to provide adequate cadres for higher education. By that time workers, peasants and communists had had two decades of opportunities to enter higher education through special preparatory faculties, and the remaining need for adult access to higher education could be met by maintaining evening and correspondence programs of higher education.

Like collectivization, the rabfaks became a part of the Stalinist system: an achievement to be boasted about, praised in history books and poetry, and shared with other fraternal socialist societies. Rabfaks were established in the Baltic states following creation of Soviet regimes there, and were utilized in Eastern Europe in the late 1940s, in China in the early 1950s, and Cuba in the 1960s. They are still considered a particularly appropriate model for the early stages of socialist construction in developing nations recently liberated from colonial domination. Even in the USSR, a variant of the rabfak with a different name reappeared in the 1960s to provide access to higher education for children from working class and peasant families (25).

The Rabfaks' Contributions

The rabfaks were part of an effort to change science by changing the social base from which scientific, technical and other specialized cadres were recruited. They achieved some success, and the related policies implemented during the Cultural Revolution resulted in a massive influx of peasants, proletarians and communists into higher education (26). The rabfaks guaranteed previously disadvantaged groups an opportunity to prepare for admission to higher education — the most one can fairly ask. The subsequent *vydvizhenie*, with its drastic lowering of standards and greater emphasis on political work, more often guaranteed not an opportunity for admission but graduation. This defies science, but is very much in the spirit of cultural revolution.

Discriminatory policies succeeded in the narrow purpose of leavening the scientific community with a large group of loyal beneficiaries of Soviet social engineering. Whatever their social origin and earlier political sympathies, the *rabfakovtsy* and *vydvizhentsy* had reaped the rewards of the new Soviet system and overwhelmingly remained loyal Party minions. They benefited not only from the educationl opportunities provided by expanded schooling and special programs, but also from the rapid career advancement made possible by the purges and the demand for personnel during rapid industrialization. They were a new elite, and developed an outlook and culture appropriate to their situation (27).

Thus, in terms of simply generating new cadres, the rabfaks and related policies had positive results. (The story of the relationship between these cadres and the existing specialists is another matter (28).) But in a broader sense the results were not what the regime intended.

Despite the purges, the dominance of Party hacks, and the rapid upward mobility of many *vydvizhentsy*, scientists of proletarian origin and communist persuasion did not necessarily prove to be so different from bourgeois scientists in seeking to protect their autonomy and scientific turf. As Kendall Bailes has so eloquently illustrated in his study of the technical specialists in the 1930s, the Stalinist experience did not put an end to political debates about policy: the debates ceased to take place in public, but they continued just as fervently within the corridors of political power (29).

Soviet leaders were only partially correct in believing that they could rapidly transform proletarians into trained specialists. They were incorrect in thinking that the new specialists would not absorb the professional culture of their disciplines, either from their educational experience or from their institutional position. Even when Stalin sought to circumvent the educational system by promoting "praktiki" to responsible positions, the institutional factor had a strong influence on their subsequent behavior. Some of the vehemence of the purges may perhaps be explained by the frustration felt when two decades of social engineering had failed to produce the desired results.

Having created an intelligentsia that was (is) both red and expert, the Soviet regime discovered that being red might insure basic political loyalty, but would not guarantee that policies from the top would be implemented automatically.

Conclusion

The Soviet experience provides a sort of negative proof that top-down efforts to transform the science system are not viable. Stalin's handling of the rabfaks and vydvizhenie demonstrated that a "top-down" transformation may perhaps be effective in the short run, if it is applied with sufficient ruthlessness. But the costs are enormous, and the distortions are likely to be severe. Bottom-up processes are generally more effective, but present their own problems. They are slower and therefore less satisfactory to groups (or regimes) in a hurry. In education, even bottom-up processes involve the state, raising a whole series of issues related to educational and social policy. And the long-term nature of bottom-up processes tends to expose them to less blatant political and social pressures over extended periods of time.

The effort to provide access to higher education for workers and peasants forced the Russian/Soviet scientific community to interact with two groups of non-specialists: the political/Party authorities and the mobility-aspiring workers and peasants. The former could extract loyalty to the regime, but could not insist on total compliance in science without destroying the scientific content of entire disciplines. This was allowed to happen in biology, but Lysenko was an exception. Most fields of science and technology avoided the worst perversions perpetrated in the name of proletarian science. The mass of workers and peasants who sought higher education could in some cases vault into the scientific community, but only if they mastered its knowledge base. In the process of acquiring this knowledge, they also were likely to be socialized by the practicing scientists. Ideology aside, does a worker who aspires to higher education really want to expend all that effort in order still to have the outlook of a worker?

The discriminatory social policies of the 1920s and 1930s did succeed in altering the class composition of the student body from which scientific and technical personnel are recruited. But the science system has proved remarkably resilient. It was possible to generate cadres whose loyalty to the regime was not in question, but these loyal cadres did not become automotons mindlessly carrying out the wishes of political leaders. The resilience of science was due in part to an inherent contradiction by which communist cadres created by the regime's policies almost by definition cannot be accused of being disloyal (except in highly unusual circumstances). In part, it is explained by the participation of these same cadres in formulating official policy. And in part, it is due to the continuing indispensability of the knowledge of scientific experts. These experts realize that real science requires them to mesh with the development of the international science system rather than the demands of megalomaniacal politicians. Even at the height of Stalinist terror and demands for Marxist biology, some scientists found ways to keep other lines of research open (30).

The resilience of science in turn has created a cycle of ebb and flow in political interference. Russian/Soviet political authorities have never maintained a consistent policy toward their scientific community, precisely because those policies that most closely correspond to the wishes of political leaders are so often precisely those that do the most damage to science, and thereby to other parts of the national political agenda. Neither the massive recruitment of new groups to higher education in the 1920s and 1930s, nor the effort to insure administration by Communist Party members in the 1960s and 1970s succeeded in eliminating the power of those with specialized knowledge.

Notes

1. Harley D. Balzer, "Education, science, and technology" in James Cracraft (ed.), *The Soviet Union Today: An Interpretive Guide*. Chicago: The Bulletin of the Atomic Scientists, 1983, pp. 233—243.
2. For recent discussions of this theme see the essays in Thomas L. Haskell (ed.), *The Authority of Experts*. Bloomington, IN: Indiana U. Press, 1984.
3. Tsentral'nyi gosudarstvennyi istoricheskii arkhiv (Central State Historical Archive), fond 25 opis 5 delo 3 list 67.
4. Robert Lewis, *Science and Industrialization in the Soviet Union*. New York: Holmes & Meier, 1979; Loren Graham, *The Soviet Academy of Sciences and the Communist Party*. Princeton: Princeton U. Press, 1967; L. V. Ivanova, *Formirovanie sovetskoi nauchnoi intelligentsii (1917—1927 gg.)*. Moscow: Nauka, 1980; and E. A. Beliaev, *KPSS i organizatsiia nauki v SSSR*. Moscow: Izdatel'stvo politicheskoi literatury, 1982.
5. Ivanova, op. cit., 1980. Note 4. And the works of S. A. Fediukin, most recently *Partiia i intelligentsiia*. Moscow: Izdatel'stvo politicheskoi literatury, 1983.
6. Basic sources on the rabfaks are Ivanova, op. cit., 1980. Note 4; N. M. Katuntseva, *Rol' rabochikh fakul'tetov v formirovanii intelligentsii SSSR*. Moscow: Nauka, 1966; N. M. Katuntseva *Opyt SSSR po podgotovke intelligentsii iz rabochikh i krest'ian*. Moscow: Mysl', 1977; Sheila Fitzpatrick, *The Commissariat of Enlightenment: Soviet Organization of Education and the Arts under Lunacharsky*. Cambridge: Cambridge U. Press, 1970; Sheila Fitzpatrick, *Education and Social Mobility in the Soviet Union, 1921—1934*. Cambridge: Cambridge U. Press, 1979; James C. McClelland, "Bolshevik approaches to higher education", *Slavic Review* **3**, 4, December 1971, 818—831; and James C. McClelland "Proletarianizing the student body: The Soviet experience during the new economic policy", *Past and Present* No. 80, August 1978, 122—146.
7. For the first years of rabfak development see Katuntseva, op. cit., 1966, Note 6; McClelland, op. cit., 1971, Note 6.
8. Fitzpatrick, op. cit., 1979, 49.
9. N. K. Krupskaia, *Pedagogicheskie sochineniia*. Moscow: Pedagogika, 1957—1963. Vol. II. p. 312.
10. The most complete account is V. I. Bessonova, "Sozdanie i razvitie rabochikh fakul'tetov v 1919—1921 gg." in *Iz istorii velikoi oktiabrskoi sotsialisticheskoi revolutsii*. Moscow: Izd. MGV, 1957. pp. 148—180, here pp. 153—155.
11. Fediukin, op. cit., Note 5. 116; Katuntseva, op. cit., Note 6.
12. McClelland disagrees, op. cit., 1971. Note 6. But cf. A. P. Kupaigorodskaia, *Vysshaia shkola Leningrada v pervye gody sovetskoi vlasti (1917—1925)*. Leningrad: Nauka, 1984.

13. A. V. Krasnikova, "Iz istorii stroitel'stva sovetskoi vysshei shkoly (35-letie dekreta 'O rabochikh fakul'tetakh')", *Vestnik vysshei shkoly* **13**, 12 December 1955, 55—60, here p. 55.
14. McCelland, op. cit., 1978, Note 6. Sheila Fitzpatrick, "The 'soft' line on culture and its enemies: Soviet cultural policy, 1922—1927," *Slavic Review* **33**, 1974, 267—287.
15. Kupaigorodskaia, op. cit., 1984, Note 12, 144—145.
16. See I. Khodorovskii, "K proverki sostava uchashchikhsia", *Krasnaia molodezh'*, No. 1, 1924, 112—116. And L. Milkh, "Itogi priema v vuzy v 1926 i problema ikh obrabocheniia", *Kommunisticheskaia revoliutsiia*, No. 8, 1927, 44—49.
17. McCelland, op. cit., 1971, Note 6.
18. Fitzpatrick, op. cit., 1979, Note 6, p. 110.
19. Ibid.
20. Kupaigorodskaia, op. cit., 1984, Note 12, p. 128.
21. V. S. Emel'ianov, cited in Fediukin, op. cit. 1983, Note 4, pp. 120—121.
22. Ivanova, op. cit., 1980, Note 4, p. 255.
23. Fitzpatrick, op. cit., 1979, Note 6, p. 188.
24. For statistics on Rabfaks in the 1930s, see Katuntseva, op. cit., Note 6.
25. Katuntseva, Op. cit., 1966, Note 6, pp. 179—191.
26. Fitzpatrick, op. cit., 1979, Note 6. Note McClelland's claim of major success.
27. Fitzpatrick, op. cit., 1979, Note 6.
28. Fediukin, op. cit., 1983, Note 4. Kendall E. Bailes, *Technology and Society under Lenin and Stalin*. Princeton: Princeton University Press, 1978.
29. Bailes, op. cit., 1978, Note 26.
30. Mark B. Adams, "The Kol'tsov Institute", in Linda L. Lubrano and Susan Gross Solomon (eds.), *The Social Context of Soviet Science*. Boulder, CO: Westview Press, 1980.

INTELLECTUALS IN SOCIAL MOVEMENTS: THE EXPERTS OF "SOLIDARITY"

OLGA AMSTERDAMSKA
Department of Science Dynamics, University of Amsterdam

On August 21, 1980, the eighth day of the strikes in Gdańsk shipyard, Bronisław Geremek, a medieval historian, and Tadeusz Mazowiecki, the editor of the Catholic periodical *Więź*, both members of the unofficial "flying" university (TKN), arrived in the shipyard to give to the strikers a letter of support signed by 64 Warsaw intellectuals (1). They met with Lech Wałesa, who not only thanked them for transmitting the letter, but also asked for more direct help from the intellectuals. Geremek and Mazowiecki, together with several other intellectuals who arrived in the shipyard during the following days (2), were then nominated as "experts" by the Interfactory Strike Committee in Gdansk (MKS) and asked to help in making sure that "'they' (i.e., the government) will not cheat us" (3). Experts were also needed to draft appropriate documents and to prepare for negotiations. During the following sixteen months of "Solidarity's" above-ground existence, the partnership between the worker-activists and intellectuals was one of the permanent features of Poland's "self-limiting revolution."

This partnership, in its origins, its functioning within "Solidarity" (including its failures and shortcomings), and its possible influence on the development of the social sciences in Poland is the subject of this paper. More specifically, I will examine how this cooperation between workers and intellectuals was possible and what it consisted of, what conditions and limitations it encountered, what roles the intellectuals played in "Solidarity," and finally, what possible consequences for the social sciences in Poland resulted from the intellectuals' commitment to and participation in the political and social changes which took place in Poland in 1980 and 1981.

Although unique in many respects, the case of "Solidarity" experts offers us not only an opportunity to examine the roles of intellectuals in

S. Blume, J. Bunders, L. Leydesdorff and R. Whitley (eds.), The Social Direction of the Public Sciences. Sociology of the Sciences Yearbook, Vol. XI, 1987, 213–245.

social movements, but also to investigate some of the more generally relevant conditions and characteristics of cooperations between scientists and other social groups. Thus, in examining the conditions necessary for such cooperations, the fact that the pre-"Solidarity" political system in Poland severely limited the possibilities of any bottom up initiatives points to the importance of political factors which might easily be overlooked in studies of cooperative ventures in Western democracies. The case of "Solidarity" makes it amply apparent that the convergence of values and interests between the relevant groups might well have been an insufficient basis for cooperation without the existence and availability of channels of communication and without political opportunities for organization and institutionalization. As several other contributions to this volume reveal, the scientists or intellectuals themselves might be instrumental in creating such channels of communication, but the social and political conditions under which this takes place deserve separate attention.

Recent work in sociology of science has emphasized the dependence of scientists on audiences other than those of their peers, and thus also the possibility that scientific research might be affected by the demands, evaluations, and resources provided by such extra-scientific audiences (4). By providing an alternative source of legitimacy and authority, such extra-scientific audiences might affect the functioning of the reputational system of control in the ideal typical academic sciences. In examining such dependence between groups of scientists and their various audiences, however, it is important to examine not only the existence of structural differences between the scientific fields, as Whitley suggests, and differences among the institutional settings in which the alternative systems of rewards might be operating, but also the political settings in which scientists are more or less able to respond to the demands of their diverse potential audiences (5). Thus, in studying the cooperation between social scientists and "Solidarity," I will focus on the existence and emergence of various alternative audiences and on the extent to which the social scientists' cooperation with "Solidarity" was shaped by the opportunities and limitations for achieving legitimacy and authority in the political arenas controlled by such alternative audiences. Differences between the relevant audiences will also be used to elucidate the distinction which must be drawn between research conducted as part of the direct provision of various kinds of expertise to "Solidarity," and the sociological and economic research which was influenced by the social and political changes taking place in Poland but not necessarily linked to specific cooperations between worker-activists and social scientists.

The Democratic Opposition: The Intellectual as Ideologist?

Discussing the theories of social movements and their applicability to the Polish "revolution," Piotr Sztompka writes:

> The third stage (according to Smelser's theory of collective behavior) is the formation and popularization of "generalized beliefs," identifying and defining contradictions, ambivalences, and tensions, providing critical evaluations, ascribing responsibility and guilt and pointing to possible ways in which they can be overcome. A special role is played here by the intellectuals who articulate such beliefs and provide them with a form of embryonic ideology or doctrine. Here one cannot avoid recalling initiatives existing prior to the August breakthrough such as the discussion group "Experience and Future" (DiP), the lectures of the Society for Scientific Courses (TKN), educational activities organized by the Church, publications outside the reach of censorship and some courageous political journalism surfacing in official periodicals, critical scientific conferences in the area of the social sciences, and open discussions in some centers connected with the Party, for example, in the Cracow "Kuźnica" (The Forge) (6).

Sztompka fails to mention the first group of opposition intellectuals who, after the workers' demonstrations of 1976, organized KOR, a committee with the initial goal of helping workers who were being persecuted for their participation in protests against price increases. The political reasons for Sztompka's peculiar omission are not relevant for our purposes, but if we are to understand the cooperation between the workers and the intellectuals in 1980-81 and to examine more critically the notion of intellectuals as ideologists, it is necessary to consider briefly the activities and the program of the Workers' Defense Committee (KOR) (7).

The initial task of KOR — bringing legal and material assistance to the punished workers and their families — necessitated not only some integration within the intellectual milieu that provided support for KOR activities, but also the establishment of direct links with the workers, their colleagues and families. Such integration and cooperation became even more important when KOR transformed itself into the Social Self-Defense Committee and broadened its sphere of activity. One of its major goals was to support independent social initiatives whose aim was to exercise or defend civil and human rights. Thus, KOR supported such initiatives as the Initiating Committees of Free Trade Unions, Student Solidarity Committees, Peasant Self-Defense Committees, and a large number of independent publishing ventures outside the reach of censorship, including the periodical *Robotnik* (*The Worker*) which was addressed directly to the workers and edited jointly by intellectuals and workers. In 1979, *Robotnik* published a "Charter of Workers' Rights" in which the creation of independent trade unions was postulated as a long-term goal of workers' self-defense.

It is not our goal here to assess the influence of all these initiatives on the emergence of "Solidarity," but the history of the democratic opposition makes it clear that the intellectuals from KOR and later also from other groups were more important in the role of catalysts (for example, by supporting the Gdansk Initiating Committee of Free Trade Unions whose members and organizers included such later "Solidarity" activists as Lech Wałesa, Andrzej Gwiazda, Anna Walentynowicz, and Bogdan Lis) than as ideological leaders, and more important for the channels of communication they established than as political leaders *sensu stricto*. This was so both because of KOR's ideology of "supporting social initiatives aiming at the realization of human and civil rights" and because of the systemic limitation on unfettered social communication in Poland.

The activities of KOR during the 1980 strikes provide a good illustration of its consciously chosen strategy of service and support. In the summer of 1980, KOR issued statements on the situation, gathered information about strikes and distributed it through expanded editions of independent publications (e.g., *Robotnik*) and through foreign journalists; but KOR was not directly involved either in the organization of strikes or in the formulation of demands. This form of assistance accorded well with the plurality of more specifically political positions represented within KOR and with its shared ideological commitment to democracy, according to which respect for civil and human rights could be achieved in Poland only through the formation of various independent social groups able to exert pressure on the authorities, that is, through society's becoming the active subject rather than the passive object of state actions.

By emphasizing the service role of KOR, I do not wish to deny its more "ideological" or "intellectual" functions. KOR repeatedly issued analyses of the social, economic, and political situation in Poland. By breaking the barrier of fear in the intellectual milieu KOR also contributed indirectly to the formation of other groups, so that in the late 1970s, analyses of the current situation in the country as well as studies of Polish history and culture were being promoted through such initiatives as the Society for Scientific Courses, which organized lecture series on a number of sociological, historical, and cultural topics, the discussion group "Experience and Future," and a growing number of independent publications (8). Given the state monopoly on information and the continuous repressions directed against all independent activities, the achievements of the pre-1980 opposition in Poland were indeed impressive, but we should not overestimate their actual social effectiveness: although the number of independent publications was growing between 1976 and 1980, even toward the end of this period they reached only a very limited number of workers. Perhaps more importantly, all these activities contributed to integration within the intellectual milieu itself, overcoming much of the ideological conflict between groups with different political genealogies. This was important to later participation in the movement, for although the divisions and animosities were by no means forgotten and could even be more directly expressed in intra-union conflicts, the union was able to accommodate a plurality of views without splitting itself from within. Even if the proposed political, social, and economic ideologies, strategies, and tactics were by no means always compatible and often led to acrimonious debates, there was general acknowledgement of a shared set of fundamental values and principles of the trade union movement: democracy, the rule of law, pluralism, self-determination, etc. At the same time, the democratic opposition succeeded in creating certain links between the intellectuals and the workers and in establishing some independent channels of communication which proved to be of importance during the summer of 1980 and during the initial period of the organization of "Solidarity." Finally, we should not underestimate the symbolic value of the intellectuals' opposition and of KOR's defense of the victims of the 1976 repressions. The ethos of KOR, exemplified in its defense of the workers, was reflected in "Solidarity's" emphasis on civil responsibility, self-organization, and social solidarity. Even if the worker activists did not know what political programs, analyses, and solutions the intellectuals proposed, they were aware of the existence of those who has spoken out against abuses or done something to break the state monopoly of information and they were conscious that among the intelligentsia they could find useful allies.

If the numerous analyses of the critical social and economic situation in Poland cannot be said to have served as a general rallying point of the striking workers, it was not because the workers had a different vision of the state of the country or of the causes of the crisis. The worsening social and economic situation in Poland and the discrepancy between the offical "propaganda of success" and the reality of empty store shelves, disorganization in production, and increasing corruption and inequality were visible to everyone. Workers interviewed by Marody in the fall and winter of 1980 argued that they had had to "take things into their own hands" because "one could no longer live like this" (9). This feeling of deprivation and wounded dignity did not need an elaborate ideology which the intellectuals could offer, in order to be transformed into action. We must remember that for years prior to 1980 Polish society had subscribed to a homogeneous and exceedingly stable system of political values. Stefan Nowak and his group found that when asked to identify the attributes of a good social system, Poles repeatedly mentioned equality of opportunity regardless of social origin, freedom of speech, the influence of all citizens on the way society is governed, the assurance of a proper standard of living for all citizens, and economic efficiency (10). For many years adherence to these values had gone hand in hand with an emphasis on individual success, the privatization of goals, and a lack of identification with any social group or organization intermediate between "the family" and "the nation"; but when the widespread frustration and irritation observed by Polish sociologists in the late 1970s were transformed into political action in the summer of 1980, the movement which emerged from the explosion of social discontent tried to put into effect precisely those values which the society at large had professed for years. In other words, if the intellectuals did not have to provide an ideology, it was for the simple reason that the values they articulated had been broadly accepted by society independently of those whose social function has been defined by sociologists as that of ideologists. In the words of sociologist and "Solidarity" expert Jan Strzelecki, the experts "were perhaps needed (to build arguments and to formulate suggested and alternative — Wałesa called them 'emergency' — solutions to various issues) but the thing they certainly did not have to do was to 'raise the consciousness of the working class' . . . People of that class already have their consciousness without us, a consciousness they got from somewhere else, from their own experience and from their own social existence" (11).

It is important to note, however, that the workers' social "consciousness" included a belief in the possibility of an alliance between the workers and the intelligentsia. In their research on the attitudes of the worker activists, Marody and her group discovered that while the activists of "Solidarity" were on the whole confident of their ability to manage union affairs, the workers repeatedly mentioned the intelligentsia as the only social group that could become their ally. At least some of the confidence expressed by the workers was based on the belief that they would be able to find "educated people" able to advise them:

an activist educated as a technician stated: "We must have a man to whom we can turn and then we simply will know. He will tell us what should be done. Not that he will direct us, but only so that he would tell us: it is like this, and if you do this then it will be like that, or this is wrong because then they can get you there . . . And when we have such a person, someone who knows the law and some regulations, we will be more effective and some matters will be taken care of sooner" (12).

Whether this trust in the intelligentsia expressed by the workers can be attributed directly to the activities of the opposition or to the general sense that these two social groups share values and interests, it created conditions which made possible the cooperation between workers and intellectuals in "Solidarity."

Without doubt, this shared though independently articulated community of values and interests was instrumental in promoting the cooperation among the various groups in "Solidarity." The official stereotype, according to which workers struck for purely economic reasons while such issues as freedom of speech or the introduction of democratic means of management and government were imposed on the movement by the intellectuals, has been amply contradicted both by sociological studies and by the chronology of events in August. Indeed, Stefan Nowak's attitude studies have found that "one important democratic value, freedom of speech, proved to be more highly prized among workers, and particularly among highly skilled workers, than it was even among people with a university education" (13); and in her interviews with the workers, Marody discovered that this value was regarded by the workers not only in instrumental terms (i.e., honest information is necessary in order to make right decisions) but also as an autonomous good since the "right to know" was deemed a part of dignified life (14). It is therefore not surprising that demands for limits to censorship and for the provision of full information about the economic situation in the country were among the 21 points formulated by the workers prior to the arrival of the experts.

During the strike in Gdańsk, most of the experts present in the shipyard believed that the main demand of the striking workers, the first of their 21 postulates, calling for the creation of new independent trade unions, was unrealistic since the authorities would never agree to make such a concession. As a result, the experts prepared two alternative documents: one relating to the creation of new trade unions, and the other describing the possibility of reforming the existing government-controlled unions. During a discussion of the alternatives with activists of the MKS, the experts were asked which of the two solutions they considered to be better. Their answer that the new trade unions would be better but had no chance of being accepted was countered by the workers' claim that it was not worthwhile to strike merely in order to reform the existing unions. This episode — related to me by one of the participants — not only demonstrates the workers' commitment to the more radical proposal of independent unions but it also defines the role of "experts" during August and within "Solidarity": the 21 points formulated by the workers were to be treated as an instruction, and the experts nominated by the MKS were to represent one of the sides in the conflict. One of the experts described this early episode as a "crisis of objective service." The experts' role was neither that of "objective" professional consultants nor that of mediators between the sides. Nor was the role of the intellectuals that of ideological leaders formulating goals and programs (15). However strongly one can aruge that the democratic opposition in Poland which developed since 1976 prepared the ground for "Solidarity," it would be misleading to see the intellectuals involved as a revolutionary vanguard. The expert was to serve with his wisdom and imagination, with his ability to articulate and argue, and with his specialized knowledge of society, economy, or law. But he was not an ideologist formulating a utopian vision of a just society, nor a professional offering "objective truth" to all interested clients. The episode described here clearly alerts us to the flimsiness of stereotypes of the ideological role of intellectuals in social movements and revolutions, as well as the stereotypes of objective professional experts indiscriminately serving all those interested in purchasing their knowledge; we must, therefore, examine in more detail the role of intellectuals in "Solidarity" and the specific forms of partnership and conflict which this social movement developed.

Intellectuals in a Social Movement: Advisors and Consultants

All the "Solidarity" experts I interviewed began by drawing a distinction between "experts-advisors" and "experts-consultants" (16). The "advisors" who functioned within the Coordinating Commission were said to have played a predominantly political role: they provided information on current issues, suggested strategies, possible compromises and alternative responses, attempting to spell out their consequences, translated the language of the authorities to the leadership of the union, and formulated acceptable answers. Their sphere of competence was not clearly defined, and they participated continuously in the activities of the Coordinating Commission, including negotiations with the government. In the words of one of the consultants, "'expert' was a convenient label for political activity." In contrast to advisors, consultants were called on intermittently to advise on specific technical issues, prepare some necessary information or documents to be presented to the Coordinating Commission and participate in negotiations with the government when these specific issues were on the agenda. Some, though not all, consultants were also coordinating and doing research within the national Center for Social and Professional Research (OPSZ). Formally, neither consultants nor advisors had any power of decision. They did not have the right to vote during the meetings of the Coordinating Commission or to make agreements on behalf of "Solidarity" in the negotiations with the authorities.

The boundaries between the two groups were, however, by no means clearly defined, and there was a complete lack of institutional differentiation. The advisory scientific council of OPSZ nominated by the Coordinating Commission was composed both of people with clearly advisory functions (e.g. Jacek Kuroń, Bronisław Geremek, Tadeusz Mazowiecki, or Adam Michnik) and of specialist consultants — professors of economics, law, or sociology whose function in "Solidarity" was limited to a specific issue or a narrow set of issues directly linked with their professional interests, such as the provision of health services, wage indexation, housing, workers' self-management, intra-union democracy, etc. (e.g., Jan Malanowski, Andrzej Tymowski, Jan Strzelecki, Jacek Kurczewski). Moreover, not all consultants or advisors were formally associated with OPSZ or with regional OBS's; some were nominated only to specific negotiating bodies, and a number of advisors stood for elections to the leadership of the union.

The drawing of a distinction between consultants and advisors makes it somewhat easier to differentiate the diffuse role of an intellectual within a social movement from that of a professional scientist offering his services to an organization. Since consultants offered technical information drawn form their specialized knowledge and skills, we would expect that their legitimacy would be based at least in part on their professional reputations. (This does not exclude considerations of trust: we must remember that consultants represented the union in negotiations in which more was at issue than scientifically established facts.) Advisors, on the other hand, would be more likely to be evaluated in terms of political rather than professional criteria, since their tasks of articulating demands, analyzing the possible consequences of various courses of action, and suggesting political strategies could be said to be based on the more general competence of intellectuals. Or, one might say that the advisors' legitimacy was based first and foremost on trust, and that they had to be trusted primarily as loyal supporters; while in order to be regarded as legitimate experts, the consultants would have to be accepted as knowledgeable professionals, and their loyalty would be less germane to their evaluation.

In broad outline, such a perspective does help us to understand why the issue of trust was of paramount importance both in the advisors' own analysis of their role and performance and in the actions and statements of "Solidarity" activists. The advisors I interviewed felt that the initial trust they received was a result of their earlier political, oppositional activities and of their clear identification with one side of the conflict already in August. They also felt that this trust later began to dissipate, particularly after March 1981, when some advisors (together with some "Solidarity" leaders) were accused of "manipulating" the Warsaw agreement and of being excessively "soft." There were accusations that the advisors were making decisions instead of the elected bodies, and that this was in violation of the democratic process within the union. A more explicit expression of this distrust came during the congress of "Solidarity," when there were pressures to formalize more strictly the position of "experts," and suggestions that they should hold only union jobs (some activists felt that employment in state institutions, such as universities, could compromise the advisors' loyalty). The advisors claim that this increasing lack of trust was a consequence of the presence of a large number of advisors representing various ideologies, of the radicalization of opinion among middle-level "Solidarity" activists, of the general lack of trust in politicians rampant within Polish society during a period of political mobilization, or of internal power struggles within the union (for example, of competition from the middle levels of the "Solidarity" leadership which felt that the advisors were blocking their possibilities of advancement within the union). All these are clearly political reasons consonant with the view that advisors were seen as political actors and evaluated in political terms.

Within this political arena, however, the advisors were clearly limited by the nature of their own position. Not only did they not have any decision-making power and their positions were not legitimized by an electoral mandate, but also their role was defined as that of providing service to the union, and this service was to promote the goals formulated by the elected bodies. Thus, although he was playing a political role, for the sake of maintaining his influence an advisor should not appear to be overtly political: he was listened to because he possessed some abilities, knowledge or skills which the activists lacked, and not because he represented a particular group or a clearly defined political program. In fact, those advisors who attempted to articulate long-term political programs or could be identified with such programs or ideologies (for example, Jacek Kuroń or Adam Michnik) were seen as particularly suspect independently of the actual content of their programs. These tensions between objective disinterested service and required political loyalty, and between the obligation to formulate a political course of action while not appearing to play a political role, were inherent in the role of the advisors and could be said to define the boundaries within which they attempted to meet their tasks. Such tensions can also perhaps serve as a partial explanation of the complaints often voiced by the intellectuals themselves that the union — staffed as it was with so many intellectuals — was unable to articulate a coherent strategy, or that there was not enough place for general programmatic discussion and political analysis of the more global institutional problems of the country. There were very few meetings of the national Coordinating Commission devoted to general strategy and tactics, or to questions such as whether there was a chance for institutional adjustment and stabilization, or what such a chance might depend on. While everybody was conscious of the fact that the union must limit its demands, there was no explicit discussion of where limits might lie and what they implied.

In drawing the distinction between advisors and consultants I have stressed the fact that the legitimacy of the consultants should have been derived from their professional expertise and that we would expect it to be less dependent on considerations of political loyalty. Given the nature of "Solidarity" as a social movement continuously involved in conflict with the authorities, it is not surprising that knowledge and objectivity was not all that "Solidarity" expected of its consultants. Although professional considerations did of course play a role in their selection, political considerations were apparently at least equally important, and an oppositional past constituted a valuable asset. Nevertheless, doubts about the trustworthiness of consultants were heard relatively rarely in comparison with the doubts raised about the advisors.

The tensions inherent in the role of the advisors were not completely different from those faced by the consultants who, after all, also took part in negotiations with the government, in which political abilities and judgements went together with technical expertise. However, because of their more limited and better defined sphere of competence, because of the often technical character of the issues involved, and because of their distance from the day-to-day activities of the union, the consultants on the whole enjoyed more trust and were less likely to be accused of political manipulation. This was so even in the case of issues in which the relevant professional expertise was not very strictly defined or issues in which members of the Coordinating Commission (which eventually decided the matter) were not completely ignorant. For example, in the negotiations over the censorship law, "Solidarity" was represented by a writer and literary critic, a journalist, a lawyer, and an elected member of the Coordinating Commission, and their compromise with the authorities (which they themselves described as not completely satisfactory) was accepted at the meeting of the Commission after a discussion in which the various provisions of the law were examined in light of the experience of the editors of union bulletins and periodicals. Occasionally, however, the consultants' loyalty was questioned. This happened primarily when the government attempted to treat a given issue as a test of political authority, so that as a result the issue became a matter of political disagreement among the union leadership. This was the case, for example, with the issue of free Saturdays, which the government attempted to decide alone without negotiations with the union. In view of the unilateral action of the authorities, free Saturdays ceased to be the technical issue of whether the economic situation would permit the introduction of a shorter work week, and became a political problem concerning the manner in which decisions of this kind should be made. As a result, "Solidarity" not only fought for free Saturdays, but economists who argued that the crisis did not permit such a reduction of working time became suspect and were accused of "not knowing what real work was like." Secondly, the trustworthiness of consultants could become an issue raised by the consultants themselves, and arguments among them (for example among the adherents of fundamentally different economic models) did on a few occasions become matters of political controversy within the union at large. Given the plurality of views represented by the consultants and the presence of a political stage on which decisions were being made, it is not surprising that consultants should compete for influence and on occasion accuse one another of professional dishonesty, currying the favor of the uninformed, or attempting to practice political manipulation. Despite these exceptions, we should emphasize again that the problem of trust was much more acute in the case of the advisors than in the case of consultants.

The relative lack of distrust directed towards the consultants was in part also dependent on their distance from the political struggles within the union, and in order to maintain this form of disinterestedness, they, like the advisors, were constrained to address and analyze specific problems rather than the more general issues concerning the political system in Poland, the possibilities of accommodation with the authorities, or the global strategy of the union. One of the consultant sociologists recounted that although he had prepared just such a general analysis of the situation and presented it at a scientific conference abroad, he felt it would be unwise to offer it for discussion in Poland.

These more general issues did arise during the preparation of the "Solidarity" program for the September Congress. Discussions on the program took place in a number of thematically organized groups composed of delegates to the Congress, advisors and consultants. The partnership between the worker-activists and their experts during the preparation of the program was evident not only in the fact that the proposals often resulted from a combination of the more theoretical knowledge of the experts with the practical experience of the workers, but also in the political differences which emerged and which in no instance reflected class differences. For example, the division between fundamentalists and pragmatists which crystallized in the group discussing the problems of the socio-political system cut across social differentiation so that both groups included intellectuals as well as workers. In the opinion of one of the chief advisors of "Solidarity," it was during these discussions on the program that a new form of partnership between the workers and the intellectuals was taking shape.

Finally, we must add that the occasional mistrust directed against the advisors (and fed by party propaganda) functioned primarily as a tool in political disputes and a means of exercising democratic control within the union. It never became acute and, for example, did not lead to dismissals. After all, advisors could be listened to or not, as the leadership chose. Ultimately, the advisors did enjoy the trust of the union and its leadership to the very end, and the underground structures of "Solidarity" to this day use some experts (17).

The Center for Social and Professional Studies

I have argued that because of their position in the union, the experts of "Solidarity" were to some extent constrained not to address general problems or become overtly ideological. This suggests that in order to understand their role we should inquire in greater detail into the actual functions and influence of the Center for Social and Professional Studies, asking what kinds of expertise it provided, how its members met their tasks, and what they felt to have been the most serious shortcomings of their work.

The formal provision for the creation of a union research center was made already in the Gdańsk agreements of August 31, 1980. Point 6 of the first paragraph of this agreement dealing with the formation of new independent trade unions reads:

> The Interfactory Committee establishes a socio-professional research center whose main function is an objective analysis of the workers' situation, the living conditions of the working people, and methods of representing workers' interests. The center will make expert appraisals of the wage and price index and suggest forms of compensation. It will also publish the results of its surveys. Moreover, the new union will have its own publications (18).

Despite this early recognition of the need for a formally organized research center, its actual organization took several months. This delay was due largely to the initital lack of clarity about the organizational structure that the new union should adopt.

Early in September, prior to the decision to organize "Solidarity" along regional lines, the Warsaw employees of educational and research institutions organized their own union, the Independent Self-Governing Trade Union of the Employees of Science, Technology, and Education (NSZZ PNTiO). In addition to organizational work among the employees of this sector, NSZZ PNTiO attempted to create "Consulting Centers" offering help to any activists interested in organizing new unions in their enterprises. Thus, the initial formal independence of NSZZ PNTiO did not signal isolation between the workers and the intelligentsia. The NSZZ PNTiO participated in all the actions of "Solidarity," intellectuals and students continued to participate in the formation of "Solidarity" structures on the national and regional levels, and NSZZ PNTiO dissolved itself in November in order to join the appropriate regional bodies of "Solidarity" (19).

The eventual decision to organize the union along regional rather than occupational lines made the access of intellectuals and various professionals to the union extremely easy. Not only the national Coordinating Commission, but also every region had its own advisors and experts and eventually its own Center for Social Research (OBS).

The professional staff of the Center working for the national Coordinating Commission was composed of economists, lawyers, and sociologists and its mandate was to provide the Coordinating Commission with information it required for its work. The OPSZ was supervised by a Scientific Council composed of 26 intellectuals while its day to day management was in the hands of Andrzej Wielowieyski and Ryszard Bugaj. While the number of employees of the Center was relatively small, many social scientists volunteered their services and performed specific jobs for the Center.

The economists constituted the largest group within the Center and the economic schools and positions they represented were quite diverse. This became apparent in the later conflicts over workers' self-management and the shape of economic reform.

Initially, however, "Solidarity" was not involved in any global economic negotiations. In the Gdańsk agreements the authorities had pledged to accelerate work on the economic reform and to define and publish its basic assumptions. The agreements did not specify what the reform should consist of beyond calling for "substantially increased independence of enterprises and real participation of workers' self-goverment in management" (20). Although the agreements also called for unions' participation in the preparation of plans for reform, "Solidarity" leadership was initially reluctant to play an active role. Thus, the "Solidarity" representatives to the Government Commission for Economic Reform, two economists, Waldemar Kuczyński and Ryszard Bugaj, had only the status of observers. Until mid-1981, "economists active in the union were often concerned with the problematics of economic reform and cooperated with a number of independent initiatives, but the union authorities did not have an understanding of the issues involved nor the possibility of working out a joint position" (21).

As a result, the economists in OPSZ (together with some sociologists) worked primarily on various social policy issues on which negotiations were being conducted or anticipated. The subjects included free Saturdays and the length of the work week, the situation of retired persons, the rationing of food, minimum living standards, the indexation of prices and wages, housing, etc. Work which involved the preparation of documents and some short-term research was divided among different groups and negotiating teams. The results were then presented to the Coordinating Commission which had the responsibility of defining the offical "Solidarity" position on the subject. Consultants participating in negotiations also reported to the Coordinating Commission on the progress of these negotations. The Center also organized scientific conferences on problems important for the union, inviting the participation of specialists from various institutions (including government or management experts), and used such meetings as a basis for the later preparation of synthetic documents concerning a given issue (e.g., housing). Work on most of these issues proceeded relatively smoothly, while preparatory discussions between union and government consultants occasionally prevented conflicts during final negotations.

The rapidly worsening economic situation and pressures from below to improve supplies to the market meant that by the spring of 1981 "Solidarity's" leadership and its economic consultants had to focus more of their attention on problems of global economic reform and on workers' self-management. In its work on these issues, OPSZ was joined by a "Solidarity" group called "Sieć" (The Network), which represented the largest enterprises in Poland. "Sieć" was the first "Solidarity" group to criticize the government draft of the law on self-management, and during the spring and summer of 1981 it prepared an alternative proposal of such a law. Later, "Sieć" published a number of documents about other aspects of the reform (22). Relations between the economists from OPSZ and "Sieć" were good, although "Solidarity's" proposals presented by its negotiating group on economic stabilization, reconstruction, and reform, were less radical than those of "Sieć." Discussions of the reform and self-management were published in "Solidarity's" press, and the matter was taken up during the Congress in September, when the union and the authorities agreed on a compromise on self-management (23). It was during these discussions on reform and self-management that differences among the economists themselves were transformed into political confrontations within the union leading to accusations of manipulation, professional irresponsibility, and a dishonest search for popularity. Despite these conflicts and despite a stormy discussion of the compromise on self-management, the compromise was finally accepted, and the reform program adopted as the basis for "Solidarity's" negotiations with the government was seen as satisfactory by the OPSZ economists. However, the conflicts over self-management and reform (and we must remember that the issue of reform was never actually taken up in negotiations with the government) can serve as a good illustration of how the consultants' involvement in broader problems, the conflicts among them, and confrontations with the authorities tended to turn into problems of trust and contests over authority within the union.

The work of the OPSZ economists was rendered more difficult by their lack of access to information about the economy which the authorities refused to release. Other shortcomings were related to the variety of needs which OPSZ was required to meet. While the work "ordered" by the Coordinating Commission had to take precedence and proceeded rapidly, long-term needs were met more slowly. Thus, the economists from OPSZ never managed to produce a regular bulletin on economic matters for the Coordinating Commission, and plans to produce a diagnostic tool for the economy were not finalized by the time the state of war was declared.

A paper by Jerzy Eysmont on the need for such a diagnostic tool is interesting precisely because it suggests that the union movement, had it been allowed to continue its activity, could have had an influence on the research of the economists who worked for it. The article begins with a telling question: "If the OPSZ of the Coordinating Commission of "Solidarity" were to undertake the preparation of regular analyses and evaluations of the economic situation, the first question which would have to be answered is: whom would these analyses and their results serve and to what ends?" (24) Eysmont lists a number of ways in which such an instrument could be useful to the union: he claims that it would allow for a more informed confrontation of opposing views, limit the possibilities of government manipulation, and serve the economic education of society. The educational function is also emphasized when Eysmont discusses the criteria which such a tool of economic analysis should meet:

> One must make sure that what will be included in the macroeconomic analysis should not diverge too far from the everyday experience of individuals. Obviously, a statistical image almost never corresponds to the selective subjective observations of individuals. It is important, however, to show why these differences exist and how they should be interpreted, so that the truth of experience does not become something altogether different from the truth of statistics, since statistics are currently regarded with excessive scepticism and distrust (25).

Claiming that the selection of the phenomena to be treated as symptomatic of the general economic situation is by no means obvious, Eysmont argues that the choices made should include "phenomena important for the fundamental functions of the trade union, that is, those which directly affect standards of living and conditions of work," as well as "phenomena which are observed directly by the people both as consumers . . . and as producers . . ." (26). Thus, in all his arguments Eysmont explicitly acknowledges that economic analyses conducted for the union should assume a special form precisely because they would be intended to serve a particular clientele and contribute to the economic education of society. What is more, he also argues that such analyses should provide a lesson is pluralism:

> Every analysis would contain commentaries written by various specialists considered to be authorities on the given issue. In this manner, the recipient would be conscious of various points of view and could convince himself of the complexity and wealth of the social and economic reality. Such an approach would have a very important influence on the formation of the economic consciousness of our society, and could increase the respect for divergent views on the same issue — even when these are supported with trustworthy statistical data (27).

The state of war declared on December 13, 1981 also interrupted the work of the OPSZ economists, and we cannot know whether regular analyses of the economic situation, had they been developed, would indeed have borne the marks of the interests they were designed to serve, nor whether they would have fulfilled the functions for which they had been intended. Here we must regard this example only as an indication of a possibility that the research of the economists in "Solidarity," their conscious espousal of its goals and interests and its pluralist character, could have had a direct influence on Polish economics.

The role of the sociologists within the OPSZ was altogether different from that of the economists. While the economists were of necessity continually involved in negotiations with the government and defined the position of the union on matters of general social interest, the sociologists by and large focussed their attention on issues internal to the union. Most generally, their task can be defined as service on behalf of democracy within the union.

Given the union's emphasis on democracy (described by some as a "democratic craze") and the belief that "Solidarity" should be fully responsive to the wishes of its membership, the numerous opinion polls conducted by sociologists in the regional Centers for Social Research served as substitutes for a system of referenda to verify and legitimate union policies. Opinion polls were conducted in the various regions after every major decision taken by the union, and they occasionally served as a tool of union activists who used their results to press for particular solutions (28). By providing for a flow of information between the cadres and the membership, opinion polls were meant to prevent routinization within the union and to reduce the danger of a separation between the union cadres and its rank and file. However, since most opinion polls were conducted on a regional basis (sociologists from the various OBS's competed with one another on this issue and, in order to avoid duplication and conflict, the OPSZ did not conduct opinion polls), whether or not their results reached the Coordinating Commission depended on the policies of the leadership in a given region (and sometimes on whether or not the results supported the views of this leadership). Thus, because opinion polls were used in political conflicts within the union, some of the cadres disliked them and resisted their use. This became especially apparent in March 1981, when a quick opinion poll conducted in the Warsaw "Mazowsze" region revealed broad support for the compromise reached by the negotiating team. Since the compromise (and especially the manner in which it was reached) was considered unsatisfactory by a number of "Solidarity" leaders and middle-level activists, the results of the Warsaw survey were not welcomed and "it was difficult to get through." The situation was apparently furhter complicated by the fact that a similar opinion poll conducted in the Wrocław region revealed popular dissatisfaction with the agreements. This situation made it amply clear that "scientific truth" is not simple; and from then on, the Coordinating Commission demanded that multiple studies be conducted on every problem.

Safeguarding democracy within the union was also the main goal of the sociologists working for the OPSZ, that is, on the national level. Accordingly the OPSZ prepared analyses of union structures, their finances and elections, attempted to maintain a register of "Solidarity" membership, studied the union press, and drafted sections of union statutes, especially those concerning the organizational structures and elections. These organizational issues were also the subject of two nationwide meetings of sociologists associated with "Solidarity" research centers. In addition, the OPSZ conducted studies of negotiations and strike mobilization, attempting to draw conclusions which could be of practical value in the conduct of union affairs

The focus on issues internal to the union seems to have involved the sociologists in relatively few conflicts. For example, disagreements about proper organizational structures (the degree of centralization and the division of responsibilities) never became open or general conflicts. At the same time, sociologists working for the OPSZ were acutely aware of their service role and of the limited character of this service. I was told by one of them that "good sociology was to be done outside, while the task of the OPSZ was to provide services." Apparently, the demand for sociological expertise as it was understood by the sociologists in OPSZ was not entirely compatible with "good sociology". But given that the two tasks were seen as contradictory, it is unlikely that the sociologists' consultancy work for "Solidarity" would have seriously affected the development of Polish sociology.

The idea that sociology was to promote democracy within the union also found its expression in the educational activities that the sociologists from OPSZ and from the "Mazowsze" (Warsaw) OBS undertook, often in cooperation with the Workers' University or with the Polish Sociological Association. Pamphlets about how union meetings should be held and how elections should be organized were published, translations of fragments of Lipset's *Union Democracy* were distributed, and seminars on these and other similar issues were organized and well attended (29).

In general, the educational activities conducted first on an *ad hoc* basis and then by the Workers' University consitute one of the most successful initiatives of intellectuals in the union. Members of DiP and TKN developed a program which included not only practical training for union activists but also lectures on history, literature, sociology, and economics. The union press, edited by intellectuals and students as well as workers, also played an educational role by publishing historical or literary articles alongside analyses of the current social, economic, and political situation.

In this context, we must mention the popularity of the demand for "truth in history." Questions about whether the new union would fight for historical truth or for changes in school history textbooks had been raised already in the first days of September 1980, during the first public informational meeting in Gdańsk, attended by the leaders and experts of the MKZ, which was to be devoted to organizational issues. The importance of the problem of historical truth for the workers was visible not only in the demand that monuments be erected to past victims of the regime, and in numerous celebrations of officially forgotten anniversaries, but also in the particularly wide appeal of the lectures of the Workers' University devoted to historical subjects.

The educational and popularizing activities of the Workers' Universities and of the union press, as well as the tasks performed by the sociologists and economists from "Solidarity" research centers, shared an attempt to be immediately responsive to the explicitly formulated demands of union leaders and members. This devotion to an ideal of service rather than one of ideological or political leadership took a number of different forms and required more or less explicit involvement in the political processes taking place with the union. We have seen that the more general the functions of the intellectuals were, and the broader the definition of their sphere of competence, the more prone they were to encounter distrust and become involved in internal conflicts. This becomes apparent not only when we compare the role of the advisors with that of the consultants, but also when we compare the kinds of services provided by the economists with those offered by the sociologists working for the OPSZ, and even comparing the role of those economists who provided expertise on narrowly defined issues of social policy with those involved in the formulation of the program of economic reform and in negotiations on that program. At the same time, given the working-class character of the movement and its democratic and pluralistic ethos, the constraints implicit in the various "service" roles of the intellectuals (who were not elected to represent specific constituencies) resulted in a virtual rejection of the role of intellectuals as ideologists. While some experts and consultants regarded this limitation as a sign of a new partnership between the workers and the intellectuals, others saw it as an abdication of responsibility. Jerzy Jedlicki, a historian of ideas, asked himself whether the intellectuals had a responsibility to steer the movement intellectually — to "discover sources of threat," to "formulate more distant goals" and to provide "information about the general state of the country" — and argued that the answer to this question should be an emphatic "yes":

> I think that the OPSZ was created also because of the conviction that formation of public opinion is possible and that we should strive for it. Up to now, I believe, the situation in this respect has been bad. That is, the numerous army of intellectuals . . . who have eagerly involved themselves in the work of the union, have played within it almost exclusively the role of experts. They serve the union interests with their professional — economic, sociological, and legal — knowledge, and execute . . . momentary "orders" of "Solidarity." I believe that the role of intellectuals should not be reduced to this alone. (. . .)
>
> What took place amounts to an abdication in the face of the elemental vitality and strength of the movement in which we are participating: we decide only to serve its interests and never — God forbid — to oppose those interests, even when from time to time we consider them irrational or even dangerous. It is a dangerous abdication. (. . .) I am thinking here about the formation of opinion in general matters (30).

Whether or not we share Jedlicki's diagnosis of the pitfalls of the "service" role, it is important to realize that the limitations of this role were implicit in the positions which the intellectuals held within the movement and in the character of "Solidarity" as a whole. We must see these limitations not as a result of a voluntary abdication of responsibility but as a function of the conditions determining the intellectuals' legitimacy and authority within a pluralistic democratic movement, one which throughout its history was ridden by the problem of reconciling its trade union character with the necessarily political role it had to play under the conditions of "real socialism."

"Solidarity" and the Social Sciences in Poland

The conscious limitation of the role of the consultants in "Solidarity" to the provision of services demanded by its leaders and members resulted not only in the lowering of the importance of the ideological functions of the intellectuals, but also in the relatively minor influence of their services on research in the social sciences. The strength of this conclusion and its applicability to other cases of the intellectuals' involvement in social movements must of course be tempered by the fact that "Solidarity" led an above-ground existence only for sixteen months, and that during this period it was involved not only in a permanent conflict with the authorities but also in the organization of its own internal structures.

The experience of "Solidarity" consultants in economics was important for research insofar as it revealed certain gaps in the knowledge of the economists and stimulated interest in such subjects as the measurement of the costs of living and their changes, the minimum socially acceptable standard of living, the indexing of wages, and workers' income and spending (31).

Other changes in economics which occurred during this period were less a result of the demands articulated during the provision of services to "Solidarity" than of the more general changes taking place in Polish society. Obviously, the economic crisis constituted the most immediate challenge for the economists, while the relaxation of censorship and the emergence of the possibility that reforms might be introduced stimulated economists associated with various institutions to formulate plans for overcoming the crisis (32). The state of war eliminated hopes for an economic reform and it is unlikely that work in this area is being pursued by more than a handful of economists. A summary of "Solidarity's" achievements in this field has been offered by Waldemar Kuczyński, one of the chief economic advisors of the union:

> What was left was a conviction that good legislation is a result of the confrontation of social forces and interests, and that good law is impossible without freedom. Also left was the idea of a socialized but not statist economy, the idea of a market economy which would not be capitalist. I believe that this was a utopian idea, a transient stage in the evolution towards an economic model resembling those of the western world. But it is impossible to exclude the possibility that it was a beginning of some new, as yet untravelled path (33).

It is not accidental that Kuczyński's summary is formulated in terms of economic policy rather than economic science. "Solidarity's" heritage in this latter field has not been very obvious thus far. In some measure we can attribute this to the characteristics of the economic community in Poland: the economists have had a long tradition of serving the government which constitutes the primary audience for the results of their work even when these results are totally disregarded in the formulation of economic policy (34). At the same time, the work of the economists was and is more directly dependent than that of other social scientists on access to government-controlled information, and the outside "market" for their expertise or an audience for their work remains very limited.

The *direct* effects of sociological consultancy for "Solidarity" on the research of sociologists also appear to have been limited. As we have seen, this was in part a result of the conscious decision of the sociologists in the OPSZ to confine their interests to the immediate task of serving the cause of union democracy. At the same time, the need not to appear overtly political effectively blocked the development of political sociology within the union (35).

But the lack of an obvious and direct link between sociological consultancy for "Solidarity" and later research should not obscure the fact that Polish sociology was very profoundly influenced by the emergence and formation of "Solidarity," by the discussions it generated, and by the social and political changes it brought about. Although an exhaustive analysis is imposible within the framework of this paper (and probably premature), I should like briefly to discuss some of the changes in Polish sociology and their causes.

The most immediate and obvious of these changes was a result of the relaxation of censorship. Prior to the formation of "Solidarity," the authorities not only prevented the publication of the results of a number of opinion polls, but were involved in the research process itself and censored questionnaires. Despite protests from the Polish Sociological Association, researchers attempting to study public opinion in Poland faced constant difficulties, and the fear that research results might be unpublishable prevented others from trying. The most natural escape was to abandon empirical research in favor of theoretical analysis. The relaxation of censorship not only permitted the publication of earlier studies on such subjects as social and political values and attitudes, or social inequality, but also allowed various institutions and individual sociologists to undertake new studies of public opinion in the hope that their results would be published (36). To some extent, this relaxation of censorship on sociological research continues today: the results of surveys conducted by, for example, the Center for the Study of Public Opinion of the Polish Radio and Television are still being published, and certain other studies prepared by university sociologists or by sociologists from the Polish Academy of Sciences appear in very limited off-set editions (100 copies) which are subject to more lenient censorship. In this context, we must also mention the continued existence of publications beyond the reach of censorship which provide an outlet for some sociological studies unpublishable in the official media (37). Although this so-called "second circuit" allows for the complete circumvention of censorship, those who decide to resort to it run considerable personal and professional risks, and scholars, especially those still pursuing their degrees, who decide to publish underground might effectively ruin their future careers. That some of them do decide to pursue this course testifies to their courage, but the "second circuit" cannot be regarded as a functional equivalent to the official (and censored) scientific periodicals.

No less important than the relaxation of censorship was the fact that the formation of "Solidarity" and the rapid — one might even say, revolutionary — changes in Polish society created a laboratory for sociological research which the Polish sociologists eagerly exploited. For example, the process of the formation of "Solidarity" has been studied by a group of sociologists from Warsaw University, while another group working in the same institution studied differences in values and attitudes among three social groups: "Solidarity" activists, passive observers, and managers (38). Large-scale surveys of the changing social and political attitudes of the population and especially of the perceptions of crises and conflicts were conducted by a large group of sociologists under the direction of Władysław Adamski from the Institute of Sociology and Philosophy at the Polish Academy of the Sciences (39). Strikes constituted another newly accessible subject for sociological studies in Poland, and while some sociologists collected memoirs of striking workers, others used participant observation to study striking students and their negotiations with the authorities (40). Analysis of strikes and the collection of strike materials was also initiated by the Polish Sociological Association. The entire 1981 congress of the Polish Sociological Association held in Łódź was devoted to subjects directly related to the changes taking place in Polish society, their causes and consequences (41). This explosion of interest in immediate events, and the interest generated *by* these events rather than by prior theoretical queries, has not disappeared after the imposition of the state of war. The experience of profound social crisis has been the subject of a number of studies, many of which were included in a volume edited by Stefan Nowak entitled *Polish Society in the Period of Deep Crisis* (42).

The use of the social movement and crisis as laboratories for social research went hand in hand with changes in the selection of theoretical problems and approaches, and in the methodology employed in the empirical studies. Perhaps the most striking feature of many of the sociological studies of the transformations occurring in Polish society in the early 1980s is their emphasis on the role of socio-psychological factors in the events. The problematics of dignity, religiosity, stress, pride and humiliation, or of the ethos of social groups, their values and consciousness are central in many of the sociological studies conducted during this period (for example those of M. Marody and I. Krzemiński). More recently there have been attempts to develop theoretical frameworks in which such factors could be used to explain social action, revolutions, collective behavior, and social movements (43). While Polish sociology has a long tradition of studying values and ethos (44), the recent emphasis on these subjects has clearly been affected by the experience of "Solidarity" and the perception that its actions often seemed determined by emotional and moral responses. (The most widely noticed example of such a response was probably the message issued by the "Solidarity" congress to workers in other communist countries which had a purely symbolic value, was certain to infuriate the authorities both in Poland and abroad, and yet was seen by the delegates to the Congress as a moral necessity.) At the same time, this interest was also suggested by government propaganda and the statements of such sociologists as Kazimierz Doktór, who asserted emphatically that workers struck in August 1980 for purely economic reasons (45). An attempt to test the validity of this claim influenced the selection of problematics on the part of at least one of the researchers in the group which prepared the extensive study *Polacy '80*.

Problems such as whether the workers were motivated to strike by economic deprivation or psychological frustration and civic powerlessness also played a role in the selection of methodologies in the studies of attitudes and beliefs. Given the lack of established dimensions within which the problem of social activism could be studied, closed questionnaires appeared to be much less adequate than open interviews. For example, it was useless to ask: "Did you join the strike because (a) your economic needs were not being met; or (b) your dignity was being insulted; or (c) because you thought the country was not being governed properly?" In order to receive some credible answer to such a query, it was much more sensible to ask simply "Why did you join the strike?" or "Why do you think "Solidarity" was created?" At the same time, open interviews are more amenable to hermeneutic or phenomenological interpretation than to quantitative analysis, so that the greater reliance on interviews brought with it also some changes in what might be called the interpretative model.

One can also detect certain changes in the theoretical interests of Polish sociologists who have turned increasingly to such subjects as the dynamics of social movements and revolutions, studies of conflicts and negotiations, revolutionary consciousness, the differentiation of linguistic codes and competences, the language of propaganda, and the problem of political legitimation (where the underlying problem could be formulated as "how can a system that society considers illegitimate continue to exist?"). While it is immediately obvious that these interests were dictated by the course of events in Poland, it would be premature to judge the theoretical innovations which the study of these problems in Poland might yet bring.

It is difficult to subsume all these changes under a single heading, but one of the sociologists I interviewed ventured the following generalization: "Before August, the social sciences in Poland fed on themselves. Most studies were theoretical or designed so as to test some theory, and science developed in relation to what was going on in science. Since then, there has been a turn from theory to reality, to the analysis of what is actually happening." Let me leave aside here the epistemological problems involved in such a claim, the question of whether it is possible to "study reality" without "theoretical preconceptions." It is certainly possible to conduct sociological research without a prior and explicit theoretical commitment; but the turn toward "what is actually happening" and away from theory-generated problems signals also the emergence of a new "audience" for sociological research. Works about recent events can obviously enjoy much more broadly based interest among the lay public than theoretical studies which by their very nature are of interest primarily to a narrowly defined circle of professional colleagues. To some extent, this change in the audience for sociological research was clearly unintentional (though once this widening of the audience becomes apparent, it may serve as a stimulus to further work); in some measure, however, it might have been encouraged by a stronger belief among sociologists that sociological knowledge might be socially useful and that "it is possible to talk to the workers." Although this might have been one of the most direct influences of "Solidarity" on the social sciences in Poland, it is suggested here only as a hypothesis. Moreover, given the limitations on free publication and social communication in effect in Poland today, this change in the audience for sociological research might be only temporary.

Throughout this paper I have repeatedly drawn attention to the important role which audiences, with their criteria of legitimation and evaluation, and their needs and demands, play in the structuring of the roles of intellectuals in social movements and of social scientists presenting their work to this movement, to the public, the government, and to their colleagues. The intellectuals who served as advisors and consultants of "Solidarity" were constrained not only by their formal position in the movement and by the complications of the political situation in Poland, but also by the need to maintain their legitimacy in the eyes of the Coordinating Commission and the membership of "Solidarity." The scope and character of their expertise was influenced both by the explicit demands of the movement and by the criteria of legitimacy and authority used in the political arena in which the cooperation was taking place. The political processes within the union provided opportunities for certain kinds of research and for novel uses of the social sciences in negotiations with the authorities and in the organization and functioning of the union, but at the same time they imposed constraints on the more broadly ideological functions of intellectuals and social scientists.

"Solidarity's" effects on the social sciences, however, cannot be measured only by the accomplishments of the direct cooperation between the movement and the experts. The social and political changes brought about by the emergence of the union also had an important influence on the research of those social scientists who were less directly involved in the work of "Solidarity." Changes in the social sciences can be traced back not only to the political liberalization and the social transformations taking place in Poland but also to the emergence of alternative audiences for research in the social sciences. The new research interests of Polish sociologists as well as their novel theoretical and methodological approaches reflected the emergence of a broader audience for sociological research which was no longer drawn exclusively from the ranks of other sociologists. Moreover, the change in the audience for sociological work might have at least retarded the return to theory in which the Polish sociologists have so often found a refuge from the censorship. At the same time, the effects of the experience of "Solidarity" appear to have been much more superficial for the economists than for the sociologists because of the greater dependence of the economists on the state and the corresponding lack of alternative audiences for their work. The suggestion that audiences play an important role in determining the work of scientists is of course not new, but we have not yet begun to understand the specific influence of different kinds of audiences and the conditions they impose on scientific and intellectual activities.

Acknowledgements

Without the cooperation and help of a number of people in Poland this article would not have been possible. I am profoundly grateful to them for sharing their knowledge with me. I also owe thanks to Loet Leydesdorff, Richard D. Whitley and Rob Hagendijk for their comments on earlier drafts of this paper.

Notes and References

1. This letter was later signed by some two hundred other writers, scientists, scholars, artists, etc., and sent also to the authorities. Similar petitions expressing support for the Interfactory Strike Committee and calling for a negotiated settlement were circulated in a number of academic centers in Poland as well as in such professional bodies as the Polish Sociological Association and the Polish Political Science Association. For an excellent account of the "Solidarity" movement see Timothy Garton Ash, *The Polish Revolution: Solidarity 1980—82*, London: Cape, 1983. The most detailed account of events is given in Jerzy Holzer, "*Solidarność*," *1980—81. Geneza i Historia*. Paris: Instytut Literacki, 1984.
2. In addition to Geremek and Mazowiecki, the group of experts to the Gdańsk MKS included: economists Waldemar Kuczyński and Tadeusz Kowalik, historian and Catholic activist Bogdan Cywiński, all three associated with TKN; Andrzej Wielowieyski, the head of the Club of Catholic Intelligentsia in Warsaw, associated also with the discussion group "Experience and Future"; and Jadwiga Staniszkis, a sociologist linked to oppositional circles.
3. For obvious reasons, some of the sources I have cited must remain anonymous.
4. See, for example, Karin Knorr-Cetina, "Scientific communities or transepistemic arenas of research? A critique of quasi-economic models of science," *Social Studies of Science* **12**, 1982, pp. 101—30; and Wesley Shrum, "Scientific specialties and technical system," *Social Studies of Science* **14**, 1984, pp. 63—90.
5. Richard D. Whitley, *The Intellectual and Social Organization of the Sciences*. Oxford: Clarendon Press, 1984.

6. Piotr Sztompka, "Dynamika ruchu odnowy w świetle teorii zachowania zbiorowego," (The dynamics of the renewal movement in light of the theories of collective behavior), *Studia Socjologiczne*, No. 3—4, 1982, p. 79.
7. The history of KOR has been written by one of its members. See Jan Józef Lipski, *KOR — A History of the Workers' Defense Committee 1976—1981*, tr. O. Amsterdamska and G. M. Moore. Los Angeles: Univ. of California Press, 1985.
8. Discussions of these independent cultural and scientific activities have been published in a number of places. See, for example, the two special issues of *Survey* from 1979 and 1980 and the works of Peter Raina, e.g., *The Political Opposition in Poland 1954—1977*. London: Poets' and Painters' Press, 1978.
9. Mirosława Marody, J. Kolbowski, C. Labanowska, K. Nowak and A. Tyszkiewicz, *Polacy '80* (Poles '80), Warsaw: Inst. Soc., Warsaw Univ., 1981; and M. Marody and K. Nowak, "Wartości i działania: o niektórych teoretycznych i metodologicznych problemach badania wartości i ich zwiazków z działaniem" (Values and actions: Some theoretical and methodological problems of the study of values and their links to action), *Studia Socjologiczne*, No. 4, 1983.
10. Stefan Nowak, "The values and attitudes of the Polish people," *Scientific American* **245**, 1981; also his "System wartości społeczeństwa polskiego" (The system of values of Polish society), *Studia Socjologiczne*, No. 4, 1979; and "Przekonania i odczucia współczesnych" (The convictions and feelings of our contemporaries) in *Polaków Portret Własny*. Cracow: Wyd. Literackie, 1979. Similarly, studies of political attitudes among young workers in Rzeszów conducted in 1977/78 revealed that "observance of civil freedoms by the authorities" ranked first among the attributes of a good social system, followed by a general and equal right to work and by democracy in government. See: Stanisław Marczuk, "Ideał ustroju społecznego młodzieży akademickiej i robotniczej" (The ideal of a social system as seen by students and young workers), *Studia Socjologiczne*, No. 4, 1981.
11. "Inna propozycia bycia obywatelem" (A different view of what it means to be a citizen), an interview with Jan Strzelecki by Hanna Krall, *Kultura*, No. 39, Sept. 28, 1980.
12. M. Marody, et al., 1981. *op. cit.*, Note 9, p. 32. Even the less educated workers expressed their confidence that experts would be available and able to assist "Solidarity," which "in their view is to be a place for a particular alliance of workers and the intelligentsia, an alliance in which the intelligentsia supplies people who will articulate the demands of the workers and who will look at what the authorities are up to: Q. 'Do you believe that the workers will manage?' A. 'The worker by himself? No. There are statements at meetings that workers by themselves will not manage. But there are people with higher education there, and they are watching out.' A similar vision of such an alliance was presented by an older unskilled woman worker: '(In "Solidarity") We have very wise people, whether workers or some educated people who are wise and understand it all and they are well able to direct it, it seems to me.'" *Ibid.*, pp. 34—35.
13. S. Nowak, *op. cit.*, Note 10. 30.
14. M. Marody and K. Nowak, 1983. *op. cit.*, Note 9.
15. Such "ideal types" are suggested, for example, by Edward Shils, in *The Intellectual and the Powers, and Other Essays*. Chicago: Univ. of Chicago Press, 1972.

16. More precisely, they drew a distinction between "experts" and "advisors," though given the fact that formally as well as in informal parlance the term "expert" was applied to all intellectuals who advised "Solidarity," I will use the terminology of "advisors" and "consultants," reserving the label "expert" for the group as a whole, in conformity with the general usage in Poland.
17. The extensive report on the state of the country prepared for the fifth anniversary of the Gdańsk agreements is ample testimony to this continuing cooperation.
18. *Protokoły Porozumień Gdańsk, Szczecin, Jastrzębie.* Warsaw: 1981, p. 2. English translation in *August 1980: The Strikes in Poland.* Munich: Radio Free Europe Research, 1980, pp. 422 ff.
19. I. Krzemiński, G. Bakuniak, H. Banaszak, and A. Kruczkowska, *Polacy — Jesień '80: Proces Powstawania Niezależnych Organizacji Związkowych* (Poles-Fall '80. The process of formation of independent trade unions). Warsaw: Inst. Soc., Warsaw University, 1983.
20. *Protokoły, op. cit.*, 1981, Note 18, p. 4.
21. Waldemar Kuczyński, "Katastrofa i przebudzenie (Polski kryzys gospodarczy w latach 1980—81)" (The catastrophe and the awakening: The Polish economic crisis, 1980—81), in *Zeszyty Historyczne* **72**, Paris: Instytut Literacki, 1985, p. 60.
22. "By the end of October, proposed legislation on financial management, corporate taxes, banks, prices, associations of enterprises, and planning, alternatives to those prepared by the government, were also ready." *Ibid.*, pp. 64.
23. A number of such discussions on the reform were published in "Solidarity's" national weekly *Solidarność*, in the *Zeszyty* of the "Mazowsze" (Warsaw region) OBS, and in *Ruch Związkowy*, the periodical of OPSZ, only one issue of which managed to appear before the state of war was declared.
24. Jerzy Eysmont, "Okresowe analizy sytuacji gospodarczej," (Periodic analyses of the economic situation), *Ruch Związkowy* **1**, 1981, p. 104.
25. *Ibid.*, p. 105.
26. *Ibid.*
27. *Ibid.*, p. 106.
28. For one example of such a poll see Nina Kraśko, "Oceny, Szanse, Niebezpieczeństwa" (Evaluations, chances, dangers), *Ruch Związkowy* **1**, 1981.
29. The Warsaw section of the Polish Sociological Association cooperated with the regional Workers' University in preparing materials and meetings on self-management. For example, Jerzy Drążkiewicz, "Doświadczenia a szanse samorządu pracowniczego w PRL (czym był a czym może być samorząd pracowniczy)" (Experiences and Chances of Workers' Self-management in the Polish People's Republic: what the workers' self-government was and what it could become) and Jacek Tarkowski, "Podstawowe założenia systemu samorzadu w Jugosławii" (The basic assumptions of the self-management system in Yugoslavia), both prepared by the Warsaw section of the Polish Sociological Association and issued as working papers by the Workers' University, NSZZ "Solidarność", Mazowsze region, 1981.
30. Jerzy Jedlicki, "'Solidarność' w polskim systemie społeczno-politycznym" ("Solidarity" in the Polish socio-political system), *Ruch Związkowy* **1**, 1981, pp. 15—16.
31. For example, L. Deniszczuk, "Społecznie niezbędny standart potrzeb bytowo konsumpcyjnych (minimum socjalne) w warunkach lat 1981—1985" (The socially

necessary standard of socio-economic needs, or the social minimum, in the conditions of 1981—85), Warsaw: Instytut Planowania, 1980 or Irena Żukowska "Zmiany w dochodach ludności Polski" (Changes in the incomes of the Polish population), Warsaw: I.G.N., 1983.

32. Kuczyński (*op. cit*, 1985. Note 21. 60.) cites six different projects published in *Reforma Gospodarcza: Propozycje, Tendencje, Kierunki* (The economic reform: Projects, tendencies, directions); Warsaw: 1981. These projects were prepared by such bodies as The Polish Economic Society, a group from the Main School of Planning and Statistics, the Economics Department of Warsaw University, the Management Department of the same university, the Economic Academy in Wroclaw, and the Main Technical Organization.
33. Kuczyński, *op. cit.*, 1985. Note 21., p. 64.
34. One indirect evidence of this greater dependence of the economists on the political authorities is the small number of the representatives of this profession in the ranks of the pre-1980 democratic opposition (for example, only 3 economists were members of TKN, as opposed to 8 historians, 7 sociologists, 9 literary critics, and 15 writers and artists). Attempts to minimize the dependence on the government necessarily led to more theoretical interests and such interests in turn would not be immediately affected by changes taking place in society.
35. The only exception might be the work of Jadwiga Staniszkis who not only served as a "Solidarity" advisor but also wrote and continues to write analyses of the experience itself, of the Polish socio-political system, revolutionary consciousness, etc. See, for example, her *Poland's Self-Limiting Revolution*, Princeton: Princeton University Press, 1982. Jacek Kurczewski's *Konflikt i "Solidarność"* (Conflict and "Solidarity"), *Zezyty* of the OBS of "Mazowsze" Region, 1981, pp. 4 ff., is much more practically oriented.
36. Among earlier studies made public for the first time we might list Z. Sufin, ed., *Społeczeństwo polskie w drugiej połowie lat siedemdziesiątych. Raporty z badań* (*Polish Society in the Second Half of the Seventies. Reports from Studies*). Warsaw: IPPM-L, 1981; or *Nierówność i niesprawiedliwość społeczna w świadomosci społeczeństwa polskiego* (*Social Inequality and Injustice in the Consciousness of Polish Society*). Warsaw: OBOP, 1981. A number of results from Stefan Nowak's continuing studies of values and attitudes were also published for the first time.
37. For example, a number of sociological studies have been published in the underground periodical *Krytyka* which began appearing in 1978 and continues publication today.
38. I. Krzemiński et al., *op. cit.*, 1983. Note 19; and Mirosława Marody et al. *op. cit.*, 1981. Note 9. Other works on the formation of union structures and workers' activism include E. Kaczyńska, "Les candidats et élus au congres de Solidarité 1981: aspects demographiques," *Sociologie du travail*, No. 3, 1982; D. Duch, "Powstanie, rozwój, działanośc NSZZ "Solidarność" Służby Zdrowia;" (The formation, development and activities of the "Solidarity" of health service workers); A. Potocka-Horerowa, "Robotnicy 1979—81. Obraz własnej osoby a aktywność w organizacjach społeczno politycznych," (Workers 1979—81. Self-image and activity in socio-political organizations); A. Bielewicz, et al., "Prohibicja sytuacyjna

— Doświadczenie gdańskie — sierpień '80" (Situational prohibition — The experience of Gdańsk in August 1980); to the best of my knowledge these last three works have not been published.

39. W. Adamski, et al. *Polacy '81: Postrzeganie kryzysu i konfliktu* (*Poles '81: The Perception of Crisis and Conflict*). Warsaw: IFiS, PAN, 1982 and W. Adamski et al. *Polacy '80. Wyniki badań ankietowych* (*Poles '80: Results of Survey Research*). Warsaw: IFiS, PAN, 1981. Fragments of these works have been published in English in 'Crises and Conflicts. The Case of Poland 1980—81, *Sisyphus* **3**, Warsaw: PWN, 1983.
40. Strike memoirs were collected by M. Latoszek, and except for the fragments which appeared in *Sisyphus* **3**, they remain unpublished; also unpublished are the studies of the student strike in Lodz conducted, among others, by A. Piotrowski and A. Bokszanski.
41. The three Ossowski Prizes awarded by the Polish Sociological Association in 1983 were given for empirical studies dealing directly with the experience of Polish society in 1980—81. These prizes went to Mirosława Marody and her group for *Polacy '80,* to Władysław Adamski and his group for *Polacy '80,* and to M. Latoszek for the collection of memoirs of striking workers.
42. Stefan Nowak, ed., *Społeczeństwo Polskie Czasu Kryzysu.* Warsaw: Inst. Soc., Warsaw University, 1984.
43. For example, Piotr Pacewicz, *Miedzy Myślą a Rzeczywistością. Rewolucja jako Zjawisko Psychologiczne* (*Between Thought and Reality. Revolution as a Psychological Phenomenon*). Wrocław: Ossolineum, 1983; M. Marody and K. Nowak, *op. cit,* 1983. Note 9, as well as a number of studies included in *Polish Society in the Period of Crisis.*
44. A tradition well exemplified in the classic works of Maria Ossowska on moral norms, and on knightly and bourgeois ethos.
45. "Inna propozycja bycia obywatelem", an interview with Kazimierz Doktór by Hanna Krall, *Kultura,* No. 39, Sept. 28, 1980. Asked about the causes of the conflict, Doktór argued that the eruption of discontent was a result of the blocking of economic needs. To the question, "Are you sure that economic issues were the only ones workers were concerned with?", Doktór answered, "Yes, I am sure. Other issues are less important or secondary, and making them into important issues is a myth of the intelligentsia."

PART IV

COLLABORATIONS AND THE EMERGENCE OF NEW SCIENTIFIC FIELDS

SOCIAL CHANGE, TRADE UNION POLITICS, AND THE SOCIOLOGY OF WORK

KATRIN FRIDJONSDOTTIR
Dept. of Sociology, Lund University

Introduction

The highly developed division and specialization of labour form one important pillar of modern culture; and the organization of work and labour is also an important issue of policy-oriented social science in modern society. This issue has been dealt with from various points of view, for example with the help of economic theory and organizational theory as well as general sociological and social-psychological theories. Anchored thus in academic social science, research on work and labour also develops by interacting with the social policy agreed upon in the area of the labour market and the work organization. The aim of this article is to illustrate certain aspects of this interaction between science and politics, using as an example the development of Swedish sociology of work, especially in relation to trade union policy, in both its broader societal context and in the disciplinary context of Swedish sociology.

The Historical and Societal Context

Whatever definition of the welfare state or welfare state society one uses, Sweden is almost always taken as an outstanding example of this phenomenon. Since World War II, Swedish development has come to represent an "ideal model" of societal development for proponents of the welfare state, as well as a principal target of criticism of reformist politics from both the right and the left. But what are the main characteristics of Swedish development? Several observers have noted some striking contradictions. On the one hand, Sweden is a technologically highly advanced welfare state as regards certain areas of collectivism: a far-reaching redistribution of income and consumption, a

S. Blume, J. Bunders, L. Leydesdorff and R. Whitley (eds.), The Social Direction of the Public Sciences. Sociology of the Sciences Yearbook, Vol. XI, 1987, 249—276.

high level of general education, an all-embracing social security system, and — even today — a relatively low level of unemployment. On the other hand, Sweden has preserved a market economy and bolstered the institution of private property far more than other countries. As a bridge between these features, Sweden has a long tradition of peaceful coexistence in the labour market, together with a tradition of conflict-solving by means of negotiations and compromise rather than in fierce conflicts and confrontations between employers and labour (1). Back in the 1930s, when the Swedish project for a welfare state was launched, one observer, M. Childe, coined the concept "Sweden — the Middle Way" to characterize Swedish development. Its successor, the "Swedish model" — or the "Scandinavian model" — describes a specific model for societal development in which the trade union movement has a powerful voice in influencing political developments, but is also a somewhat abstract concept in need of clarification.

The essence of the Swedish (Scandinavian) model is a contract between labour and capital based upon the political hegemony of the reformist state and the economic hegemony of capital: (a) neither part tries to subjugate the other; (b) labour guarantees capital favourable conditions of economic growth; and (c) in return, capital accepts that resources will be allocated for a comprehensive welfare state policy. Thus, in Scandinavia, and especially in Sweden, the labour movement, acting through the social democratic parties, has gained strong representation in the state and in politics, and thereby a decision-making power on overall development that differs radically from the power of labour unions in other Western countries. At the same time, the exercise of this power has respected certain self-imposed limits. The framework of the model also draws a line between economy and policy, between work and welfare. The social policy of the Swedish welfare state was designed so as not to interfere with the basic mechanism of capitalist development: management's authority to organize production and the work process, and the prime role of price mechanisms in clearing the labour market (2). Trade unions are allowed to press for a higher level of wages. When low-profit firms go bankrupt as a result and workers are sacked, labour power is transferred to high-profit firms or to the public sector. Swedish labour market policy thus includes the creation of a nation-wide and almost free-floating labour market, partly with the help of a structural rationalization of industry and the re-training of labour. The expansion of industry has thus been made easier, and the course of development has also proceeded in harmony with the national policy of economic growth and the wage policy of the trade union movement. Different interests could thus be satisfied within the framework of the overall political and economic strategy of the Social Democratic Party, which has been in power most of the time. Within this strategy, while economic growth is encouraged and its effects more or less equally distributed among the citizens, more and more sectors have become objects of socialization. The role of policy-oriented social science is important in this process.

Some inherent characteristics of the "historical compromise" underlying the Swedish model are of special importance in those areas of policy-oriented social science dealing with matters of interest to trade union policy. Under this 'historical compromise' the trade unions accepted the employers' right to manage and assign work and to direct a policy promoting productivity and growth, in return for which workers were given a fair share in wages of the gradually increasing GNP-pie via negotiations and a governmental policy that promoted a more equal distribution of income, at least among the wage-earning strata. With this distinction between welfare and work, trade unions accorded their highest priorities to wage claims and work safety, priorities that also showed up in their expectations of what scientists could do for the union. For a long time, trade union interest in science was limited to questions of whether, and how, research could provide concrete solutions to the practical problems faced by the unions in the fields of wage policy and work safety.

The Swedish model functioned rather well until the mid 1960s. Before then, the rapid growth of production, profits, real wages, employment, and public income and expenditure had eased the strains between capital and labour and made it possible for Social Democracy to siphon off resources for welfare. But in the course of the 1960s, Swedish industry experienced stiffening competition on the world market, the demand for iron, steel, and wood-products stagnated, and profits declined. As a result, firms were closed down or amalgamated, structural unemployment rose steadily, and labour came under greater pressure to increase productivity. Wildcat strikes broke out. The cracks in the wall disclosed some urgent problems to be solved. These were not primarily claims for higher wages; many of the problems had more to do with bad working conditions and with increased demands for the participation of workers themselves in decisions on how to organize their work. These events also provide the historical and societal background for increased cooperation between trade unions and sociologists, for the cracks in the wall had also revealed new and interesting questions to be studied empirically by social science. During this period Swedish sociology was in the midst of a phase of reorientation, manifest mainly in the actions and writings of its radical students, but also affecting the discipline as a whole.

Following the dramatic events of the late 1960s, new questions and new claims on behalf of the trade unions have reflected a heightened awareness of the negative consequences of technological development, together with demands for the improvement and development of the organization of research policy related to trade union interests. Sociologists have taken a positive attitude towards this development, contributing with empirical studies and theoretical elaborations to the reformulations of the policy of trade unions. These elaborations have, for example, concerned the concepts of the working environment and of democracy and participation at the work place. The sociology of work, as a research specialty, has grown and developed in this process. In addition to research under the joint auspices of the trade unions and employers carried out in research institutes or university institutions, several research projects have been initiated involving direct cooperation between researchers and local trade unions (and sometimes their central boards). This partnership has not necessarily been without open conflict. Indeed, as I will show later, cooperation has sometimes been accompanied by controversies and expressions of rather divergent views on the role of scientific knowledge. At the same time, the relevance of sociology of work and working conditions has generally been recognized and accepted by the trade unions (3).

The Disciplinary Context and the Development from Industrial Sociology to Sociology of Work

Few tasks can be more important than obtaining reliable knowledge about group formation by workers and the attitudes and sentiments that characterize this group formation, not only because job satisfaction is highly valuable in itself, but also because it is vitally important for our future that these groups should function and produce. (From the Foreword to the first Swedish industrial sociological study by Segerstedt & Lundqvist, 1952)

Swedish research on working life is firmly rooted in the efforts of employers to raise productivity by means of modern technology, developed primarily in the areas of psychology and business administration. As an institutionalized branch of research, the sociology of work and industry is a child of the post-war period. Before that time, the rather modest social-psychological and psychological studies of problems of work organization had mainly the status of a helping hand for industrial administration (4). The primary domain of sociologists in this area of research, in the beginning termed industrial sociology, is also intimately connected with the early development of the discipline of sociology in Sweden.

The expansion of the social sciences in Sweden took place during the post-war period in parallel with the construction of the welfare state. The social sciences were seen as capable of contributing directly to the promotion of social justice, and at least indirectly to the acceleration of economic growth, through the knowledge they were expected to develop about society. One of the main incentives behind their expansion was thus of rather instrumental nature (5). These claims were met and handled differently in different social science disciplines, partly because of their different institutional and cognitive premises, and partly because of the nature of the new roles expected of them. Of importance might also have been the then current view in sociology as to what was the most fruitful research orientation as a tool to handle the roles in post-war society. This seems to have been especially clear in the case of sociology, which as a new discipline also had to decide what (international) research traditions it would follow and make "Swedish". From the very beginning it was affirmed that sociology in Sweden was to be mainly an empirical and policy-oriented discipline. The dominant empirical tradition in sociology at that time was that of the USA, and its empirical tradition was seen as especially useful in Swedish sociology. When the decision was taken to make sociology an independent social science discipline, in the late 1940s, the following statement was issued of the desirable orientation of the new discipline on Swedish ground, as compared to sociology elsewhere:

If one is schematically to characterize this field (sociology), it has partly a strong speculative orientation, partly a pronounced empirical orientation. Among the representatives of speculative sociology we find a number of well-known French and German scholars. Empirical sociology, on the other hand, has been successfully developed in the United States. There is no doubt that it is sociology as an empirical science with the objective of studying modern societal conditions which has been favoured in Swedish discussions on the need to establish sociology as a scientifc discipline (6).

The first generation of Swedish sociologists, however, were recruited from the field of social philosophy, and had initially made their way into sociology by way of the theories and basic assumptions of their former subject. As an independent academic discipline, Swedish sociology was established in 1947 when the first professorship in the subject was created at Uppsala University. Previously it had been possible to study sociology within the discipline of social philosophy in Lund and Uppsala. A number of young students of philosophy in Stockholm, partly through contacts with German refugees and the Frankfurter School, had also developed an interest in social psychology, above all in the work of Adorno (7). Social psychology (which never formed an independent discipline in Sweden) and sociology in general fairly soon became oriented almost exclusively towards American-inspired empirical studies. Sociological research concentrated at the "micro-level" on small groups and sometimes individual behaviour, and at the "macro-level" on studies of social behaviour and attitudes and other measurable phenomena amenable to social engineering. An important factor behind this development seems to have been simply the possibility of obtaining grants for empirical investigation and studies within the new discipline (8). The theory of norms and normative systems elaborated by the first professor of sociology, Torgny Segerstedt, a former professor of social philosophy, was also important in the early development of theoretical orientation. Segerstedt's theory of norms as a uniting factor in social groups and group formations bears some resemblance to theories of the "middle-range" within the Structural-functionalist School in American sociology. As deployed in early Swedish industrial sociology, it serves as a good example of a marriage of the two strands in the early development of Swedish sociology (9).

One of the first interest groups in Swedish society, outside the state administration, to appreciate the potentiality of the new discipline, was private business; and as a result the first major field of research was opened up. The interest of industrial enterprise was aroused mainly by visions of higher productivity through management-initiated social reforms at the workplace, and here the inspiration came from the USA and social reform in American business. The theory corresponding to such enterprise in the USA was the "Human Relations School" associated with the works of Elton Mayo and others. The key concepts in the research of the Human Relations School were adjustment and satisfaction at work, as measured by workers' attitudes. The main focus of interest, therefore, was not on how different working conditions affected the workers, but rather on how the workers perceived and reacted to these conditions. The relation between the workers' adjustment to work and their satsifaction with productivity was based on the claim that well-adjusted workers would produce more. Some studies within the Human Relations tradition had also corroborated this. Improvement in human relations at the workplace, by informing the workers and by improving the conditions of work groups, would therefore be seen, at least indirectly, to raise productivity (10). The central thesis of the Human Relations School was also in accord with the Segerstèdtian theory of norms as a coherent factor in social groups, small and large. Standing out as something of a milestone in the early history of Swedish sociology, presenting data at that time unique from the every-day life of workers, the message and implications were however clear: the spirit, or the "symbolic milieu" of the workplace — as a factor amenable to change — was more important to study than the working milieu understood in more materialistic terms (11).

The Human Relations approach had its limitations, which were soon pointed out by the main recipient of its results: the business community. The decreasing popularity of the school internationally was important here, as was the growth and strengthening of coordinated research under the aegis of private enterprise (12). Other branches of practical knowledge more directly related to the domains of psychology, business administration, and economics, increasingly gained ground within the research perspective of private industry (13). The early honeymoon with business, however, had important consequences for Swedish sociology. Some sociologists found jobs in industry as researchers and thereby adapted to the changing expectations of business. Of even greater importance, however, were the effects of industry on academic sociology: virtually all of the first generation of Swedish sociologists, including those who were to occupy the first professorial chairs, had in one way or another worked in industrial sociology. As formulated by Gunnarsson: "For an outsider, it could seem as if Swedish sociology was for a while synonymous with industrial sociology" (14). Most of this early generation of sociologists left the area of industrial research to follow other interests,. and for them the “laboring man” seems to have been something of an incidental research interest, besides offering training in empirical work. Others, however, would develop the subject within their discipline and would soon shape a distinct sociological tradition, which after a while came to be termed the “sociology of work”.

Three names stand out in this process: those of Edmund Dahlström, Bertil Gardell, and Bengt Rundblad. They all joined the newly-established Department of Sociology at Goethenburg University in the early 1960s. Gardell continued the tradition of more micro-oriented and social-psychological studies of working conditions and adaptation to work. Rundblad analyzed the effects of the mobility-stimulating efforts of Swedish regional policy, and established thereby a more macro-oriented research tradition for the sociology of the labour market (15). Dahlström, as a professor and head of the new department, came to act both as a stimulator and catalyst in this area, whilst carrying on his own research involving some reformulation of the subject. In various works Dahlström discussed the importance of sociological research into work and working conditions, as well as the importance of broadening the scope of studies towards sociology of work in general, including studies of societal changes related to technological changes and changing working conditions (16).

Along with these changes in cognitive definitions, a new interest group came to the forefront: the trade unions.

Trade Unions and Social Science

Interest in research policy is of quite recent origin in the Swedish trade unions and did not become a topic of official discussions and programs until the 1970s (17). One pioneering report was that of the white-collar union (TCO) published in 1970, which stated that research was far too important a societal activity to be “left to the scientific elite to decide upon.” Consequently, the TCO report argued for increased political and union representation at all levels of science policy, including the research foundations (18). The report was widely discussed and seems to have influenced the ensuing discussions of research policy in Sweden. The first official programme of the Trade Union Confederation (LO) was presented at the union congress in 1976. One result of this congress was the formation of a committee on research policy at the LO, with the aim of investigating further trade union research policy interests (19).

This new-found interest in research policy questions does not mean, however, that science, or different branches of science, as well as the research results from scientific research, had played no part in earlier union policy and the struggle for better working conditions. At various congresses of the trade unions and their central body, the LO, especially in the late 1950s and 1960s, questions had been raised about how to encourage economic research to deal with union problems. Various specialities in medical science and toxicology — so far as they could contribute results relevant to the unions — were also seen as important fields of knowledge for possible use in the struggle. Thus on one occasion in the 1950s, contacts were initiated with medical scientists, who contributed test results showing how asbestos affected the lungs of employees handling the material at work. These results were used with some success in negotiations with employers over the physical aspects of the working environment, although to begin with, not even the workers were prepared to accept changes that might reduce their wages (20).

In social science, and perhaps academic economic science in particular, matters seemed to be rather complicated. There have been two themes of special importance to union policy in the domain of economic science: first, the question of economic democracy and the co-determination of economic policy first formulated concretely in the early 1920s by the Social Democrats (21); and second, the problem of wage-formation.

Neither of these questions has been dealt with in academic economic science from a union point of view. One reason for the strained relations between academic economic science and trade unionism might therefore lie in the dominant theories of economic science and in the strength of its intellectual tradition (22). The questions of economic democracy and co-determination were to reappear in political and economic debates, in the 1950s in discussions of the comprehensive State Pensions Funds (ATP reform), and in the 1970s in demands for co-determination and Wage Earners' Funds. In both cases, with very few exceptions, the objectives and policies of trade unions in these matters have been met with scepticism and even strong criticism by the academic community of economists.

Outside and alongside academic economic research, trade unions created their own Department of Applied Economic Analysis (the LO's utredningsavdelning) in the 1950s in order to develop the theoretical and practical arguments for their policy. The department mainly analyzed problems of immediate relevance to policy, but the theories it elaborated (mainly by the economists Gösta Rehn and Rudolf Meidner) were in fact accepted by the Social Democratic government for use in their labour market policy. The main point of these theories emphasized that inflation could not and should not be fought by curbing union power. Instead, labour market policy was activated insofar as unemployment created by high wage increases was mopped up by transferring the unemployed to growing firms and areas (23). Later, in 1984, the LO started its own Institute of Economic Research to be able to carry out a more comprehensive research programme. This institute can be regarded as a consequence of the work of the research policy committee of the LO.

This history suggests a general mistrust on the part of the trade unions towards academic social science, at least economic science. But this mistrust does not perhaps hold for all branches of social science (and not all social scientists), nor is it relevant for all branches of trade union policy. Following the increased professionalization of trade union politics, the unions needed their social scientists, and they also needed to keep up with the results of social science in various areas of interest to policy. One such area was industrial sociology, where the trade unions had followed research for some time with considerable interest. The direct engagement of trade unions in organizing research on work, using sociologists of work, began, however, as joint research under the auspices of both local trade unions and employers. This specific trait of the Swedish labour organization — the considerable strength of trade unions not only at the central level, but also at the *local* level — had resulted in various arrangements for negotiations between the parties on the issues of working conditions and organization.

This local system of negotiations between the parties of the labour market thus became the "gateway" for sociologists collaborating with the trade unions. Some of these joint projects sought to evaluate the very system of local negotiations, and showed that democracy behind the factory gate left much to be desired (24). These studies sometimes led to controversies, not only between employers and sociologists but also between trade union officials and sociologists; but they also paved the way for direct cooperation between the unions and the scientists. The development of sociology, and dramatic happenings at the societal arena, also supported this cooperation.

Sociological Research in Relation to Trade Union Interests: The Development of Sociology of Work and Working Life

Swedish sociology has been characterized as rather positivistic an developing along rather "untheoretical" lines during the 1950s and early 1960s. The Finnish sociologist Erik Allardt has characterized Swedish sociology during these years as a sociology "without conflicts and without surprises" (25). Surprises were, however, yet to come. In their empirical research the Swedish sociologists contributed to recipes for welfare state policy at various intermediary levels (social policy, educational policy, housing policy, etc.). Sometimes they managed, with the aid of their empirical data, to lay bare some of the "dilemmas" of welfare state policy itself (26). Other "surprises" came from the theoretical import of marxism and the applications of marxist theories to Swedish development. When direct contacts between trade unions and sociologists were established, Swedish sociology was about to enter a phase of reorientation. Above all, "surprises" of specific relevance for the sociologists of work were created by the objects of study: the workers themselves.

In the late 1960s, the early interest of the trade unions in sociology was mainly related to two issues: the effects of technological developments on working conditions and the contents of work, and how to explain wildcat strikes. The background to this was the growing dissatisfaction of workers with a deteriorating work environment, together with severe troubles in the labour market.

These concerns within the trade unions could in turn be met by the sociologists. A group of work sociologists under the direction of Dahlström presented a report at the Congress of the Trade Union Confederation (LO) in 1966; and this report would be used as background material for the following LO official report on union policy and technological development. Besides the complication of technological development, the sociologists' report also dealt with co-determination at work (27). The work of Dahlström in 1966 is also interesting because it signals a new orientation in Swedish sociology of work compared with the earlier studies in industrial sociology. Technology is here brought in as an important factor in working conditions, and in this report a criticism of the more traditional organizational theories used in industrial management is developed in relation to a concrete analysis of working conditions (28). This work is theoretically inspired by the work of Blauner, and solutions were accordingly sought to a great extent in technological development itself: automation would reestablish the qualifications and freedom of work that had characterized former craft work but had been destroyed by mechanization (29). In this sense, technological development could be viewed as leading to an "upgrading" of qualifications in general. This, moreover, agreed rather well with the then dominant trade union view of technological development as generally leading to a structure of work organization characterized by a steadily growing number of skilled jobs and a decreasing number of unskilled ones. In these years, the trade unions saw the main problems concerning the quality of work as related to the physical aspects of working conditions. In social science the corresponding view is the "upgrading thesis" put forward in the influential work of Blauner and followed up in much contemporary sociology of work, not only in Sweden.

Later on a counter-thesis was brought into the discussions on work and working conditions, namely that of Braverman and Mendner, who advocated a kind of "downgrading thesis" to explain the effects of technological development on industrial work (30). The import and application of these latter ideas, however, also followed the general reorientation and radicalization of sociology in the early 1970s. In the meantime, much had also changed in Swedish society and in its labour relations.

Negative and disturbing consequences of the rapid structural changes in the Swedish economy and labour market became clearly visible in the late 1960s. One aspect of these problems was related to the implementation of the regional policy and the labour market policy and programmes for those people (especially in the North) who could not (or refused to) leave their community for new areas where jobs were available. Another aspect had to do with deteriorating working conditions and discontented workers, culminating in a series of strikes in the winter of 1969/1970 at the state-owned mining enterprise at Kiruna in Northen Sweden, which had previously been taken as a model of human relations management (31).

The struggle for improved working conditions often met with greater resistance from employers than direct wage struggles. When the Volvo workers went out on strike in 1969, they issued two major demands: for improvements of hourly wages and a slower speed on the assembly line. The wage demand was met almost immediately, but not the demand for a slower assembly-line speed. A similar distinction was made in the demands put forward during the miners' strikes in Northern Sweden. It was, however, in the interest of both employers and trade unions to find solutions that would satisfy both sides. A study by Walter Korpi, initiated in collaboration with trade unions, is entitled: "Why do the workers strike?" (32). This was of course a very urgent question for both employers and trade unions, and one which interested other sociologists too (33). In their studies, both Korpi and Dahlström noted the discrepancy between the rights of the employers and employees (34). The studies of the strikes thus inspired far-reaching discussions on the issue of power relations in the employment sector, as well as furnishing arguments for increased participation of employees in decision-making at the workplace and discussions on how these rights could be enforced by law. According to Dahlström, the analysis of the strikes and related problems in the late 1960s was perhaps the most important single contribution of social science to later reforms of working life (35).

Of considerable importance for the further development of the sociology of work was the general reorientation of sociology following the student radicalism of the late 1960s. Parts of the left-wing student movement became more interested in collaborating with groups like trade unions (36). In addition to more theoretical analysis of this development, direct contacts were established at some universities between trade unions and social scientists. In Lund for example, the "AHUF circles" were initiated as the result of a growing interest in collaboration with trade unions on the part of the social-democratic student movement. Contacts were in this instance established with the central body, the LO, after which local contacts grew and were organized into circles. One aim of these circles was to test ideas and new forms of collaboration between local trade unions and universities (37).

In sum, the sociology of work became more and more related to the problems of workers and their unions. In its further development, the sociology of work in Sweden increasingly addressed its research to union policy, both outside and inside academia. Alongside developments in sociology as a discipline, which remained relatively "open" to the demands and expectations of outside groups like the trade unions (and, following the radicalization of sociology, groups like trade unions became of *specific* interest), the development was also supported by the *research policy* of the Swedish state.

The Strengthening of Sociology of Work and Working Life: Professionalization and Pluralism in Theory and Method

Favourable funding, the creation of institutional settings, and other research policy steps have all contributed to the growth and strengthening of the field of sociology of work and working life since the 1970s. Of significance in this development was the creation of the Swedish Work Environment Fund (Arbetarskyddsfonden) in 1972, financed by a special payroll tax. From the beginning, the fund's objective was defined as financing research in the areas of occupational safety and health, rather than the social aspects of the working environment in the broader sense usually assumed by sociologists. However, following the Act on Co-determination at Work, new guidelines for research were also initiated in more broadly defined fields of working life and working conditions. The Swedish Center for Working Life (Arbetslivscentrum) was established shortly afterwards, the objective of which was to stimulate and conduct interdisciplinary research on working life either under the joint guidance of the partners in the labour market or in direct cooperation with trade unions. Studies at the Center have dealt, for example, with structural changes in production, the impact of new technology on work organization, economic democracy, etc.

The research financed by the Work Environment Fund and research in cooperation with the Center for Working Life grew in importance also at the universities. Some local arrangements were also to follow, such as the above-mentioned AHUF circles at Lund University. In the beginning, the AHUF aimed at setting up courses for union members at the university level to deal with various aspects of working life and labour market policy. In 1978, as a following-up of these courses, several research circles and small research projects were arranged, including local union members and social scientists at the university. The topics have ranged from problems of work organization to regional planning (38).

Along with the developments in funding research, several research posts in the area of working life and labour market policy have been created in the social science departments of the universities. As a result, there has been a considerable strengthening of these fields, with ensuing specialization and professionalization in sociology and other social science disciplines. These developments at the universities should also be viewed against the background of a more general re-orientation of university research and research education in the 1970s and partly as a consequence of the policy of sectorial funding, which also has led to the creation of specific research posts related to sectorial policy interests. As a result, many researchers and doctoral students tend to specialize at an early stage of their career, often in an area that can be fitted into the existing structure of financing research.

This situation, taken together with the general strengthening of the field theoretically as well as financially, has encouraged the establishment of working life studies in various sociology departments. As mentioned above, research is also carried out in collaboration with the Center for Working Life, and the structure of finance for research as well as the sphere of knowledge utilization is approximately the same for both types of settings. The research settings themselves, of the Center and the university, are, however, different. The research at the Center is meant to be interdisciplinary and also rather strongly influenced by the interests of parts of the labour market. University-based research in sociology of work is perhaps more "free" in this regard but also "bound" to its disciplinary structure. These differences seem to have played some role during the late 1970s when some division of labour seemed to develop between university-based sociology and sociological research at the Center. A group of social scientists at the Center developed a more or less distinctive research tradition: action research. Action research in this field has been characterized as a close interplay between action and reflection in a process of change (39). The research process starts with practical and concrete problems, as these problems are conceived and experienced by workers. The major objective of this approach is to further the employees' influence and their awareness of technical and organizational change. The role of the social scientists is to help articulate the problems and propose means of dealing with them. At a later stage, the social scientist can then work out the experiences in a more theoretical form, reflecting the processes of change (40). One example of a project of this kind is the DEMOS project, which supported and studied four local efforts to influence the process of change in technology and work organization (41). Another action research project, the PAAS project, dealt especially with computerization, and the researchers' role was to equip workers with sufficient knowledge to enable them to consider alternative courses of action (42).

Action research in Swedish sociology of work is seldom conducted by university social scientists, with the possible exception of the AHUF circles at Lund University (43). This is partly a result of the financing pattern: action research usually requires studies over a longer period of time, involving several researchers, preferably recruited from a number of different disciplines. University research in the area of working life is mainly financed by project money on a short-time basis, involving a single researcher, or at best a small group of researchers. The disciplinary structure of the university includes an academic reward structure that emphasizes the importance of theoretical work limited to the discipline; of necessity, university research develops in a dialogue with general academic sociology. But the reward structure of academia also affects research being done in the same field outside, opening the way to various types of conflicts, as was the case in the action research projects. Of importance were also different views on science within the union and the group of researchers, respectively.

The researchers complained about the lack of time and opportunity to develop the lessons of action research into a more coherent theoretical structure. The viewpoint of the union is of course important in action research at workplaces, since direct cooperation between the scientists and union members is a prerequisite for this type of research. The image of the "the scientific role", however, did not always fit with the expectations of trade union members, which caused frequent discussions in the above mentioned DEMOS project, for example. According to the participating scientists, the union people expected more immediate results for use as arguments in ongoing discussions concerning the application of new technology. Against this, the researchers argued for the need for more comprehensive knowledge (44). These differences can also be illustrated with the following description by one of the union members helping to organize results:

> . . . many problems of cooperation originate in different types of ethos in trade union organization and research organization, respectively. One may call these the ethos of the shop-steward (ombudsmannamoral) and the ethos of science. The first gives priority

> to loyalty to the organization, and a person with this background tends to evaluate facts from the viewpoint and standpoint of the organization. He is prepared to accept what is in tune with the politics of the organization and to fight against what contradicts it. . . . Against this stands the ethos of science, which is a kind of search for truth and questioning that requires a total lack of loyalty to established organizational principles and standpoints. For a scientist it is even rewarding to question the current policy consensus and to test data against different hypotheses. To the trade unionist this kind of attitude belongs rather to the culture of the employer . . . (45).

Action research is, however, only a small part of the research conducted by sociologists of work. Significant for developments in the 1980s is the growth of more theoretical studies, at both the Center and in the university. This development is also characterized by a pluralism in approaches, often through the combination with other specialities within social science, such as political sociology, economic sociology, women's studies, social studies of technology, etc. (46). Still, the sociology of work is to be regarded as a separate field of studies and as such viewed as one of the most vital subfields of Swedish sociology. As I have discussed earlier, the "roots" of this development were there from the very beginning, when sociology was established as an independent discipline in Sweden. What factors shaped this development, including the reorientation of the field? I will now discuss some aspects of this.

The Processes Shaping the Field of Sociological Research in Work and Working Conditions

Swedish research in the sociology of work and working conditions would probably not be what it is today without its confrontation with practical union policy. It is by means of these contacts, together with later theoretical and methodological developments, related in turn to the development in sociology in general, that the sociology of work as a specialty has grown and developed and been encouraged by the overall research policy of the Swedish state. Partly, this development, including the marked reorientation of industrial sociology, has, as I see it, to be explained by reference to the nature of the "contact field" between sociology and trade union politics; partly it has to be explained by reference to the nature of sociology as a discipline.

This brings us also to the question of "bottom-up" processes, defined as "processes initiated by scientists who wanted their disciplines to be more socially relevant" (47). How do these processes function in social science? "Bottom-up" processes, if they are to rise and grow in science, probably need both a flexible disciplinary context and a favourable utilization context, with the latter promoting professionalization within the framework of the discipline. For social science in modern society, this utilization context consists among other things of the prevailing societal policy and the interests of the different actors shaping this policy.

The contacts between the different spheres, that of scientific knowledge production and the utilization of knowledge, is, however, not a "direct" one. Even urgent social and economic problems are not necessarily always the most obvious and immediate "raw material" for which an enlightened social science must seek solutions. Such themes and issues have to be filtered through the cognitive and social organization of science at a particular time and in a particular context, in order to become feasible "research problems" (48). The different spheres of policy and the interests of different actors may also be more or less "obvious" for different areas of social science. I have mentioned that the knowledge-utilization area of Swedish sociology lay particularly in various intermediary levels of welfare state policy, and in contributing with various concrete recipes for this policy. This may have had some importance for its development; probably reforms of work organization were seen as an appropriate area for sociologists.

What is, however, rather clear is that before the 1960s the trade unions were not obvious partners for direct cooperation within any of the academic social science disciplines. Sociologists had, however, already been involved in a few local developmental projects, jointly commissioned by both parties of the labour market and aiming to improve the organization of work. This participation seems to have functioned as a "gateway" for the direct cooperation that was to follow. One area of particular interest to sociologists working in the area of industrial sociology was the system of negotiation between the parties (49). This was also of disciplinary interest since the possibilities of studying the local system of negotiation furnished arguments for criticism of then-dominant organizational theories and offered possibilities for profiling the sociological arguments towards other social scientific explanations and recipes for work organization.

An important factor behind the initiatives taken by the trade unions in the mid 1960s, offering sociologists of work possibilities to develop their arguments, was the already developed professional competence in sociology to perform research into the conditions of work: a competence which, however, had mainly been developed in cooperation with the employers. A "change of partners" followed a diminished interest on the part of employers, and also paralleled shifts in perspective and new orientations in the sociology of work internationally. The newly emerging perspectives and theories were in turn applied to trade union oriented research. One precondition for this success was that the new ideas should fit rather well into the new orientation in trade union politics toward problems of the workplace (50).

The radicalization of sociology was, in parallel and especially later, also certainly of importance, since it prepared the ground for new approaches in research and also led to a growing interest in labour market problems, notably among younger scholars. One of the effects of the new-left discussions in the field of higher education and research was a new awareness of the political dimensions of science. Among the catchwords of the new movement was: "More research in the interests of the working class." These discussions were most highly developed in sociology and had more profound effects there than in other social science disciplines, and of course also influenced other areas beyond the sociology of work. Sociologists might be — as a tribe — more "radically" oriented than other social scientists; but the development of "bottom-up" processes in sociology, as for example the one described here, might also indicate something about the nature of sociology as a social science discipline.

Compared to acedemic economics, the discipline of sociology seems — at least from the standpoint of the trade unions — to have been a somewhat more accessible discipline when it came to responding to questions of relevance to trade union policies and struggles. The report leading to the decision to establish a separate trade union institute for economic research in the mid 1980s includes the following statement:

> The institute is being created because of the massive threat to full employment and the equitable wage policy. An institute is needed that in the long run can function as an active instrument for a valid strategy of economic policy to put questions of full employment in focus. The need for economic research focusing the analysis — from the union point of view — on societal change (. . .) is obvious, and this need cannot be satisfied within the existing system of economic research in this country (51).

Academic research in economics seems, at least as compared with sociology, to have developed much more strictly according to the "hard-core" rules of the discipline. This holds also for research of interest in trade union policy. Main-stream economic research has tended to highlight and explain the problems of labour market policy mainly from the vantage point of received neo-classical economics, which dealt with these problems in terms of market efficiency. One example of a macro-economic model within this school is the explanation of unemployment as the result of a wage level that is higher than the marginal productivity of labour. Similarly, the micro-economic approach to the study of the behaviour of the partners in the labour market has been based primarily on the assumption of a rational, utility-maximizing individual actor. From this perspective, structural unemployment is at least to a certain extent the natural outcome of the behaviour of workers as rational actors searching for new jobs, e.g., weighing their discounted future income from a more lucrative or satisfying job against the costs of searching and the reduced living standard from unemployment benefits during the search period. Economic literature has therefore paid relatively little attention to actual institutional formations in the labour market, and perhaps even less attention to trade unions as specific societal actors (52). Since the relative strength of trade unions in forming political and economic policy is one outstanding trait of the Swedish welfare state, the question arises why academic economics seems to have developed a kind of "theoretical immunity" to such crucial aspects of societal development.

A. W. Coats has described a distinctive trait in the discipline of economics: the high value of the "hard core" of theory compared with questions of empirical validity and relevance. Coats explains this trait with reference to the specific hierarchical organization of work and control in the field of economics (53). Coats's explanations for this development in economics are historical and sociological, and also resemble those of Richard Whitley on the social organization of science. According to Whitley, academic scientific work can be described as organized into several reputational units, i.e., disciplines or research specialities, whose organizational traits strongly influence the direction of research (54). Coats's and Whitley's descriptions of academic economics as characterized by a strong hierarchy and bureaucratic organization seem to me to be fairly apt. The results of my own research into the different disciplinary "cultures" of Swedish academic science indicates also that theory and theoretical work (according to the hard core) play a supreme role in academic economics as the "top" of the hierarchy and the main factor in reputational development (55).

But this is also the case in academic sociology (also according to my studies). The difference between these two fields of academic inquiry might lie no only in the "uni-paradigmical" structure of the one discipline and the "multi-theoretical" structure of the other, but also in how "external" reputational structures can enter and affect the basic structure of disciplinary work. The organizational structure of sociology not only allows room for many competing structures of reputation, but is also relatively "open" at its core(s), so that these structures may influence the theoretical orientation of knowledge production, with fateful effects in its policy orientation too.

Even though economic science can be seen as following more its rules of hard-core science, and therefore its development might look more like the development of an "untouchable" and even non-empirical abstract social science; economics might also be in its policy-orientation relatively "open" to political and ideological shifts relevant to its potential cognitive contributions. Important differences between these two social sciences might also arise over the question of precisely which policy spheres and actors the different disciplinary areas of research consider significant and are therefore "open towards". Compared with sociology, economics may also have had relatively more "stable" partners of knowledge-utilization in private business and in state policy when it sought to promote conditions favourable to private business.

In sum, the cooperation between trade unions (in their policy interests) and sociologists (at least partly in their professional interests) can be described as resulting from several cooperative processes, some of them rooted inside the discipline and some of them outside, but all leading to a situation in which the interests of both parties could be met. This happened in a certain context, and the policy issues raised at that time, the manner of their formulation and the solutions that were seen as politically possible, seem to have accorded rather well with the disciplinary interests and policy-research assumptions of sociology. It has also been suggested that a preconditions for this "happy marriage" between science and politics were found both in the nature of the "contact sphere" between social science and its societal utilization, and in the organizational features of social science. As concluding remarks, I will now discuss briefly how this analysis could be fitted into a more general discussion of the role of social science as well as "bottom-up" processes in shaping the development of social science in its interaction with societal developments.

The Welfare State and its Social Science

I started this paper with a short description of some of the characteristics of the "Swedish model" of societal development, and mentioned in particular the role of trade unions. To begin with, the role of trade unions was mostly limited to wage struggles. Consequently, many areas of potential conflict relating to questions involving authority in the production process or problems of work control were defined outside the domains of direct union struggle. This does not mean that there did not exist potential conflicts between capital and labour; but these conflicts could be eased and dampened as long as a high growth rate made possible arrangements from which both parties profited. The "contract" underlying the Swedish model satisfied both parties, not by denying underlying conflicts but by defining a path of development that gave each party a stake in adhering to the contract. The role of policy-oriented social science in a welfare society has been to formulate parts of this agreement, to legitimize it, and also to revise and reform it as necessary.

One example of how social science has contributed to the revision of the "contract" is offered by the development of the sociology of work in the late sixties and early seventies: the conflict-ridden joint experiments and analysis, the later analysis of wildcat strikes, and the contribution of sociologists to the reformulation of the law on co-determination at work. The strikes and incidents of conflict can be seen as examples of the underlying differences which exist between labour and employers on the question of how to define problems in working life and how to search for solutions. But these incidents also demonstrate the role of social science in making visible such differences of interest and in probing the cracks in the wall of agreement between capital and labour in a mixed economy. Initially, the problems perceived by employers were restricted to individuals and groups of workers and their adaptation to the work process. From the viewpoint of the trade unions, problems came gradually to be defined as structural and therefore much more complicated. Apart from actual events forcing the trade unions to act — dramatic happenings in the labour market in the late

1960s — social science also played an important role in elaborating and broadening the concept of the work environment (so as to include, for example, the connection between social and technological factors). The development between science and policy is, however, reciprocal.

"Bottom-up" processes in social science in modern society may be regarded as closely connected to revisions of parts of the underlying societal contract. They are thereby related to the "need" seen by some actors to strengthen their position in the policy-making process. But a corresponding "need" is also felt on the part of the social sciences to consider the actors of change and action. Such processes, to be followed up in science, require not only a certain openness and responsiveness to questions of relevance for policy, but also the possibility of translating these problems into the disciplinary and reputational structure of the science. This was obviously the case with sociology of work in its development.

Acknowledgements

Gudmund Larsson and Enar Ågren contributed their insights into the problems of trade union research policy. I would also like to thank Edmund Dahlström, Lars Gunnarsson, Bo Gustafsson and Richard Whitley for their helpful comments. This work was supported by a grant from the Swedish Board of Higher Education.

Notes and References

1. Bo Gustafsson, "Conflict, confrontation and consensus in modern Swedish history," in L. Arvedsson (ed.), *Economics and Values*, Stockholm: Almqvist & Wiksell, 1986.
2. See Gösta Esping-Andersen, "Work, employment and the welfare state: Pattern of welfare state entry" (Berlin: Wissenchaftscentrum, 1985; mimeo). The literature on the "Swedish model" and its inner development is immense. Works of immediate relevance to the present topic include Walter Korpi, *The Working Class in Walfare Capitalism: Work Unions and Politics in Sweden*. London: Routledge & Kegan Paul, 1978; Ulf Himmelstrand et al., *Beyond Welfare Capitalism: Issues, Actors and Forces in Societal Change*. London: Heinemann, 1981; and Gösta Esping-Andersen, *Social Class, Social Democracy, and State Policy*. New Social Science Monographs E 8, Copenhagen: C.U., 1980. For interesting discussions of the general development, see also Gustafsson, *op. cit.*; Göran Therborn, "Social-demokratin träder fram," *Arkiv för studier i arbetarrörelsens historia*, No. 27—28, pp. 3—71; and Gunnar Olofsson, "Den svenska socialdemokratin — en rörelse mellan klass och stat," *Ibid.*, pp. 84—96.

3. The general development of the research field is dealt with in the following works: Edmund Dahlström, "The role of the social sciences in working life policy: The case of postwar Sweden," in Hans Berglind et al. (eds.), *Sociology of Work in the Nordic Countries*. Oslo: The Scandinavian Sociological Association, 1978, pp. 75—100. A slightly updated Swedish version is Dahlström's "Samhällsvetenskapernas roll i svensk arbetsmarknadspolitik efter andra världskriget," in Hans Berglind & Lars Tunvall (eds.) *Arbetssociologi i de nordiska länderna*. Stockholm: Liber, 1982. A somewhat different account of the development of sociology of working life in Sweden is Åke Sandberg, "From Satisfaction to Democratization," a paper presented at a symposium on the roles of sociologists in relation to industrial management and conflict at the ISA Conference, Mexico City, 1982. An updated version in Swedish of this paper is "Från trivsel till demokratisering," in Heine Andersen & Christian Kundsen (eds.), *Videnskabsteoretiske grundlagsproblemer i konomiske discipliner*. Copenhagen: Akademisk Forlag, 1984. See also Åke Sandberg, "Trade union oriented research for democratization of planning in working life — Problems and potential," *Journal for Occupational Behaviour*, No. 4, 1983, pp. 59—71. For more general overviews on the development of social, behavioural, and medical studies of working life, see *SOU 1973:55. Beteendevetenskaplig arbetslivsforskning*. Stockholm: Industridepartementet, 1973; and Rolf Gustavsson & Anders Kjellberg, *Beteendevetenskaplig arbetsmiljöforskning*. Stockholm: Arbetarskyddsfonden, 1983; Rapport 1983:6).
4. Gustavsson & Kjellberg, *op. cit.*
5. See also the contribution of Björn Wittrock in Georg Thurn et al., "The development and present state of public policy research. Country studies in comparative perspective". Berlin: Wissenschaftscentrum, 1984; mimeo.
6. SOU 1946:74 Samhällsvetenskapernas ställning Stockholm, 1947, pp. 73.
7. The early development of Swedish sociology is dealt with in Katrin Fridjonsdottir (ed.), *Om svensk sociology*. Stockholm: Carlsson Förlag, 1987. See especially the contributions by Torgny Segerstedt, Bertil Pfannenstil, and Georg Karlsson.
8. Edmund Dahlström has suggested that this early orientation towards positivistic and empiricistic research in Swedish sociology, including mainly surveys and related methods, not only mirrored the general methodological limitations (following American sociology), but was also directly conditioned by the demands and expectations of industry concerning individual-oriented research. (E. Dahlström, *Samhällsvetenskap och praktik. Studier i samhällelig kunskapsutveckling*. Stockholm: Liber, 1980, p. 124). But the development was of course not specific to Swedish sociology. For the US.-influence on German sociology in the post-war period, see Horst Kern, *Empirische Sozialforschung*. München: Beck, 1982. For an account of the empiricist dominance in Swedish sociology as seen through the eyes of a foreign visitor, see Alvin Gouldner, "Personal reality, social theory and the tragic dimension of science," in *For Sociology*. London: Allan Lane, 1973.
9. Torgny Segerstedt & Agne Lundquist, *Människan i industrisamhället. Vols. I—II*. Stockholm: SNS, 1952—1955. Segerstedt's theory of norms is presented in Torgny Segerstedt, "The Uppsala school of sociology," *Acta Sociologica*, No. 1, 1956, pp. 85—119. For a more comprehensive treatment see Torgny Segerstedt, *The Nature of Social Reality*. Stockholm: Scandinavian University Books, 1966.

10. The most famous studies and experiments were undertaken at the Western Electrical Company in Chicago. For an overview of these studies see George Homans, *The Human Group*. New York: Harcourt Brace & Co., 1950, pp. 48–80. The development of the Human Relations School can be regarded as a reaction to the then-dominant Tayloristic view of work organization. In stressing the importance of viewing the worker as a social being, the Human Relations approach was certainly a reaction against the more mechanical view of Taylorism; and as such it also paved the way for sociological studies of industry.
11. In the first Swedish sociological study of industry, Segerstedt & Lundquist made a distinction between "spirit" and "satisfaction at work," spirit being understood as the more basic, broader concept, covering e.g. the degree of the workers' identification with the company. This, in turn, was in accordance with one of the key concepts in the Segerstedtian scheme: the symbolic milieu.
12. For the development of Swedish industrial sociology, see Lars Gunnarsson, *Att förändra arbetsprocessen*. Lund: Department of Sociology, 1980, and also Torsten Björkman & Karin Lundqvist, *Från MAX till PIA. Reformstrategier inom arbetsmiljöområdet*. Malmö: Arkiv, 1981.
13. In the late 1950s the president of the Industrial Council for Social and Economic Studies (SNS) remarked: "The Human Relations School gained popularity in business because of the assumption that there was a distinct correlation between improved human relations (or increased 'satisfaction at work'), and increased productivity. This is a naive assumption, as naive as the assumption that productivity can be increased indefinitely by means of technical rationalization. We have thus reached the point where we can conclude that there is a conflict between the goal of increased productivity and improved human relations" (Hans Torelli, cited in Björkman & Lundqvist, *op. cit.*, p. 38).
14. Gunnarsson, *op. cit.*, p. 30.
15. Bengt Rundblad, *Arbetskraftens rörlighet*. Göteborg, 1964.
16. See Dahlström 1978 and 1982.
17. For an overview of the development of the research policy programs of the various unions see Rune Premfors, *Facklig forskningspolitik. De fackliga huvudorganisationernas program och aktiviteter*. Stockholm: Froskningsrådsnämnden, 1981.
18. TCO: *Forskning och utveckling*. Stockholm: TCO: 1970; Congress report.
19. Similar organizations had already been set up in Norway in 1972 (Landsorganisationens Forskningsutvalg) and in Denmark in 1973 (Fagbevaegelsens Forskningsråd). For the program of the Swedish LOFO, see LOFO: *Forskning, en fråga för fackföreningsrörelsen*. Stockholm: LO, 1978.
20. Personal communication with Enar Ågren, the former head of LO's committee on research policy (LOFO).
21. The question of economic democracy was above all heralded by the late Minister of Economic Affairs, Ernst Wigforss. This part of the original Social Democratic Program was shelved in the post-war creation of the Swedish model.
22. For an interesting discussion of how this intellectual tradition seems effectively to have ruled out alternatives to main-stream views on economic and societal developments, especially those of relevance for trade union politics, see Villy

Bergström, "Nationalekonomerna och arbetarrörelsen," Herlin & Werin (eds.), *Ekonomisk debatt och ekonomisk politik*. Stockholm: Norstedts, 1977.

23. Gösta Rehn & Rudolf Meidner, *Fackföreningsrörelsen och den fulla sysselsättningen*. Stockholm: LO, 1951.
24. Edmund Dahlström, *Information på arbetsplatsen* (1954). This study showed that the local system of negotiation did not result in much real democratization at the work place. According to Dahlström this study, and the "negative picture" it presented of joint efforts, was not received with much enthusiasm by either party. For a while Dahlström (as well as his co-researchers) were treated like "personae non grata" by both parties and not engaged as researchers in working life problems until later on in the sixties, and then by the union (personal communication with Edmund Dahlström).
25. Erik Allardt, "Swedish Sociology," *International Journal of Sociology*, No. 3, 1973, 50—71.
26. Besides discussing conflicts in the labour market in the latter half of the 1960s, sociological studies of the distribution of welfare also came to illustrate some surprises if compared to the offical image, both outside and inside the workplace.
27. E. Dahlström et al., *Teknisk förändring och arbetsanpassning*. (Stockholm: Prisma, 1966) and LO, *Fackföreningen och den tekniska utvecklingen*, Stockholm, LO: 1967.
28. Dahlström, *op. cit.*
29. R. Blauner, *Alienation and Freedom*. Chicago: Chicago University Press, 1964.
30. H. Braverman, *Labor and Monopoly Capital*. New York: Monthly Review Press, 1974. J. Mendner, *Technologische Entwicklung und —Arbeitsprocess*. Frankfurt am Main: Fischer, 1975. For a discussion of the acceptance of Braverman's thesis in Swedish social research see, Christian Berggren, "Braverman — och sedan?" *Sociologisk Forskning* **1**, 1982, 2—25.
31. E. Dahlström et al., *LKAB och demokratin*. Stockholm: Liber, 1971.
32. W. Korpi, *Varför strejkar arbetarna*? Stockholm, 1970.
33. G. Therborn et al., *LKAB-strejken*. Halmstad: Zenit, 1970; Dahlström, *op. cit.*
34. This was based on a central agreement (1906 § 23, later § 32) giving the employer the sole right to hire and fire workers as well as to decide on how to distribute work tasks among them.
35. See the discussion in Dahlström, 1982 (1978), and also E. Dahlström, "Efficiency, satisfaction and democracy in work: Conceptions of industrial relations in post-war Sweden," *Acta Sociologica*, No. 1, 1977. For a more detailed historical analysis of the development, see Bernt Schiller, "LO, paragraf 32 och företagsdemokratin," *Tvärsnitt. 7 forskningsrapporter utgivna till LO:s 75-års jubileum*. Stockholm: Prisma, 1973.
36. As to direct cooperation with the trade union, the leftist student movement differed in its attitudes; the "New Left" left of social democracy rather tended to regard the theoretical struggle as a more proper struggle in the interest of the working class, and the organized social democratic students were more oriented towards direct cooperation with trade unions.
37. L. Gunnarsson & M-L Perby, *Forskningscirklar — en metod i facklig kunskapsuppbyggnad*. Lund: Arbetslivscentrum, 1981.
38. *Ibid.*

39. Pelle Ehn et al., "Demokratisk styrning och planering i arbetslivet," *Nordisk Forum* **4**, 1975, 30—46; Åke Sandberg, "Kunskapsuppbyggnad och aktiviering. Om forskning och facklig praktik i DEMOS-projektet," in Sandberg (ed.), *Forskning för förändring*. Värnamo: Akademilitteratur/Arbetslivscentrum, 1981; Åke Sandberg, "Från aktionsforskning till praxisforskning", *Sociologisk Forskning*, No. 2—3, 80—89; and Sandberg, 1984.
40. Å. Sandberg, 1984.
41. Å. Sandberg, 1981.
42. Bo Göranzon (ed.), *Datautvecklingens filosofi*. Malmö: C & J Bokförlag, 1983. Bo Göranzon et al., *Datorn som verktyg — krav och ansvar vid systemutveckling*. Lund: Studentlitteratur, 1983.
43. One example of action research at the university is reported in Marianne Svenning, *Tjejerna på tvätten*. Stockholm: Liber, 1984. A group of employees at a laundry in Lund, together with a sociologist at Lund University (Svenning), dealt with problems of the working environment and discussed ways of increasing the participation of the employees themselves in the decision-making process.
44. Sandberg, 1981.
45. Contribution by O. Hammarström in TCO: *Facklig forskningsplanering — rapport från ett seminarium*. Stockholm: Liber, 1982, p. 43.
46. For an overview of recent publications and ongoing research in the area of working life studies and related fields, see M. Olovsson, "Swedish research on work: Some examples related to the themes of the symposium 'Work in 1984 — Emancipation or Derogation?'", *Economic and Industrial Democracy*, No. 1, 1985, pp. 121—134. An informative discussion of the development of Swedish studies in the area of work and technology is given in Boel Berner, "Sociology, technology and work," in Ulf Himmelstrand (ed.), *Sociology — From Crisis to Science*? London: SAGE Publ., 1986. Vol. II.
47. Stuart S. Blume et al., *Social Direction of the Public Sciences: Causes and Consequences of Collaboration Between Scientists and Non-Scientific Groups*. Amsterdam, 1984; mimeo.
48. Helga Nowotny, "Marienthal and after: Local historicity and the road to policy relevance," *Knowledge*, No. 2, 1983, 169—192. See also the contribution by S. S. Blume in this volume on the the university-industry relationship, and the contribution of P. Wagner on social science development in France, Italy, and West Germany.
49. Dahlström, 1982 (1978); see also note 25.
50. Notably as to the view on the effects of technological development. The hot topic during the mid-1960s was, furthermore, co-determination at work. See Dahlström, 1977, and Schiller, *op. cit.*
51. LO/LOFO, *Utredningen om ett fackligt institut för ekonomisk forskning*. Stockholm: LO, 1984; mimeo, p. 40.
52. For an overview of recent contributions in the area of economics of the labour market, see Lars Calmfors, *Trade Unions, Wage Formation and Macroeconomic Stability*. Stockholm: Institute for International Economic Studies, 1985.
53. A. W. Coats, "The sociology of knowledge and the history of economics," in W. Samuels (ed.), *Research in the History of Economic Thought and Methodology*. London: JAI Press, 1985.

54. R. Whitley, *The Intellectual and Social Organization of the Sciences*. Oxford University Press, 1984; R. Whitley, "The structure and context of economics as a scientific field," Manchester Business School, October, 1985 (mimeo).
55. *Culture and Community of Research*. This project studies the different "cultures" of disciplinary research, and also the interplay between different kinds of users of scientific knowledge.

SOCIAL SCIENCES AND POLITICAL PROJECTS: REFORM COALITIONS BETWEEN SOCIAL SCIENTISTS AND POLICY-MAKERS IN FRANCE, ITALY, AND WEST GERMANY

PETER WAGNER
Wissenschaftszentrum Berlin für Sozialforschung

Introduction

In the post-World War II history of the social sciences in France, Italy, and West Germany, a distinct period can be detected in which a policy orientation was (re-)introduced into these disciplines. In this period major research efforts were undertaken, either on specific policy areas or on the politico-administrative processes themselves, with a view to improving policy-making by putting it on a "scientific" or "more rational" basis. These processes took place in the 1960s and early 1970s and thus followed previous developments in the USA where the term "policy sciences" (1) was coined, but they differed significantly from the US experience. The specific nature of the confrontation between the innovative approach and the established national science traditions and the interaction of the emerging policy researchers with actors in different politico-administrative systems had a significant influence on the shape the process took (2).

The key actors were groups in the political and economic elites who saw the need for an encompassing societal modernization process, including the introduction of rational planning procedures and the formulation of reform policies which would enable all social groups to participate in economic and societal progress (3). Social scientists sometimes joined outright reform coalitions with these groups, shared their basic political convictions, and saw their own task in designing the required political innovations on the basis of social science expertise. In Germany and Italy, the scientists' interlocutors were mainly members of groups in the major political parties, which had not yet reached power

S. Blume, J. Bunders, L. Leydesdorff and R. Whitley (eds.), The Social Direction of the Public Sciences. Sociology of the Sciences Yearbook, Vol. XI, 1987, 277–306.

positions; in France social scientists were more oriented to the process of "*planification*", which as such took a more long-term perspective on social developments and seemed more open to heeding scientific advice and to initiating social reform (4). These intense interactions between scientists and non-scientists, however, proved to be of a short-term nature. Recognizing political and scientific deficiencies in their approach to a science-politics interaction, in a learning process which was speeded up by social developments, researchers turned away from these coalitions. Some then started to search for new social actors to engage in research-based political action.

The following is an attempt to analyse these developments by tracing the social processes which brought about the new orientation in the social sciences. To do this, an analysis at the macro-level — some concepts of the state of the discipline and of socio-economic change influencing disciplinary perspectives — will have to be linked with one at the micro-level dealing with the behaviour of particular actors and their strategies. In this regard, the three-country comparison may serve to elucidate the relative importance of certain "macro-structural" determinants as compared with the micro-level characteristics specific to each situation.

As in the case of university-industry relations analysed by Stuart Blume in his contribution to this volume, the emergence of a policy-orientation in the social sciences can be characterized as "the construction of a research programme in relation to a particular sort of social network, which consciously embodies the goals, needs, interests or aspirations of actors within this network" (5). The following analysis is devoted to such coalitions between social scientists and political actors, "common social projects" in Blume's terms, and to their demise and the related emergence of new coalitions and research programmes with different political orientations. As an analysis of the social sciences it thus deals with struggles over the production and imposition of legitimate representations of the social world, and will have to pay particular attention to political actors and their positions within the different societal contexts (6). The importance of the political context will already become clear from a short description of the situation of the social sciences before the policy-orientation emerged.

The Social Sciences in an Early Phase of Institutionalisation

The processes under analysis here, through which a policy orientation emerged in the scientific fields of sociology in France and Italy and political science in West Germany, took place mainly between the late 1950s and the early 1970s. In the 1950s the social sciences in these three countries were characterised by a low degree of academic institutionalisation, by strong links to philosophy and to normative thinking and by lively debates about the consequences to be drawn from the first experiences with modern, methodologically rigorous empirical research techniques.

In France, sociology formally received early recognition as an academic subject when Emile Durkheim became the first regular professor in social science at the philosophy faculty of Bordeaux University in 1896, and more significantly, when in 1913 his later chair at the Sorbonne in Paris was renamed "education science and sociology", thus recognising the notion of sociology. This success in terms of institutionalisation, however, must be ascribed mainly to Durkheim's personal endeavours. He had purposefully worked to secure his scientific approach institutionally (7). Although his school remained moderately influential in different fields in the inter-war period, no further steps towards full institutionalisation could be made. In the mid-1950s in the whole of France there were only four university chairs in sociology plus three chairs in ethnology (8). The post-war debate over the central importance of the social sciences in rebuilding French society had raised high expectations, which, however, materialised only to a small extent. The most important organisational innovations were the creation of the Centre for Sociological Studies (*Centre d'études sociologiques, CES*) in the National Centre for Scientific Research (*Centre national de la recherche scientifique, CNRS*) and of the Sixth Section of the School for Advanced Studies (*Ecole pratique des hautes études, EPHE*, later the *Ecole des hautes études en sciences sociales, EHESS*), which were soon to become the most important social science centres in France (9). These academic institutions remained outside the universities, which is a typical feature of the French research system but was also significant for the low academic reputation of the discipline. A decisive change resulted only much later from the full recognition of sociology as a university subject, fixed institutionally by the introduction of a specific full degree, the "*licence*", in 1958.

Compared with the low level of development of sociology as an academic discipline in France, the situation in Italy before the 1950s was even less impressive. Neither the intense empirical research of early positivist sociologists, such as Enrico Ferri or Cesare Lombroso, nor the more theoretically-oriented works of Roberto Michels, Gaetano Mosca, and Vilfredo Pareto had any lasting impact on the academic organisation of the social sciences. Before World War II there was no university base for social science research except for some teaching in social philosophy and some social psychological research at the Catholic University of Milan (*Università Cattolica del Sacro Cuore di Milano*) — if we exclude the Statistical Institute under the directorship of Gini and the faculties of political science created at a number of universities during the fascist regime for the purpose of training future diplomats and high-level administrators to deal with social facts on the basis of an elitist ideology (10). In the early 1950s there was just one university chair in sociology in Italy, in Florence, and not until the mid-1960s was sociology — after long disputes — fully recognised as a university subject leading to a separate degree.

The academic tradition of sociology is much stronger in Germany, with the early inquiries of the Association for Social Policy (*Verein für Socialpolitik*) around the turn of the century, the foundation of the German Sociological Society (*Deutsche Gesellschaft für Soziologie*) in 1910, the important research institutes in Cologne and Frankfurt in the Weimar period, and about a dozen chairs in sociology by 1932 and again by 1955 after the reestablishment of the discipline (11). Political science had a predecessor in the Cameral and State Sciences which were designed to fulfill analytical functions for the absolutist state. But the rise of liberal ideology and industrial capitalism caused this quasi-inter-disciplinary approach to decay as political aspects lost their importance in favour of legal ones.

After the Second World War, West German political scientists were influenced by the double experience of witnessing the break-down of parliamentary democracy and the establishment of a dictatorial regime on the one hand, and on the other, by witnessing in American exile the capacity of a different political system to deal with enormous economic, social, and military problems without abandoning its institutional order. Consequently, priority was placed on the study of institutional systems of government, in particular Western ones, with special regard to the stability of formal democracy and with an explicitly normative orientation in this direction; while the study of the history of political thought was considered a precondition to a debate on democratic values. Philosophical reflection on political developments was seen as the specific task of the discipline, one which would distinguish it from the other social sciences and from law and would turn it into a somewhat superior subject (12).

A high degree of political involvement and strong philosophical orientations were also characteristic of French sociology after 1945, but in a very different fashion. Because of its historical links with the philosophy faculties, sociology in France had a permanent point of reference in the ongoing philosophical discourse. From this perspective, the orientations of the intellectually dominant philosopher-sociologists of the immediate post-war period, such as Raymond Aron, Maurice Merleau-Ponty, or Jean-Paul Sartre, can be regarded as a revolt against the routinised and institutionalised sociological approach of the Durkheim school (13). Historicity, class consciousness, and commitment (the consecrated words of the philosophical semantics of the time, to use a phrase of Pierre Bourdieu and Jean-Claude Passeron) could not easily be linked either to the Durkheimian tradition of thought or to modern research techniques. In this intellectual climate, sociology proper had little chance of attracting the most enterprising minds; this attitude "thereby helped to hold back the development of the human sciences and especially the social sciences" (14). The attitude of the "*intellectuel engagé*", which doubtlessly had been furthered by the experience of the Résistance, anchored the intellectual debate firmly in the political environment of the time and thus in some sense prepared the ground for a more practical orientation of the social sciences than was the case during the inter-war period. Given the presence of a strong communist party as the major force of opposition in society, intellectuals were under a virtual obligation to clarify their position towards Marxism and Communist politics.

In this latter respect, the intellectual atmosphere in post-war Italy resembled the French one. Many intellectuals were attracted to a strong Communist opposition, which was soon almost totally excluded from political power and from the process of societal restoration. In this environment, the rejection of modern sociological approaches was so strong that the word "sociology" was used with the pejorative connotation of an American imperialist instrument to secure bourgeois domination. This attitude was in accord with the still culturally dominant idealistic philosophy mainly represented by Benedetto Croce, whose influence had even increased through his anti-fascism. Throughout his life, Croce had opposed empirical social science research as unable to "understand" society, which he believed was possible only through historical-philosophical intuition (15). Sociology gained its first foothold in post-war society in intellectual environments which were both non-Marxist and anti-idealistic: groups in and around the Catholic University of Milan and in the modern industrial city of Turin.

It was mainly in Turin that the first generation of sociologists found an opportunity to do empirical research supported and financed by industrialists such as Olivetti. Industrial work organisation — and to a certain extent also urban and regional planning — became the early thematic focus of sociological research related to topics such as modernization, rationalisation, and technological progress, and their consequences for societal development. Similarly, French sociologists at about the same time began to engage in empirical research on demand and with the support of non-academic actors. The research in industrial sociology pursued by Georges Friedmann, Alain Touraine, and others, although partly based at the CES and the EPHE, was financed by sources such as the European Productivity Agency or the European Coal and Steel Community, and was encouraged and made possible by industrial companies such as the Régie Renault (16). In the same period, Jean Stoetzel, who was strongly influenced by Paul F. Lazarsfeld, tried to spread the use of quantitative empirical methods at the CES and in French sociology in general (17). In Germany as well, the debate over the relevance and necessity of sophisticated modern methods and, following this, of the relation between theory and empirical facts, was led primarily by sociologists who had more experience in this area than political scientists. Normative attitudes and the emphasis placed on the study of ideal-type workings of institutional systems of governments, rather than the empirical distribution of power and influence, made the latter a marginal issue in political science until the 1960s.

In spite of all the differences in the national situations of these social sciences in the 1950s, there emerges a common picture of these disciplines as hardly possessing secure legitimate roots in academic institutions; as strongly linked to philosophical discourse, at least partly as a result of the lack of consensus on basic conceptual matters; as disputing imported methodologies with which they were as yet unable to cope; and as firmly tied to developments in the political environment as a point of reference for their normative intellectual self-image.

Social Transformations

By the end of the 1950s the initial room for economic expansion — which had resulted from reconstruction needs, the influx of labour, and changed consumption patterns — had been filled in the three countries. Comparatively high rates of employment had been reached; internal migration or immigration increased, which led to social tensions; inflation rates rose rapidly; and growth rates, though still high, showed a tendency to decline. The response to these signs of crisis in the political and economic system was to strengthen those forces which sought to modernize both the economy and the politico-administrative apparatus, which should then become capable of supporting economic restructuring as well as the planning and implementing of reform projects to prevent or mitigate social tensions.

One of the early topics of this emerging modernization debate was the need for a science and technology policy. Public funds for research, which had traditionally been considered a consumption expenditure, were now seen increasingly as a national investment, the returns from which would come from improved international economic competitiveness based on technological progress and increased productivity. The creation of science policy institutions, such as ministries, research councils and commissions, dates from the mid- and late 1950s in all three countries (18). The social sciences, however, were rarely the subject of science policy debates in these early years. A similar debate concerned the need to improve and expand the educational system in order to increase the qualifications of the work force. However, this issue had a second and equally important political focus in the discussion over social reform and the equality of educational opportunities.

This "reform mood", common to all three countries in these years, found different expressions in changes in the political majorities. However, in all three cases, reform policies were based on the concept of sustained economic growth, harmonised and regulated by government intervention and linked to a set of social reforms through which the increase in welfare would be more evenly distributed in society and the external costs of growth would be diminished. The government apparatus had to be adapted to this new and more demanding style of policy-making. Such measures as ministerial reorganisation, the introduction of improved techniques for monitoring administrative activities and efficiency control, and the creation of new institutions were designed to allow for purposeful and effective intervention into economic and social developments (19).

In the following section the development of the social sciences will be sketched against this background of a pronounced "reform mood" in the political system. As we shall see, the interactions between social scientists and actors in the political system were not part of a "top down" social science policy: this did not yet exist. Moreover, although the non-scientific actors involved in these interactions can doubtless be considered part of a larger societal elite, most of them were in the political opposition and strong advocates of change.

The Policy-Oriented Social Sciences: The Rush to Relevance

Remaining almost completely outside the academic institutional framework, Italian sociology since the 1950s had developed a strong social position (20) closely linked with the emerging so-called Centre-Left politics, the entry of the socialists into government, which was previously dominated by the Christian-Democrats. Theoretically, the American functionalist approach provided a comprehensive and coherent framework under which social phenomena could be subsumed, and it allowed the steering capacity of specific political actors to be made the central point of reference in theoretical reasoning. It can be plausibly argued that the absolute reliance on private funding sources with specific policy interests, namely on private and public enterprises, was crucial in establishing the predominance of the functionalist approach in Italian sociology which was stronger and imported more easily than in other European countries (21).

Modernization was the thematic and conceptual focus of almost all sociological research. The basis of the modernization process was seen in technological innovation; the major task of sociology was to analyse the impact of technological change on work organisation and productivity, on social mobility and the class structure, on urban and regional developments and, more generally, to focus on the possibility of harmonising and controlling the social processes set in motion by technological development.

One year before the long-awaited Centre-Left coalition would shape politics at the national level, in 1961, the Italian Association of the Social Sciences chose as the topic for their meeting in Ancona a problem of central concern to most sociologists seeking to become socially relevant: "Sociologists and the Centres of Power". The main thrust of the Ancona debates can be summarized by saying that most of the social scientists felt that their field had now been consolidated; they possessed the analytical tools to interpret social developments and were now at the point where they wanted to translate their findings into practice (22).

In fact, the period of Centre-Left dominance brought sociologists increased importance in defining reform policies. In disciplinary terms, these developments even increased the dependence of sociology on a policy model, since academic institutionalisation proved to be a much slower process, beginning in 1962 with the establishment of the Institute for Advanced Studies in Sociology at the University of Trento (which was also politically motivated), but bringing full recognition of the subject only in 1968.

Italian sociology outside academia was thus influenced by a politico-scientific constellation to such an extent that one may speak of the absolute predominance of a functionalist approach in scientific terms, and one oriented toward modernization and reform in political terms. By contrast, in French sociology, with its moderate but more secure institutional basis, changes in the socio-economic environment led to divergent approaches in the sociological community rather than a complete and unanimous orientation towards reformist policy-making.

This latter position was clearly represented in France by Michel Crozier, who explained his view in terms of a profound change in the role of the intellectuals from the immediate post-war period to the early 1960s (23). Following Raymond Aron's version of the "end of ideology" theorem, which emphasises steady economic growth, internal stability, and cumulative progress in the sciences and in technology as factors allowing for a new form of rationality and for knowledge eliminating the need for force, Crozier argues that intellectuals have to move closer to action, and their thought must be much more pertinent and applicable in a direct way. In political and intellectual developments beginning with the late Fourth Republic government of Mendès-France, he traces the emergence of a new elite which brings the whole sphere of political and social action and the strategy of reform into the intellectual domain. For the social sciences, this will lead to increasing the influence of those groups who "form a sort of link . . . between the intellectual tradition seeking to reform itself and the world of action which is trying to renew itself through a process of more scientific reasoning" (24).

Seeing himself as a representative of this new type of social scientist, Michel Crozier had already contributed to this approach by elaborating an analytical model for the study of bureaucracies. His book on "the bureaucratic phenomenon" (25) generated a French school in the sociology of organisations at the *Centre de Sociologie des Organisations*, which was set up in 1966 under Crozier's directorship and has been innovative in opening up private and public administrations to empirical research. Reluctant to give policy advice in this book, Crozier in his later publications came closer to the new role of the intellectual as he conceived it, and offered more direct policy recommendations. During the post-1968 government he took up the role of informal policy advisor to Prime Minister Chaban-Delmas.

Basically sharing Crozier's view on the desirable future position of sociologists as policy intellectuals, Raymond Boudon voiced a different perspective on a strategy for the development of the social sciences (26). In his opinion, the way to achieve both a greater utility of sociology for societal practice and greater scientificity lay in methodological advances and, concomitantly, in the extension of the research infrastructure which this would necessitate. In spite of profound differences, Crozier and Boudon can be grouped together with some other sociologists under the heading of a politics-science interaction model, in which the role of the social sciences is to provide knowledge for the socio-technical steering of societal development by government and administration.

In contrast, the sociological approach of Alain Touraine, although it also seeks to break with traditional philosophical discourse and to be more empirical, tries to maintain a greater distance from politics. Touraine acknowledges the legitimacy of the call for societal relevance and recognises the pressures to produce knowledge instrumental to political and administrative needs. But his reaction is a refusal to make the analysis of the functioning of social systems his primary research objective. Instead, his "sociology of action" analyses the commitment and engagement of individuals and collective actors, by which they create situations and establish meanings (27). Consequently, he sees the possible application or utility of sociological knowledge not necessarily and not primarily in the politico-administrative system, but in terms of other social actors, such as social movements. He was later to elaborate this approach in greater detail.

A third group of sociologists around Pierre Bourdieu were not centrally concerned with the potential practical use of their sociology. Linking up with the Durkheimian tradition and to structural analyses of society, their work on the educational system and on intellectual milieus tries to relate micro-level analysis to macro-sociologically conceived developments. Refusing a "voluntarist orientation towards knowledge application" (Michael Pollak), their analysis of the reproduction of social structures in and through the educational system (28) has in fact played a part in the gradual disillusionment with the possibility of social reform through educational policy.

Compared with developments in Italy, a rather similar socio-economic environment in France did not lead to an identical reaction among sociologists. Strong incentives to produce politico-administratively relevant knowledge accompanied the emergence of policy-oriented sociological approaches, but at the same time brought about different "non-governmental" reorientations in French sociology. Comparable processes took place in the West German political science "community", although along different lines of debate.

A debate on the need to reform the governmental apparatus began in West Germany in the early 1960s, when the political system was increasingly considered as malfunctioning, in particular with regard to harmonising economic and social development. As earlier in France and Italy, this debate centered on the issue of modernization understood in terms of adjustment to worldwide technological and economic change. Unlike the other countries, however, the German debate concentrated on the issue of restructuring the politico-administrative system and did not really widen to include societal modernization in general (29).

Around 1965 the issue of political reform began to be discussed in the social sciences. In the following years a number of different programmatical positions emerged besides the traditional administrative/juridical one, which proved unable to encompass the empirical analysis of policy-making. In addition to the sociological approach of Niklas Luhmann and an attempt to widen the scope of the traditional administrative sciences (e.g., W. Thieme and F. Morstein Marx), at least three different conceptions emerged within the framework of political science (30).

Very much in the tradition of normative political philosophy, Wilhelm Hennis sketched the task of a "*Regierungslehre*" (government science) as the study of "the manner in which given the challenges to the modern state the steering, governance and coordination of a policy can be effected" (31). In a similar way, but more willing to deal with the concrete policy issues of the time, Thomas Ellwein argued for studying the processes of political decision-making and their transformation into policy programmes. Thirdly, a rudimentary alternative research programme was emerging as a reaction to the government-centred approach. At its core was the analysis of the entire political process, including the way in which changes in the economic system pre-structure and define the government's capacity to act in an anticipatory way (32).

The breakthrough from this programmatical debate to a broad range of empirical research projects occurred at the end of the 1960s and was made possible by the Volkswagen Foundation, a major social research funding institution, and by a whole set of research contracts issued by the newly-formed project group on governmental and administrative reform and mainly given to the "reform university" in Constance.

This emerging policy-orientation is best exemplified by the work of Fritz W. Scharpf. His understanding of administration in terms of political actors, who thus need to be studied within the framework of political science, is sketched in a paper published in 1971 in which he explicitly refers to the advancing policy research in the USA and advocates this approach as a promising new direction in political science (33). His publication (together with Renate Mayntz) on the "organisation of planning" (34) collects a number of papers written on political demand for the reform of the Federal politico-administrative apparatus, and can be considered the centrepiece of the early phase of policy-oriented research in Germany. The emphasis of the authors on intra-administrative organisational factors provoked the criticism of a number of researchers who considered themselves politically to the "left" of social democratic reformism, and who argued that Scharpf and others failed to recognise the decisive importance of external, mainly economic factors which limited the effectiveness of governmental reform policies and the scope of reformism itself. From this position emerged a "critical" approach which agreed on the need for empirical studies of policy processes. One strand of this approach sought to integrate the viewpoint of the citizens concerned, while the other, a much stronger neo-Marxist current in political science, concentrated on determining the extent and manner in which external contraints limit the autonomy of the political sphere (35).

Following this brief sketch of the scientific positions which developed during the 1960s, the task of the next section will be to trace the politico-scientific interactions which lay behind the successful appearance of new research programmes.

Political Innovativeness as a Topic: Interactions between Social Scientists and Policy-Makers

The changes which occurred during, broadly speaking, the 1950s and the 1960s and which amounted to the emergence of a policy orientation were conceptually least significant in Italian sociology. With no real change in theoretical and political assumptions, the shift from industrial relations to policy issues is mainly to be explained by a growing debate in this period on the need for public intervention into various policy fields. To understand why political influences could reorient the entire scientific discipline so strongly, one has to remember the low degree of academic institutionalisation and recognition. "Sociologists constitute a reference group of yet low relevance and cohesion, and thus Italian sociologists address themselves only in part to a public of scientific colleagues, but value much more highly the reference group formed by the larger cultural environment" (36).

Around 1960 a striking convergence of interests emerged between social researchers and policy-makers in the "cultural environment" of the North, especially in Milan and Turin. Left-Catholic, socialist and modernization-oriented lay groups transcended party boundaries on the basis of the common idea of politically regulated and planned social change. Social scientists offered their support as scientific experts able to analyse social problem constellations and to indicate the necessary political interventions: "In these years, sociology identified completely with the social project of changing Italian society" (37). Political and intellectual elites from the North joined forces to achieve this project. A number of initiatives flourished in the late 1950s providing contacts between people from different professional groupings and political orientations. The cultural and political association *Il Mulino* in Bologna was one of the most important of these initiatives, starting mainly as the editorial committee of a journal and later on expanding to include one of the most important social science publishing houses and a research centre. Another one was the *Centro nazionale della prevenzione e difesa sociale* (CNPDS) in Milan which gave birth to the most important sociological research institute of the 1960s, the *Istituto Lombardo per gli Studi Economici e Sociali* (ILSES). The CNPDS was instrumental in founding the Italian Association for the Social Sciences in 1958 and in organising and financing its first conference. In 1960, debates at CNPDS led to the idea of setting up a research centre after a congress on the impact of technological progress on society. Informal talks in Milan with leaders of the new Centre-Left coalition sufficed, and the ILSES developed into the place where almost all sociologists of the first and second post-war generations worked and which gave an enormous impetus to sociological research in Italy. To grasp fully the close relation between science and politics, it is useful to quote Alessandro Pizzorno, one of the founding fathers, describing the project in retrospect: "It was a study centre of the anticipated Centre-Left government, where political forces met with a group of independent scholars. The ILSES was to be the symbol of a new politics which should be based on analyses and research" (38). This concept was transferred to the national level after 1962, when the Centre-Left alliance was formed in Rome and again brought about a number of ambitious political reform initiatives and the setting up of policy-oriented research institutions.

In her analysis of the relation of sociology and politics in Italy, Diana Pinto asserts that the interest in modernization was translated in this period into a recourse to American political, economic, and social models, and as such, into an interest in American social science. This science model, on the other hand, was a precondition enabling sociology to become

> the new *lingua franca* of the Centre-left, capable of bridging the historical and cultural gaps which separate the christian democratic tradition from that of lay socialism. Ahistorical and international in scope, the social sciences smooth out any major conflict between the two political traditions, primarily by treating Italy's problems as the product of qualitatively new social and economic transformations, for which past solutions (and therefore conflicts) would be of little use (39).

Much of this analysis would also hold for Michel Crozier's political and sociological ambitions. His article on "the cultural revolution," cited above, can be read as an attempt to present his conceptions of the political advances required in French society and the role of the intellectuals engaged in these processes (40). But in France it proved much more difficult to get this position accepted in and against a much stronger and more self-conscious intellectual life. Thus, sociologists intending to adopt the prescribed new role of the intellectual had to look outside for support and acceptance.

Personal contacts were vital, and the main reference group was high-level administrators and managers, both in the public and the private sectors. As in the Italian case, a major catalytic influence can be traced to the private associations which organised around journals or simply as meeting places such as the *Club Jean Moulin*, the *Association d'études pour l'expansion de la recherche scientifique*, or the journals *Esprit* and *Prospective*. Unlike in Italy, these groups could not yet refer to a specific emerging political project, but in a less clearly defined way they assembled members of the political, economic and intellectual elites (41).

Looking back, Jacques Lautman has recently recalled the importance of the "personal connections which played a capital role in the formation of the spirit of state patronage" (42) for social science around the Fourth Plan. He notes that Alain Touraine maintained close relations with Claude Gruson, the head of the statistical institute INSEE between 1962 and 1967. In fact, Gruson, who argued in 1964 for the integration of sociological expertise into the planning process — though on a position subordinate to economics — seems to have offered Touraine the post of director of a new sociological section to be created at the INSEE, but these plans did not materialise due to a lack of financing (43). Michel Crozier's research profited from the patronage of Jean Ripert, at that time a high functionary in the Planning Commission, and was conceptually oriented strongly towards the sociological analysis of the planning process (44). Touraine himself now regards the constellations around the colloquium at Caen in 1956, one of the major science policy initiatives of the Mendès-France government, as a confused coalition of all modernization-oriented elements. "This coalition of the Mendès-type for a number of years constituted the universe in which the social sciences developed" (45). Its strength even increased with the return to power of de Gaulle and the reformism which then set in; after a great number of more or less informal meetings, one major event in this phase of intense interaction between social scientists and modernization-oriented administrators was the 1965 congress of the French Sociological Society, in which a group of high-level bureaucrats participated, among them Gruson, P. Massé, the director of the Planning Commission, and R. Grégoire, who was active in economic policy-making.

The orientation and behaviour of individual social scientists must be seen against the background of a changing socio-economic and political situation in which a strong demand for certain types of social knowledge exercised a major influence on the social sciences. But they have to be understood also as individual strategies to achieve or secure a reputation in a scientific field which was undergoing rapid transformations. The importance of the traditional Parisian intellectual community diminished, and political recognition gained in significance. This development has been described as the emergence of a second pole of reference for social scientists, the pole of power as opposed to the intellectual pole (46).

The social processes which lay behind the emergence of a policy-orientation in German political science seem to suggest that here the pole of power predominantly structured the field in which different scientific approaches competed (47).

Without doubt, the launching of the initial research projects and the opening up of the administrative apparatus towards empirical research were closely linked with the political intentions of the social democrats who became minority partner in government coalitions in 1966 and majority partner in 1969. A key figure in this process seems to have been Horst Ehmke, himself a professor of law, state secretary in the Ministry of Legal Affairs, and minister in the Chancellor's Office after Willy Brandt, a close friend of his, became Chancellor. Ehmke was among the programmatic thinkers of the SPD, who wanted to develop long-term perspectives for social democratic government, and who saw a primary task in restructuring the politico-administrative apparatus to increase rationality and efficiency in political processes. His preparatory work led, after some political compromises, to the formation of a reform cabinet which soon created the Project Group on Governmental and Administrative Reforms, still under the Great Coalition government in 1969 (48).

The inter-ministerial group, together with the Department of Planning established in the Chancellor's office by the subsequent social-democratic government, drew a great many social scientists into the policy process as advisors. In a second stage, and largely due to this advice, a demand was voiced for social science analyses to be pursued in the ministries. Many of these research tasks were contracted out to the newly-formed "group for comparative administration research" at the University of Constance. Given the political interest, it is not surprising that the research group followed the "government-centred" approach closely, dealing with issues such as internal control in planning administration, medium-term budgetary planning, or reform of the ministerial organisation.

However, this was only one step towards changing the political science agenda. In 1969, similarly influenced by the political debate on reform needs, the Volkswagen Foundation raised the issue of funding concentrated research efforts in public administration. After long discussions about the appropriate conceptual framework, a research committee was created to review funding applications. This group included representatives of the traditional administrative sciences, the reformist political scientists, and the emerging neo-Marxist current. In this way, a funding source came to exist whose decisions were not dominated by purely political criteria (49).

Both the reformist proponents of an "active policy-making" (R. Mayntz/F. W. Scharpf) and the neo-Marxists presented their research at meetings of the political science association. Given the state of the discipline, which was not homogenous, no integrative effects could be expected. Rather than the search for a common research programme or paradigm, a polarisation of the discipline occurred.

The emergence of scientific currents critical of the policy orientation which for some years either was or appeared to be dominant is a common feature in the social sciences of the three countries analysed here. Guido Martinotti suggests for the Italian case (and with some qualifications this holds true also for France and West Germany) that the social scientists were for some time guided by a mythical figure, the "innovative policy-maker" (50). A subsequent disillusionment and the occurrence of social revolt against the policy model they had advocated led to a reorientation not only of political conceptions but also of scientific approaches.

Political Crisis and Social Revolt: Social Scientists in Search of a New Coalition

In the changed political environment of the late 1960s and early 1970s a process of rethinking conceptual approaches started or accelerated among social scientists. The revolts of 1968/69, the abandonment of economic planning in Italy, disillusionment about the capacity to steer economic and social developments in France, and emerging financial and political restrictions on the reformist capacity of government in West Germany all helped to destroy the political and "epistemological optimism" and brought about or strengthened less directly policy-centred conceptualizations.

In Italy, the sociologists had begun to withdraw from the modernization project in the early 1960s, but this movement became generalised at the end of the decade with increased working-class militancy, the student rebellion, and the observation of political impotence in the face of an aggravation of social problems. Of great significance was the concept of action research, *conricerca*, as developed by the *Quaderni Rossi* group. *Quaderni Rossi* was a leftist political group organised around a journal which criticized the traditional workers' organisations for their integration in the capitalist system and their class collaboration. In the first half of the 1960s — six issues of the journal appeared between 1961 and 1967 — the concepts developed by this group had considerable influence on strategic debates in the reviving workers' struggles. Central to these discussions was the notion of autonomy, which indicated not only a disassociation from the established organisations for the representation of workers' interests, but also a rejection of the entire capitalist state, society, and way of life, and an attempt to organise self-determined spheres of work and life.

The concept of *conricerca* was considered by *Quaderni Rossi*, which consisted mainly of intellectuals, as an essential instrument to establish contact with workers on the shop-floor and to intervene in industrial relations. The group was aware of the American origins of the idea of action research, but intended to redefine it politically (51). Drawing on Marx's workers' inquiry worked out in 1880, the aim was to use such sociological techniques for a number of purposes. First, the responses would allow an assessment of the development of capitalist work organisation and thus help to refine the theoretical analysis of the capitalist system. Second, it would make the workers aware of the exploitative conditions under which they were living and working, and would initiate or advance their reflections about counter-strategies. Third, it would establish or improve contacts between revolutionary groups and the workers (52). The actual research pursued by members of *Quaderni Rossi* hardly reached the high theoretical and methodological levels which were sought; but these reflections were important in preparing the ground for a reorientation in the sociologists' work concerning both the users of research findings and theoretical developments. In a cultural environment where in particular the (communist) left was suspicious of modern sociology as an instrument of class domination, this approach reconciled Marxism and sociology by defining Marxism as sociology (53).

Among the first generation of Italian sociologists, who were already established professionally in the early 1960s, the impact of the rising political and social tensions can best be exemplified by the work of Franco Ferrarotti. Acting as a sort of spokesman for Italian sociology, he proclaimed as early as 1966: "It is necessary to advance from sociology as a palliative to sociology as a means of participation" (54). In the early 1970s, a programmatical essay on "the alternative sociology" summarized his critique of sociology, and he published the findings of a study of the victims of modernization in the shantytowns around Rome (55). An organised expression of irritation and the desire for a new conceptual orientation among sociologists was the 1971 conference on "Sociological Research and the Role of the Sociologist" in Turin, which focused exclusively on attempts to clarify past errors, assess the development of the discipline in Italy and abroad, and define a new role for the profession in society (56).

The problem of an adequate interpretation of the notion of "society" now emerged as a central conceptual issue. Society in the early 1960s had been seen in a very restricted sense as a "force to be moulded and regulated by planners and technicians from above" (57), whereas through the revolt it had come to constitute itself as an actor in its own right. Similarily, at the Second Congress of the French Sociological Society in 1969, Robert Castel argued that the methodologies adopted in sociology had strengthened the importance of empirical statements, of formal schemes of organisation, and of superficial equilibria, and had tended to misconceive the social significance of change and of social conflicts. This was thought to have led to similar concepts of political action now reduced to the management of social order and the manipulation of subjects to adapt them to this order. Alain Touraine summarised the significance of the social crisis for sociology: "The intellectual problem which is posed to the sociologist is the political problem which is posed to society; and the political awakening of society . . . or at least the calling into the question of its orientations and its forms of organisation allow the sociologist to rediscover the unity of the object he studies and thus of his own approach (58)." In his own work since 1968 Touraine has continuously tried to draw conclusions from this diagnosis. The approach which he had set out in "la sociologie d'action" in 1965, seeing social actors as transforming society by creating instead of merely responding to situations, is further developed by his turning to social movements as the principal agents of history. In this analysis of the French May revolt (59) he not only chooses to study different social phenomena but links them to a concept of sociological intervention. The interventionist sociologist, as an analyst in interaction with social actors, seeks to stimulate the autoanalysis of groups acting in social movements. This autoanalysis, it is assumed, enables the actors to give meaning to confrontation and their position in it, and thus contributes to advancing the movement as a whole. In more recent years Touraine has applied this approach in sociological interactions to the student movement of the 1970s, the anti-nuclear movement, and the Polish workers' movement around the "Solidarity" union (60).

Politically motivated action research was also a topic in German sociological debates in the wake of the student revolt. Compared to Italy and France, it was, however, seen less as a new and alternative political and scientific approach than as a critical reorientation, complementing and adjusting existing research programmes and reform strategies (which were still being pursued by the social-liberal government at that time) (61). In the field of political science, which is traditionally focussed on institutional analysis, the latently counter-institutional approach of action research had no influence. The politicization of the research environment strengthened the "leftist" position in political science, but very few studies tried to advocate a participatory approach. Most research efforts were meant to counter the "policy-oriented" researchers on the system level by developing the concept of systemic limits to state action in capitalist society based on a reconstruction of the Marxian critique of political economy (62). In its extreme versions, this neo-Marxist current virtually denied all relevance to intra-organisational factors in the politico-administrative system for the outcome of (reformist) policy-making. By contrast, the proponents of "active policy-making" laid almost exclusive emphasis on the study of these intra-organisational factors, thus, at least by implication, placing a high value on their importance. This dispute can partly be described in terms of a too restrictive conceptualisation of the political system, and especially of the notion of "policy", which had been stripped of its historical and societal context. As with the influence of functionalism on Italian and parts of French sociology, this approach in German political science was shaped by the American understanding of policies as institutionally discrete and temporally exactly defined actions pursued by an independent political system. This concept easily lent itself to the support of voluntarist reformism. During the 1970s, the experience of increasing difficulties and the partial withdrawal from reformist policy-making on the one hand, and the exhaustion of increasingly barren attempts to determine purely theoretically the restricted nature of the state's autonomy under capitalist conditions on the other, led to a convergence of both approaches on the field of an empirical and theoretically more open study of policy processes.

After these periods of high engagement and dispute, inner-scientific discussions have been less conflictual in recent years in all three countries (63). This may be in large part due to the fact that such intense alliances involving the merging of political and scientific goals no longer attract large groups of social scientists. But that the question of how to develop the interaction of social science and politics is still high on the agenda can be seen for example from the recent debates at a round table of the German Political Science Association on the topic of "policy studies and traditional political science". At its centre was the fear of a fragmentation of the discipline due to the rising political demand for specialized analyses in different policy fields (64). Similarly, at the National Colloquium in January 1982, which was to inaugurate a new and better era in science policy in socialist France, a strong resistance on the part of the social scientists could be observed, rooted in their memory of the tendency towards subordination to political interests in the 1960s and 1970s.

Conclusions for a Political Sociology of Science

This paper has described and analysed the emergence of an intense collaboration between social scientists and political actors, a reform coalition (65), its demise, and the search for a new coalition, a new definition of social relevance on the part of social scientists. Coalitions are joined for certain goals. In very broad terms, the aim of the reform coalitions discussed here, their "common social project", can be described as modernization and social reform (66), while the goal of the action research projects which followed was support for the victims of capitalist modernization and the construction of a societal alternative. The following remarks explore some of the implications of these processes in terms of the interaction between scientists and non-scientific groups.

The social demand for knowledge, and in the case of the social sciences, also for interpretations of social reality which are adequate and necessary for some actors' interests, can be considered as a general condition for such forms of collaboration. However, the movement of social scientists towards engaging in politically useful work cannot simply be seen as a "selling out". To me it seems important to note that since the end of the Second World War the "mythical promise of social renewal through the social sciences" (67) has been one of the most important motives for becoming a social scientist. This promise was not only unfulfilled by 1960, but became even more urgent, as the political restoration had left many social questions unanswered. The pressure on the social sciences to become societally relevant stemmed from a larger socio-political environment and must have been internalized by many social scientists. Their identification with the political project thus went far beyond the mere acceptance of a social task, and can only be explained in this larger context.

But the social scientists also had an academic project. In all three countries, although more in France and Italy than in Germany, the degree of institutionalisation of the social sciences was insufficient for expanding research activities and for consolidating the disciplines. The promise of immediate political utility could be used to claim resources and to facilitate academic establishment. In France and Germany the phase of intense interaction between social scientists and policy-makers paralleled expansion and the full institutionalisation of the social sciences at the universities. In Italy, delays in a planned university reform due to resistance by the representatives of established academic interests meant that the full recognition of sociology as an academic subject did not occur before 1968; but with the "Committee for the Political and Social Sciences" (COSPOS), other institutional solutions could be found to bridge this gap. From the viewpoint of disciplinary interests, then, the interaction with influential non-scientific groups served this aim of achieving full academic institutionalisation.

A third consideration concerns the role of the disciplinary community as a reference group for the individual scientist and the implications for innovative strategies. In West German political science the normative commitment to Western parliamentary systems was initially shared by almost all the members of the community, since it had almost been a constitutive element in the emergence of the discipline. The policy-orientation of the late 1960s introduced an unfamiliar perspective into the disciplinary debates: the focus was now on the empirical study of substantive policy issues, and to a number of the "traditional" political scientists, this "modern" viewpoint overemphasised technical issues and disregarded the central political problems. It can thus be argued that without external support providing research funds and influencing science policy decisions (besides providing the necessary access), the new approach would have faced much greater difficulty in gaining acceptance.

In France, the intellectuals in Paris doubtless formed a sort of community in which every member had to refer to the works of other members and to write and act according to certain standards. These intellectual circles, however strongly they may have shaped the re-emergence of sociology, did not, however, themselves form or intend to form a scientific discipline. Thus, much of the sociological work of the 1960s bears signs of the tensions between the strong grip of the traditional reference group and the attempt to build up one's own position by distinguishing oneself from the philosophies of the past and by establishing a real "science".

The issue of "scientificity" as a precondition for becoming a discipline had the least relevance in Italian debates about sociology in the 1950s and early 1960s. The sociological approach accorded almost perfectly with political reformism, and not much care was taken to emphasise the differences between science and politics. For sociologists it could be as important, or even more important, to establish close links with relevant Centre-Left party politicians than to engage in intense discussions about scientific progress in the field.

In this last case it is obvious that the degree to which a self-referential disciplinary community existed was connected to its degree of external orientation. While in Italy the absence of a scientific community was connected with the extreme importance assigned to external political processes, the two factors are not necessarily identical. In France, on the contrary, it was a key feature of the intellectuals that while being self-referential in their interactions they were also highly engaged in commenting on social developments and taking political stands, their ideal being that of the "*intellectuel engagé*". Direct and utilitarian contact with political, economic, and administrative elites was considered a breach of the traditional ideal, endangering the critical function of intellectuals by reducing their distance from power. It seems in large measure to be the definition of their relation to society that emerged in the 1960s which to this day distinguishes the different "schools" in French sociology, which are mainly grouped around dominant individuals.

The interference of these disciplinary and individual considerations in the macro-politically influenced desire to conduct socially relevant research, in my view considerably modifies the notion of "shared social purpose" which Stuart Blume assumes to be the basis for such interactions between scientists and non-scientists. In part, the different actors use a common situation for their *different* purposes.

One such feature in these coalitions — one might call it the "honeymoon syndrome" — is at the root of their demise. Coalitions are entered into with high expectations, and programmatic statements at the beginning of a collaboration are usually enthusiastic. Regarded in retrospect, both the political and the scientific outcomes are highly exaggerated, and sometimes it seems as if a more sober contemporary might well have recognized this overstatement. But exaggerating the opportunities of the coalitions might have served as a useful way to strengthen the position of the collaborators in both fields. However, once an innovative scientific approach has been established with the help of external support, and once a group of political actors has come to power proclaiming rational policy-making on the basis of social scientific advice, the different logics in the two fields reassert themselves and interests begin to diverge (68).

In the above analysis it has been suggested that the demise of the modernization-oriented reform coalition led to the emergence of the critical action research approach. This shift, of course, should not be regarded as a clear sequence. As can be seen most clearly from the discussions about *conricerca* in Italy, this conception emerged while modernization-oriented sociology was still at full strength. But a second differentiation has to be made regarding the success and the stability of such coalitions. The policy-oriented social sciences addressed themselves to identifiable actors who would ultimately dispose of large resources; the interaction between social scientists and policy-makers might become institutionalised in social science policy and thus achieve a certain continuity and stability (69). The action researchers, by contrast, had much greater difficulty in identifying their relevant public and, even when they could, in maintaining a stable relationship of communication and interaction. Their engagement very much depended on the existence of active social movements or other social groups having a definable political perspective and interest. Moreover, even when these conditions were met, continuity could only be expected if institutional settings were available which allowed researchers to concentrate on these issues. In the absence of these preconditions, the action research approach as a broad movement of social scientists proved to be very short-lived, though it survives in other forms.

Finally, the problem of the epistemological consequences of the coalitions between social scientists and non-scientists must be mentioned briefly. This problem cannot be fully analysed here on the basis of the previous analysis, but our discussion of the changing concepts of "society" and "policy/political system" clearly suggests that significant epistemological shifts in the social sciences can be observed depending on the type of actor-orientation which is dominant at a specific time (70). Although these shifts cannot be measured against any baseline (of purely internal scientific progress), the concept of "epistemic drift" introduced by Aant Elzinga (71) seems to be a valuable one. The task of a political sociology of science would thus be to specify in detailed analyses the character of specific "drifts" by studying both the socio-political environment and the potential audience of a discipline and the structure of its scientific field as a precondition to understanding the behaviour of specific actors in this field.

Notes and References

(All translations are by the author.)

1. Daniel Lerner, Harold Lasswell (eds.), *The Policy Sciences.* Stanford: Stanford University Press, 1951.
2. For an attempt to analyse the long-term development of problem-oriented social science research see: Georg Thurn, Peter Wagner, Björn Wittrock, Hellmut Wollmann, "The development and present state of public policy research. Country studies in comparative perspective", mimeo, Berlin 1984, an interim report from a research project presently pursued at the Wissenschaftszentrum für Sozialforschung Berlin and the Freie Universität Berlin in cooperation with the Swedish Collegium for Advanced Study in the Social Sciences. In the sections analysing the developments in Germany this paper partly draws on Hellmut Wollmann's analysis in this project. For further references on social science developments in Germany as well as in France and Italy see this report.
3. For an attempt to describe related processes in Austria in terms of a conflict between traditional and modernizing political forces, see Bernd Marin, *Die*

Paritätische Kommission. Aufgeklärter Technokorporatismus in Österreich. Vienna: Internationale Publikationen, 1982, in particular part III.

4. This difference may in part explain why in Germany political science was more involved in these processes, while sociologists played a major role in France. However, other factors enter into this, and I shall return to them below. The rationale of choosing different disciplines for the following analysis — sociology in France and Italy, political science in Germany — may be questioned, but this comparative approach is in my view the most appropriate, since the interaction between actors in the science system and actors in the political system is to be stressed. Some unevenness in the argument concerning the comparison of disciplinary developments must therefore be accepted. For similar reasons the three countries were chosen for this analysis, as they were all marked by short periods of particularly intense debate about modernization and the concomitant interaction of social science and politics.
5. Stuart S. Blume, in this volume.
6. Katrin Fridjonsdottir seems to adopt basically similar perspectives in this volume in her analysis of the development of the sociology of work in Sweden.
7. See Robert L. Geiger, "Die Institutionalisierung soziologischer Paradigmen", W. Lepenies (ed.), *Geschichte der Soziologie*. Frankfurt/M: Suhrkamp, 1981, and Terry N. Clark, *Prophets and Patrons: The French University and the Emergence of the Social Sciences*. Cambridge, Mass: Harvard University Press, 1973.
8. Alain Drouard, "Réflexions sur une chronologie: le développement des sciences sociales en France de 1945 à la fin de l'année 1960", *Revue française de sociologie* **23**, 1982, 66f. It may be argued that in the inter- and early post-war periods sociological thinking in France had found a place in the disciplines of ethnology and history. In the latter case, the *Annales* school also succeeded in institutionalising its approach early in the post-war years.
9. Michael Pollak, *Gesellschaft und Soziologie in Frankreich*. Königstein/Ts.: Hain, 1978. 30.
10. See Orlando Lentini, *L'analisi sociale durante il fascismo*. Naples: Liguori, 1974; Franco Leonardi, "Italian sociology within the framework of contemporary sociology", *Contemporary Sociology in Western Europe and America. Proceedings of the First International Congress of Social Sciences of the Luigi Sturzo Institute*. Rome: XXXXX, 1967; Paolo Mieli, "I Machiavellini", *L'Espresso* **29**, 1983, 66f. The strong ideological heritage of political science in Italy surely played its part in preventing an earlier reemergence of the discipline after World War II and in leaving sociology the first rank in the demand for societally relevant knowledge.
11. M. Rainer Lepsius, "Die Entwicklung der Soziologie nach dem 2. Weltkrieg 1945—1967", Günther Lüschen (ed.), *Deutsche Soziologie seit 1945*, special issue of the *Kölner Zeitschrift für Soziologie und Sozialpsychologie* **21**, 1979, 33. See for a historical sketch Horst Kern, *Empirische Sozialforschung*. Munich: Beck, 1982, and the report by Hellmut Wollmann in Thurn *et al., op. cit.*, 1984. Note 2.
12. See Hans Kastendiek, *Die Entwicklung der westdeutschen Politikwissenschaft*. Frankfurt/M.: Campus, 1977, pp. 185ff.
13. On the origins of this attitude in the 1930s, see Johan Heilbron, "Les métamorphoses du durkheimisme, 1920—1940", *Revue française de sociologie* **26**, 1985, 225ff.

14. Pierre Bourdieu, Jean-Claude Passeron, "Sociology and philosophy in France since 1945: Death and resurrection of a philosophy without subject", *Social Research* **34**, 1967, 171ff., quotation from p. 179.
15. For a short characterization of Croce's methodological writings see Franco Ferrarotti, *Introduzione alla sociologia.* Rome: Riuniti, 1981, pp. 149ff.; for a description of the intellectual climate in the post-war period see Diana Pinto, "La sociologie dans l'Italie de l'après-guerre", *Revue française de sociologie* **21**, 1980, 234ff.
16. See Claude Durand, "Les ouvriers et le progrès technique: Mont-Saint-Martin vingt ans après", *Sociologie de Travail* **22**, 1980, 5, and Alain Drouard (ed.), *Le développement des sciences sociales en France au tournant des années soixante.* Paris: CNRS, 1983, pp. 89f. and 126ff.
17. See his one-sided account: Jean Stoetzel, "Sociology in France — An empiricist view", in Howard Becker and Alwin Boskoff (eds.), *Modern Sociological Theory in Continuity and Change.* New York: Holt, Rinehart and Winston, 1957.
18. For a comparison of French science policy institutions with those in the Netherlands and the U.S.A., see Ronald Brichman and Arie Rip, "Science Policy Advisory Councils in France, the Netherlands and the United States", *Social Studies of Science* **9**, 1979, 167—198.
19. This ultra-brief description can obviously not account for the complexity of specific historical situations in three countries and — for the purpose of this paper — almost unavoidably tends to neglect important differences between the countries. For more detailed analyses see for example: Joseph Lapalombara, *Italy. The Politics of Planning.* Syracuse, N.Y.: Syracuse University Press, 1966; S. S. Cohen, *Modern Capitalist Planning. The French Model*, Cambridge, Mass., Harvard University Press, 1969; Elmar Altvater, Jürgen Hoffmann, Willi Semler, *Vom Wirtschaftswunder zur Wirtschaftskrise.* Berlin: Olle und Wolter, 1979; Jack Hayward, Michael Watson (eds.), *Planning, Politics and Public Policy. The British, French and Italian Experience.* Cambridge (UK): Cambridge University Press, 1975.
20. Diana Pinto uses the phrase "cultural centrality" in: "Sociology, politics and society in post-War Italy", *Theory and Society* **10**, 1981, 676.
21. A strong argument in this direction is made by Carlo Guido Rossetti in: "Sur la sociologie italienne vue par Diana Pinto", *Revue française de sociologie* **23**, 1982, 284f.
22. See Renato Treves, *Sociologi e Centri del Potere.* Bari: Laterza, 1962.
23. Michel Crozier, "The cultural revolution: Notes on the changes of the intellectual climate in France", *Daedalus*, Winter 1964, 514—542.
24. *Ibid.*, p. 537.
25. *Le phénomène bureaucratique.* Paris: Le Seuil, 1963.
26. See, for example his contributions to the debate reprinted in the collection: *La crise de la sociologie.* Paris: Droz, 1971.
27. Alain Touraine, *La sociologie de l'action.* Paris: Le Seuil, 1965.
28. Pierre Bourdieu, Jean-Claude Passeron, *La reproduction.* Paris, 1970.
29. Reasons for this may lie in the traditional "*Staatsfixierung*" (centrality of the state) in German social thought, or quite differently, in the smaller need for such a widening given the high competitiveness of the German economy. But whatever

the reasons are, this political focus may have been decisive for the turn toward political science instead of sociology to deal with this problem.

30. For a description of the development of the debate see for example Thomas Ellwein, "Verwaltungswissenschaft: Die Herausbildung der Disziplin", Jens Joachim Hesse (ed.) *Politikwissenschaft und Verwaltungswissenschaft, Politische Vierteljahresschrift*, special issue 13, 1982, 34ff.; and Joachim Hirsch and Stephan Leibfried, *Materialien zur Wissenschafts- und Bildungspolitik*. Frankfurt/M.: Suhrkamp, 1971. 236ff.
31. Wilhelm Hennis, "Aufgabe einer modernen Regierungslehre", *Politische Vierteljahresschrift* **6**, 1965, 424.
32. Joachim Hirsch, *Parlament und Verwaltung*, Vol. 2. Stuttgart, 1968; see the discussion in: Rolf-Richard Grauhan, "Politikwissenschaftliche Forschung zur Verwaltung", *Die öffentliche Verwaltung* **23**, 1970, 591.
33. Fritz W. Scharpf, "Verwaltungswissenschaft als Teil der Politikwissenschaft", reprinted in : Scharpf, *Planung als politischer Prozess*. Neuwied: Luchterhand, 1973.
34. Renate Mayntz and Fritz W. Scharpf, *Planungsorganisation*. München: Piper, 1973.
35. For useful reader in English see John Holloway and Sol Picciotto (eds.), *State and Capital. A Marxist Debate*. London: Arnold, 1978, For a summary of the critique of the approach taken by Scharpf and others, see Wolfgang Fach, "Verwaltungswissenschaft — ein Paradigma und seine Karriere", in: Hesse, *op. cit.*, 1982, Note 30, pp. 55—73.
36. Guido Martinotti, "Il condizionamento della ricerca", Pietro Rossi (ed.), *Ricerca sociologica e ruolo del sociologo*. Bologna: Il Mulino, 1972, 143f. In this section of a paper written in 1971 Martinotti rejects on these grounds the proposition that Alvin Gouldner's critique of sociology can be applied in similar terms to the Italian case.
37. Guido Martinotti, *L'istituto superiore di sociologia di Milano*, mimeo, Milan, 1984, p. 3.
38. Quoted from : Silvia Giacomoni, *Miseria e nobiltà della ricerca in Italia*. Milan: Feltrinelli, 1970, p. 105.
39. Pinto, *op. cit.*, 1981, Note 20, p. 680. Franco Ferrarotti saw as one of the insidious dangers for Italian sociology in the mid-1960s its being considered as a "*deus ex machina* for evidently complex social problems" ("Changement social et sciences sociales en Italie", *Revue française de sociologie* **7**, 1966, 29).
40. From a different standpoint Lucien Goldmann wrote: "Future historians will probably identify the years 1955 to 1960 as the sociological turning point in France between crisis capitalism and organised capitalism, accompanied by a transition from philosophical, historical and humanistic sociology to the a-historical sociological thinking of today." (*Sciences humaines et philosophie*, preface to the new edition. Paris: Gonthier, 1966, p. 6.)
41. See e.g. recently Drouard, *op. cit.*, 1983, Note 16, p. 68; see also Bourdieu, Passeron, *op. cit.*, 1967, Note 14, p. 187; and Pollak, *op. cit.*, 1978, Note 9, p. 54.
42. In: Drouard, *ibid.*, p. 78.
43. Claude Gruson, in: *ibid.*, p. 147. See also his programmatical paper: "Planification

économique et recherches sociologiques", *Revue française de sociologie* **5**, 1964, 435—446.
44. See his: "Pour une analyse sociologique de la planification française", *Revue française de sociologie* **6**, 1965, 147-163.
45. A. Touraine, answer to a questionnaire, published in: Drouard, *op. cit.*, 1983, Note 16, p. 180.
46. Michael Pollak, "L'efficacité par l'ambigüité", *Sociologie et sociétés* **7**, 1975 p. 36.
47. In fact, such an interpretation has been given in an early assessment by Hirsch, Leibfried, *op. cit.*, 1971, Note 30.
48. See Heribert Schatz, "Auf der Suche nach neuen Problemlösungsstrategien: die Entwicklung der politischen Planung auf Bundesebene", Mayntz, Scharpf, *op. cit.*, 1973, Note 34, pp. 29ff.
49. For a similar view see Erhard Friedberg, Pierre Gremion, *La recherche administrative et le réformisme politique*, mimeo, Paris: CSO, 1974, pp. 8ff.
50. Martinotti, *op. cit.*, 1972, Note 36, p. 146.
51. See Dino de Palma, Vittorio Rieser and Edda Salvatori, "L'inchiesta alla Fiat nel 1961", *Quaderni Rossi*, No. 5, March 1965.
52. Dario Lanzardo, "Intervento socialista nella lotta operaia: l'inchiesta operaia di Marx", *Quaderni Rossi*, No. 5, March 1965.
53. "Marxism . . . originates as sociology, what is 'Capital', as a critique of political economy, if not sociology? . . . In my view . . . sociology is not a bourgeois science, . . . we can use and criticise sociology in just the same way as Marx did with classical political economy." (Raniero Panzieri, Contribution to the seminar "Uso socialista dell'inchiesta operaia", Turin 1964, published in *Quaderni Rossi*, No. 5, March 1965, quoted from the reprint in: Claudio Pozzoli (ed.), *Spätkapitalismus und Klassenkampf*. Frankfurt: Europäische Verlagsanstalt, 1972, pp. 105 and 108).
54. Ferrarotti, *op. cit.*, 1966, Note 39, p. 31.
55. Franco Ferrarotti, *Una sociologia alternativa*, Bari: De Donato, 1972, subtitled "From sociology as a technique of conformism to critical sociology", and *Roma da capitale a periferia*. Bari, 1971.
56. See Rossi, *op. cit.*, 1972, Note 36.
57. Pinto, *op. cit.*, 1981, Note 20, p. 680; see also Rossetti, *op. cit.*, 1982, Note 21.
58. Quoted from: Riccardo Scartezzini, "Il dibattito metodologico in Francia", in: Rossi, *op. cit.*, 1972, Note 36, p. 257; see also Alain Touraine, *La société post-industrielle*. Paris: Denoël, 1969.
59. Alain Touraine, *Le mouvement de mai ou le communisme utopique*. Paris: Seuil, 1968.
60. For recent critical assessments of his approach see: Remi Hess, *La sociologie d'intervention*. Paris: Presses Universitaires de France, 1981, pp. 140—149; and Peter Kivisto, "Contemporary social movements in advanced industrial societies and sociological intervention: An appraisal of Alain Touraine's pratique," *Acta Sociologica* **27**, 1984, pp. 355—366.
61. Early contributions were: Werner Fuchs, "Empirische Sozialforschung als politische Aktion", *Soziale Welt* **21/22**, 1970/1971, 2—17; and Fritz Haag, Helga Krüger, Wiltrud Schwärzel and Johannes Wildt (eds.), *Aktionsforschung. Forschungsstrategien, Forschungsfelder und Forschungspläne*. Munich: Juventa, 1972.

See also Klaus Horn (ed.), *Aktionsforschung: Balanceakt ohne Netz?* Frankfurt: Syndikat, 1979; and recently Horst Kern, *op. cit.*, 1982, Note 11, pp. 261—272.

62. The unexpected revival of Marxism both in the social sciences and in society, which occurred also in France and Italy, would require a separate analysis.
63. The debates on the role of the social sciences in society during the 1970s and early 1980s cannot be dealt with in the framework of the argument proposed here. To contribute to the still inadequate analysis of this period of the "social scientification" of politics is the intention of the above-mentioned study of the development of policy research; see Thurn *et al.*, *op. cit.*, 1984, Note 2.
64. See the proceedings in: Hans-Hermann Hartwich (ed.), *Policy-Forschung in der Bundesrepublik Deutschland*, Opladen: Westdeutscher Verlag, 1985.
65. The term "reform coalition" in this context has been used by Hellmut Wollmann, "Policy analysis — Some observations on the West German scene", *Policy Sciences* **17**, 1984, 44. I heard the much nicer term "honeymoon" from Guido Martinotti and rediscovered it in Katrin Fridjonsdottir's contribution to this volume. Its advantage (or disadvantage) is that it may be considered to contain the implicit assumption of an early end.
66. For an evaluation of the German experience in these terms see: Ulrich Beck, Wolfgang Bonß, "Soziologie und Modernisierung", *Soziale Welt* **35**, 1984, 381—406.
67. Lentini, *op. cit.*, 1974, Note 10, p. 21.
68. As a profound discussion of the idea of logics — unitary, diverse or analogous — shaping social science-politics interaction, see: Björn Wittrock, "Social knowledge and public policy: Eight models of interaction", Helga Nowotny and Jane Lambiri-Dimaki (eds.), *The Difficult Dialogue between Producers and Users of Social Science Research*. Vienna: European Centre for Social Welfare Training and Research, 1985, pp. 89—109.
69. For the German (and American) development in political science, this relative continuity contributed during the 1970s to keeping researchers in the same logic as the policy-makers by shifting emphasis from the processes of policy formulation and planning to implementation analysis and policy evaluation. But this relatively stable relationship to the political system by which larger and larger shares of the total research budget were being allocated can also be considered as responsible for the loss of a critical perspective and the potential for a fruitful intra-disciplinary dispute on conceptual and theoretical issues. For a more detailed analysis see: Thurn *et al.*, *op. cit.*, 1984, Note 2.
70. For a suggestive historical account, which, however, obviously lacks a detailed analysis, see Paul Diesing, *Science and Ideology in the Policy Sciences*. New York: Aldine, 1982.
71. Aant Elzinga, "Research, bureaucracy, and the drift of epistemic criteria", Björn Wittrock and Aant Elzinga (eds.), *The University Research System: The Public Policies of the Home of Scientists*. Stockholm: Almqvist and Wiksell, 1985.

ATTRACTING AUDIENCES AND THE EMERGENCE OF TOXICOLOGY AS A PRACTICAL SCIENCE

PETER GROENEWEGEN

Science and Society, Chemistry Department, State University of Groningen

The central focus of this article is on the strategies scientists adopt in presenting their work as relevant to the solution of significant social problems. Such problems may provide material for scientists to base their research on, and may serve as focal points for collaborations with groups of non-scientists (which I shall refer to as *audiences*). Audiences interested in particular social problems sometimes provide a market, e.g., they provide funds for scientific research. In order to win for themselves the attention of relevant audiences and thereby to obtain resources for research, scientists may attempt to exploit the "market potential" in areas of current social concern. But the *manner* in which scientists solicit the interest of these audiences, and the kinds of audiences involved by them, are assumed to be mediated by their own institutional locations, their previous experiences, and their own motivations. The point is, as we shall see, that these audiences may be given a role in the subsequent stabilization of a field of scientific research.

The interaction of scientists with their environment is a theme as old as the sociology of science. Recently a number of contributions to the field have stressed the role of scientists as entrepreneurs in establishing links with outside groups. For example, in an attempt to overcome the limitations of "laboratory studies", Latour has tried to show that the interests of social groups are not "given", but can be actively constructed by scientists. In his analysis of the development of Pasteur's microbiological research Latour argues that interests are not stable:

> Their interests (in Pasteurs' research) are a consequence and not a cause of Pasteur's efforts to translate what they want or what makes them want. They have no a priori reason to be interested at all, but Pasteur has found them more than one reason (1).

S. Blume, J. Bunders, L. Leydesdorff and R. Whitley (eds.), The Social Direction of the Public Sciences. Sociology of the Sciences Yearbook, Vol. XI, 1987, 307—328.

Similarly, in his study of the development of meteorology Friedman explores the strategy used by Bjerknes in seeking new audiences (2). Bjerknes established a new conceptual foundation for atmospheric science as the basis for a system of weather forecasting required by emerging commercial aviation. In order to profit from a more favorable institutional climate than in his native Norway, Bjerknes moved to Leipzig, where he could link his scientific goal to the practical needs of aeronautics. In the German situation aeronautics provided an important audience, which was less developed in Norway. In Norway following his return in 1917, the prediction of weather for agricultural purposes provided a significant social goal and a significant new audience. Friedman shows that these audiences were allotted a substantial role in the emergence of scientific meteorology.

Both Latour and Friedman suggest that the actions of Pasteur and Bjerknes can be seen as attempts to interest significant audiences in their research. By focussing on the actions of individuals, Latour and Friedman pay little attention to the constraints deriving from the fact that modern science is an organized and structured activity. It is as if outcomes are to be understood in terms of the insights, motivations, and cunning strategies of individual scientists. The organized and structured character of the sciences, as discussed by Whitley (3), and the way in which this may contrain the possibilities of "cunning strategies", are ignored.

Shrum takes the character of the sciences as an organized activity into account, and his concept of sciences as a technical system addresses some of the issues involved in the interaction between scientists and non-scientific groups. According to Shrum, technical systems contain the following features:

(1) a goal-oriented emphasis on the production of scientific information relevant to the resolution of a broad technical problem of social concern;

(2) direct involvement of the state in coordinating research activities concerned with the production of information;

(3) research activity from a relatively large number of actors, as compared with the number typically involved in a single scientific specialty; and

(4) contributions from sectorally diverse organizations (e.g., public, private) and from various occupations (scientists, engineers, administrators) as well as diverse scientific disciplines or specialties (4).

It is worth noting that technical systems are characterized by Shrum in terms of "administrative hegemony", and by this definition may be described as "centralized" communication networks, in which decision-making is concentrated within the public sector consisting of government agencies and government laboratories (5). In contrast to Latour's discussion of the mobilisation of interests, Shrum's central concern, the pattern of communication, involves a consideration of the roles of institutional positions in the establishment of external networks.

In this article I want to make use of two central elements drawn from the literature cited above. The studies of Friedman and Latour suggest the crucial role of individual motivations in the establishment of audiences interested in a certain research area. Both approaches suggest that attention to non-scientific interests and problems plays a crucial role in the strategies by which research areas or theories are (re)constituted. Shrum alerts us to the role of structured communication and interaction mediated by a variety of institutions. In what follows, I want to examine the enlistment of audience interest, and the way this is constrained (and attention is focussed) by institutional structures, in relation to the development of a new scientific specialty.

The development of toxicology in the Netherlands provides us with a good case for the exploration of the formation of links between science and audiences in different institutional locations, and the influence of these audiences on research. Although Johnston discusses the development of risk assessment science from a different angle than we will use here, his description of toxicology suggests that non-scientific audiences play an important role in the development of toxicology:

> There can be little argument over the extent to which this research is constrained and directed by the political issues associated with risk determination and control. It is then, *par excellence*, a field in which organisational pressures and political objectives shape the selection, production and evaluations of scientific knowledge and its study can contribute to the development of more comprehensive theories of the interaction between political and intellectual aspects of the social reality which constitute science (6).

Moreover, in an analysis of the regulation of two carcinogenic pesticides Gillespie and Johnston suggest that the relation between toxicology and audiences is different in different national contexts:

> While toxicology and its practitioners have generally been 'politicized' by their involvement in government decision-making (. . .) the respective styles of government have politicized them in different ways. In the US, toxicologists have been enlisted by

conflicting social groups, especially industrialists and environmentalists, to support their arguments, and, as a result, have been drawn into adversarial political processes (7).

In this analysis of the development of toxicology in the Netherlands, two questions will be addressed. First, what is the relationship between the kinds of toxicological problems addressed by scientists and changes in the pattern of collaboration between toxicologists and audiences? Second, to what extent and how have differences in institutional organization affected these processes of collaboration between toxicologists and their audiences?

1. The State Institute of Health

The State Institute of Public Health (RIV, Rijksinstituut voor de Volksgezondheid), part of the Ministry of Health, was where preventive toxicological research originated in the Netherlands (8). During the first decades of this century, the Ministry of Health became involved in regulating and enforcing regulations with regard to food; and research on food was initiated in the 1930s. During the 1950s the RIV worked largely in response to questions put forward by the Ministry of Health and the Health Inspectors. The Institute's research was mainly determined by this service function, and was concerned with problems such as advising on accidents involving public health, the standardization of vaccines, the production of vaccines, the standardization of vitamin preparates, etc. Not until the 1960s would more attention be given to the scientific credibility of the Institute.

Although investigations into intoxications had been part of the work of the RIV from early in its existence, systematic attention to toxicology originates in the period around the Second World War. The aim of early work in the divisions of Pharmacology and Chemical Pharmacy was the determination of chemical agents responsible for human diseases or deaths. Experience in this work played a role in the origins of systematic toxicological research, as did animal research on the effects of vitamins. During this early period, other questions were being put to this unit in addition to those relating to acute toxicity. A new subject for the RIV was the evaluation — at the request of the government — of the safety of alternatives to common food (Ersatz products) during the German occupation in the Second World War. For example, animal feeding experiments were performed in order to determine whether flower bulbs were safe for human consumption (9). So in addition to acute toxicity questions new research questions were gradually entering the purlieu of the scientists at the RIV.

Directly after the War, the increasing number of problems relating to food additives and pesticides spurred the RIV to organise a separate unit for toxicology. The relevant questions now concerned the safe (for humans) application of food additives and colorants in the food industry and of pesticides in agriculture (10). Among the issues drawing attention was the amount of pesticide residue left on produce when it was ready for the market. Most of the questions were a direct consequence of the increased use being made of chemicals in food production and conservation. Such questions about safety for human consumption were quite different from those of acute toxicity, for which clinical expertise was relevant and where direct responses were addressed to the medical authorities.

Van Genderen, then a young biologist trained as an animal physiologist, became the first head of the toxicology unit. His previous working experience within the RIV had brought him into contact with the problems of acute toxicity as well as animal food experiments. He sought to deal with the new questions by organizing systematic toxicological research. A lack of experience with preventive toxicological testing was a reason for Van Genderen, supported by a fellowship from the WHO, to visit the Food and Drug Administration in Washington in 1949, where he spent three months in the Pharmacology Division. This Division had developed into the principal governmental laboratory doing toxicity testing directly before and during the Second World War, and by 1949 it was one of the main sites of experimental toxicological research in the US. As an institute of the federal government it was involved in the development of testing procedures and guidelines, as well as the investigation of problems caused by pesticide residues and food additives (11). On his return Van Genderen began preventive toxicity research and adopted the FDA procedures. The RIV toxicology unit became a focal point for safety questions related to the regulation and testing of chemicals. Preventive testing of chemicals proved to be a new area in which research experience and regulatory structure were lacking. The toxicologists tried to fill these lacunae.

In order to strengthen the capacity of the RIV to intervene in potential health problems caused by pesticides, the toxicologists worked with other government scientists. Together these differently located scientists organized an informal committee, the Phytopharmacy Committee. Van Genderen participated in this committee together with scientists from the quality control bureau and the plant protection service (12). This committee evaluated health hazards of pesticides and agreed upon spraying rules and pesticide residue levels. The committee was active in an area that cut across the boundaries of Ministries dealing with health and agriculture. The scientists from the institutes acted on the belief that it was for them to take the initiative in dealing with pesticide problems: a view incidentally which was not appreciated by the civil servants of the two Ministries. These latter held the opinion that the committee had to work formally, and within a legal framework. Thus it was deemed essential by the civil servants to issue a law on pesticides in which the Phytopharmacy Committee was given official authority to evaluate the safety of new pesticides (13). Some civil servants became members of the committee, and as Van Genderen observes:

> The functionaries saw toxicology as a way to deal with hazards by doing a few routine experiments. This attitude changed through the influence of the cooperation in the Committee on Phytopharmacy. Some of the civil servants developed a keen insight into the necessity of developing toxicological research. This new attitude made it possible to obtain sufficient means for toxicity testing and toxicological research (14).

Toxicologists used other means to interest the government apart from participating in regulation (15). The attempt of the RIV toxicologists to develop contract research for industrial enterprises provides a good illustration. In the 1950s toxicologists did toxicity testing for a number of Dutch industries: a peculiar arrangement at a time when contract research by government institutes was rare in the Netherlands. Contracts with industry provided the scientists with some financial means for their research, and it was used for animal testing facilities and some specialized apparatuses. More important however, was the contribution in the form of research problems. Chemicals from industry provided material with which the toxicologists could gain insight into the intricate details of safety testing. This testing experience offered "real" problems on which fundamental research could be based. The links with industrial toxicologists resulting from the contracts were then used to impress on government officials the importance of health problems relating to chemicals. In other words, the relevant authorities were not yet in need of the research the toxicologists wanted to undertake. Most of the work the authorities had to offer to the toxicology unit consisted of an increasing number of questions on acute problems, and there was no policy with regard to longer-term research. However, by virtue of the problems of industry the interest of the government was aroused.

> The consciousness of the government with regard to toxicological research was raised and the number of relevant questions from the government grew. The contract research with industry was therefore less essential (. . .) (16).

The toxicologists at the RIV used the opportunity provided by the problems of industry to develop methods of toxicity testing and to explore research possibilities. Thus, the contracts with industry allowed the toxicologists to increase their scientific credibility. This new development is also relevant for our question about audiences. Industrial interest in preventive toxicological testing was used as a lever to interest government, and at the same time to signal an increasing need for a more systematic approach to questions of toxicology. This would be necessary in order to deal with scientific questions of safety evaluation at the same level as was done by industry and international bodies. The formulation of a research need to identify new problems served as a method for extending the role of the toxicologists within the RIV even further. This effort, however, was hampered by the large work load that the other tasks of the toxicology unit demanded (17).

2. Biological Toxicology at CIVO

The modernization of Dutch society also played a role in the founding of the second institute for experimental toxicological work, the Biological Toxicology department of the TNO Central Institute of Food and Nutrition Research (CIVO) in 1961 (18). The TNO (Organisatie voor Toegepast Natuurwetenschappelijk Onderzoek: the Organization for Applied Scientific Research) was founded in the thirties to stimulate the application of scientific research to problems of industry and government in the Netherlands. A large portion of the money was provided by the government as a founding subsidy, but research contracts with industry were envisioned.

Within the TNO a number of different institutes were organized, some operating in economic sectors such as the chemical, rubber, or food industry, or agriculture; while others were founded to do research in sectors of general social concern like health care and defence. In 1940 one of the institutes, CIVO, started food research. Its first Director, Van Eekelen, came from the RIV and brought with him an interest in vitamin research (19). Initially, CIVO conducted animal research on food and nutrition problems. During the war food problems similar to those at the RIV were addressed. After the war CIVO policy, influenced by Van Eekelen and later Engel, strongly emphasized the necessity of adopting a commercial attitude and listening to industry (20). This policy led to good contacts with industry in the food sector and contributed directly to the origins of toxicological research.

In 1960 a large Dutch multinational, Unilever, suggested to CIVO that an effort in toxicological testing would be appreciated. At the time, Unilever was confronted by a public scandal (the so-called "margarine disease") and was of the opinion that another group doing toxicological research in addition to the RIV would provide a useful diversity of expertise. Unilever was prepared to support this effort by financing contract research (21).

In order to acquaint CIVO with the possibilities of this type of research, a decision was made to learn at first hand about recent developments in the US. Thus in 1960, A. P. De Groot, a biologist and food nutrionist working at CIVO, on leave to do research at a nutritional institute (22), was commissioned by the directors of CIVO to learn more about toxicity testing in the US. As a result, he visited the FDA and other groups active in toxicity testing. Upon his return he advised the Directors of CIVO to introduce protocol toxicity testing along with the directives of the FDA in CIVO research.

Initially, De Groot organized this research by following the FDA guidelines strictly. He believed that this imitation was necessary as a means of gaining sufficient experience with the actual testing before more fundamental studies could be started. The first large contracts made good use of previous CIVO experience in feeding experiments on animals. For example, British Petroleum, working on the development of single cell proteins, was looking for an institute that could perform safety testing on proteins (23). The toxicological evaluation of single cell proteins not only provided CIVO during its starting period with a generous sum of contract money; it also turned out to be an extremely difficult research area. This led to a number of scientific publications in the area of toxicology and eventually to a world-wide reputation for CIVO and De Groot.

After the initial period which was dominated by research on food-related chemicals and food components, other areas of testing were developed. The development of these areas was strongly related to potential possibilities for obtaining contracts with industry. Broadening the range of available methods played a key role. After making a start in oral toxicity, facilities for inhalation toxicity testing were organized when a grant could be obtained from the Scientific Council on Health and Smoking (24).

Foreign industries were and are prominent among the customers of CIVO, although there were also relations with Dutch industry. Such contracts with foreign industries could only be obtained if CIVO closely followed standard practices of toxicity testing and research. But the close relationship with industry meant that little scope was available for research being published in international journals; most of the output of the institute was in the form of confidential reports. Research subjects are typically pursued for a number of years, then dropped; only the methods remain stable. These apparent discontinuities reflect industrial interests: they are a consequence of obtaining contracts with industry. For example, organophosphorous pesticides were studied only so long as they were of commercial interest to the pesticide industry. Collaboration between researchers and industry can be initiated by industry, or a scientist can try to sell the industry a new test. But once contacts are established, the reputation of the institute is at stake:

> The value of toxicological research increases with the reputation of the research-institute. This reputation increases with the fact that foreign firms also award contracts and thereby recognize the work of the institute (25).

The emphasis that was put on obtaining contracts from industry prevented the development of a strong basic research tradition in CIVO (26). When the customer is content, new contracts will result and the scientist who did the previous testing will be approached to do new research. During the sixties the need for toxicity testing was so great that no diversification of audiences was necessary. This has changed slowly in the 1970s (27).

3. Characteristics of Safety Evaluation Research

In the 1950s and 1960s RIV and CIVO were dominant in Dutch toxicological research and toxicity testing: a dominance moreover assured because of the lack of any industrial toxicity laboratories. A number of large companies, including Shell and Unilever, did start to organize toxicology laboratories in the fifties in the Netherlands. However, both these companies were also involved in toxicity testing in the UK, and eventually decided to locate their main testing facilities there instead of conducting toxicological research in the Netherlands (28). Since industrial laboratories were insignificant, toxicology did not become dominated by industry experts, as was the case in, for example, Britain and the United States (29). Rather, close ties existed between industry toxicologists and the RIV and CIVO. Thus, the first phase of Dutch toxicology was in all respects linked to the needs of the government and emerging economic sectors. The prime interest of both Van Genderen and De Groot was to establish research related to the concerns of government and industry in that period. The relations among toxicity testing laboratories, industry, and the state in the Netherlands suggests that at least during this period toxicologists at the RIV, CIVO, and in industry had the same central goal, namely to establish procedures for evaluating the health risks of chemicals.

During this initial phase, techniques from foreign toxicological institutes were much relied upon. Problems of chemical toxicity were not yet seen in a broader perspective by government and health inspectors. Whilst the RIV toxicologists in particular signalled the need for more toxicological research as a central goal, this goal was not yet shared by the state. The urgent need for toxicologists to establish the knowledge necessary for the safety evaluation of chemicals dominated this period. The toxicologists addressing this need saw toxicology as a separate area of multidisciplinary research which — for example when compared to pharmacology — was no more than a "stepchild of science". In general, the orientation of these institutes reveals a stabilization of relations between research and one principally relevant audience. The toxicologists at CIVO worked for industry, an audience that at the same time financed their research. Here (in contrast to the RIV), the possibilities for developing research depended on the interests of industry. The collaboration of the RIV toxicologists with industry had an ulterior motive, being based on the necessity to attract the attention of the government to toxicological problems.

Although toxicology was considered a scientific endeavour, there was limited scope for pursuing its further development as a scientific discipline within the RIV and CIVO. The official tasks of the RIV scientists were too great a constraint on their actions to build such a discipline. This provides a background against which we can locate the change brought about by Van Genderen in 1961, when he became professor of toxicology at the State University of Utrecht, and thereby established toxicological research in a third type of institutional environment.

4. The Academic Institutionalization of Experimental Toxicology

Dutch universities in the sixties were still largely organized according to traditional academic principles. This meant that the state paid all personnel and did not formulate any specific science policy with regard to university research. The sciences were organized within faculties, which functioned largely as educational and administrative units. Professors were responsible for research and the contents of education within a given area. The administrative and educational tasks at the faculty level were the collective responsibility of the professorate in that faculty (30). The sixties were a period of exponential growth in student population and university budgets, and this growth afforded ample room for the scientists, especially the professors, to organize research according to their own aims and insights.

Toxicology subsisted only on the margins of academic medicine and pharmacy in a number of universities. Whilst someone in the medical or pharmaceutical faculty often taught toxicology, there was no systematic research effort. The research that was done was related to intoxication brought to the attention of hospital clinicians. Also some research was done in connection with the work of Zeldenrust, a forensic doctor of the Ministry of Justice, by a small laboratory working on poison cases. A small amount of biological research, for example on snake poisons, that could be termed toxicological was going on, but this was not directed toward questions of chemical safety.

The internal dynamics of related displines provided no stimulus for the development of toxicological research. Some toxicologists argue that in most cognate specialities toxicological research can be seen as academically less rewarding than working on the central research questions of the speciality (31). Although toxicology might have developed as a medical specialty, the medical fields related to preventive medicine were themselves underdeveloped (32). As a consequence, there was no obvious academic location within which toxicology could develop. Toxicity testing in the state institutes was generally regarded as a boring and simple form of applied research by university biomedical or medical scientists. The existing toxicity laboratories were growing, but employed only a small number of toxicologists. Moreover, these laboratories did not seek trained toxicologists so much as they sought pathologists, haematologists or biochemists who could learn toxicology during the research process itself (33). No job market existed that could have stimulated toxicology as a separate academic educational specialty.

5. Institutional Development of Toxicology in the University

The Institute of Veterinary Pharmacology and Toxicology — in the Faculty of Veterinary Medicine at the State University of Utrecht — was the first group to develop experimental toxicology. The intention to develop the subject originated with Van Genderen, who was appointed as professor of Veterinary Pharmacology in 1961, after leaving his former post as head of the toxicology and pharmacology department of the RIV. His forerunner De Jong had occupied the chair of Veterinary Pharmacology for only two years before leaving it for a more favorable position on the Medical Faculty at the University of Amsterdam. Van Genderen agreed to accept the chair on condition that the description of his tasks be extended to include Biological Toxicology in addition to Veterinary Pharmacology (34). His experience at the RIV gave him the idea that pharmacology and toxicology belonged together in one group. The duties of the new professor included lectures and practical courses in pharmacology for veterinary students, and pharmacological research in support of this teaching seemed logical and necessary. But Van Genderen also used the opportunity offered by the chair of Veterinary Pharmacology to institutionalize toxicological research in Utrecht.

He decided to try to expand the fundamental basis of toxicological and pharmacological research, an area left open by laboratories doing toxicity testing. The organization of fundamental research had been difficult in the RIV as a consequence of an overload of short-term practical problems. Van Genderen's new laboratory provided an opportunity for such an effort. Another restriction on research within the RIV had been its strict limitation to the estimation of the health risks of chemicals with human exposure. Van Genderen, like other government scientists, had been worried about the threat of pesticides and other sources of pollution in the environment. They considered research into this subject to be urgent, although prior to Van Genderen's appointment there were no institutional opportunities to do such research. Van Genderen's personal motivation played a role when he initiated research based on a somewhat broader conception of toxicology. He saw the effects of pesticides on wildlife as a signal of side-effects in general. This central concern with a broad concept of toxicology is formulated in his inaugural address, "The Toxicological State of the Environment". Here he discusses the central issues of toxicology in terms of their relation to the domestication of nature in general; he describes the adaption of nature to the benefit of mankind as a means of finding ways to handle the toxic character of the environment (35). Van Genderen concludes that research on the most important food additives and pesticides had by then already been done. This, however, was not the case, he argues, in regard to toxicological knowledge relating to the conservation of nature, especially of wildlife, or the role of pesticides. While the overall tone of his address was very practice-oriented, Van Genderen also stressed that fundamental biological and biochemical research would in the long run be necessary to solve toxicological problems.

From the start, research was structured according to the interests discussed above. Projects were organized in veterinary pharmacology and biological toxicology. Both subjects were considered directly related to practical questions. Cooperation with biologists was sought but was difficult to obtain because of faculty boundaries (36). To support both practical subjects, neurotoxicology and biochemical toxicology were main issues. Later, cooperation with veterinary pathologists was sought and eventually a separate pathology working group was established. The way these fundamental projects were detailed was closely related to problems of pesticides in relation to wildlife. Research in biological toxicology was embedded in an interest in pesticide toxicity in general; in these first years it was especially concerned with the problem of pesticide poisoning of wild birds. The fundamental projects that were organized dealt mainly with chemicals originating in the biological toxicology group (37).

Biological toxicology proved to be a research theme of significance for the outside visibility of the institute. Biology students with an interest in the themes of the biological toxicology group were attracted by this research, and intrigued by a professor who was doing research on the problem of poisoned birds falling from the sky (38). The central research theme was probably made more attractive to these students by the growing public interest in environmental problems and by the lack of interest in such problems among the existing biological faculty. The influx of biology students is cited by Van Genderen as one reason why biological toxicology grew faster than pharmacology. More important, however, was that biological toxicology offered an opportunity to do research that was significant for environmental problems. The resulting research was organized in terms of its social impact rather than its scientific priority (39). Toxicologists in the biological toxicology group organized contacts with other scientists and with — as they say — more "emotionally" involved nature conservationists. Van Genderen saw the role of scientists as providing facts, whereas conservationists wanted to regulate without recourse to the scientific facts:

> As scientific researchers, we are in the first place obliged to search for the truth. Once this truth is formulated our motivation may come back, and from this motivation we can argue for policy decisions or try to influence management by other means. The basis, however, has to be strictly objective (40).

Links with nature conservationists were never formalized, although cooperation arising from shared interests in the issues was possible. Nature conservationists brought poisoned birds to the Utrecht group or communicated their observations of bird populations. One typical research project resulting from such cooperation studied the death of the Great Terns in the Wadden Sea (41). This subject was brought to the attention of the group by a bird watcher who was worried by the deaths of seemingly healthy birds. In this case, the Utrecht group found dieldrin and aldrin to have been the substances involved. In the actual research, bird watchers cooperated with the scientists in field observations and to obtain samples. There was also cooperation with the Dutch Society of Nature Conservation (Vereniging tot Behoud van Natuurmonumenten). In general, amateur and professional conservationists functioned as a link for field research done by the biological toxicology group.

Much of the research of the Biological Toxicology group in that period was written down only in reports and Dutch publications intended for use in debates within the government about regulation. The effort was deliberately directed to the preparation of reports intended to have a high impact on political decision-making, rather than to publication in international journals (42). This orientation towards local audiences and lay people is prominently visible in the productions of the Biological Toxicology group until the 1980s. The publication profile of this group within the Institute was significantly different from the other four research groups: the pharmacology and fundamental groups wrote articles for international scientific journals from the beginning (43). This strategy provided the whole Institute with relations to audiences that secured it a position in the field of university research.

The development of environmental toxicology in the Utrecht group was an extension of the area of toxicological research. This extension was related to its orientation towards new audiences which became interested in the work of the scientists. The new audience was composed of the nature conservationists, nature managers as well as amateur scientists, and reflected a greater flexibility in the development of research interests in the university. To summarize: the marginal position of toxicology in the universities offered opportunities for a new type of research. The formation of the first institute interested in the environmental aspect of toxicology took place at a time when such concerns were becoming common but could not be translated into a research strategy elsewhere. The attention of the group was drawn toward a new area in which the interests of others were aroused and could be used to stabilize the research effort.

6. The Committee on Side-Effects of Pesticides

The group of Biological Toxicology at the University of Utrecht formed together with other researchers a committee to study the side-effects of pesticides. In terms of the vocabulary of this paper, this committee can be seen as a structuration of audiences. Nearly all the institutes and industries these scientists came from were involved in some aspect of the pesticide problem, and most of the scientists involved were acquainted with one another through pesticide regulation. The general idea that "something should be done" was alive in a wider circle, but the scientists initiating the committee had a kind of emotional commitment in addition to their scientific interest:

We met at the house of Van Genderen to discuss the contents of Rachel Carson's book *Silent Spring* and what her description would mean for nature conservation. Most of us were members of the NJN (44) and were biologists. We were worried about the degradation of quality in nature and discussed possibilities of taking action in our professional capacities. However, the reaction of the biology faculty was only lukewarm and Van Genderen's chair was the only one to which an emerging toxicological interest could be attached without creating political problems in the faculty. The group that met was organized formally as a committee of the TNO, the committee on side- effects of pesticides. This was done to give it some official status (45).

In 1965 the committee, consisting of university, state, and industrial scientists, was installed in the TNO. The CNB (Commissie Nevenwerking Bestrijdingsmiddelen, the Committee on Side-Effects of Pesticides) acted as a forum to provide advice to the TNO on research to be conducted. It also received a small budget from the TNO for research purposes (46). The motive involved in starting the CNB was to organize research on the issue of environmental chemicals, which were regarded as a social problem:

The main motive of the CNB was that while agreeing on the necessity to use pesticides in modern agriculture, it also would be necessary to know the effects of such chemicals. Pesticides should be used with the best available technical means. You should spray when necessary, but also discuss whether some damage done by insects could be accepted (47).

The CNB initially focussed on pesticides, and working groups on soil, wildlife, and occupational exposure were formed. Interest in the side-effects of pesticides functioned as a starting point, but the research that was discussed in the framework of the CNB gradually expanded to encompass environmental toxicology in general. This was a new area of concern that was different from the methods and research oriented towards the human safety evaluation of chemical exposure. Subsequently, the occupational exposure group left the CNB, while the committee changed its name in 1969 to include effects related to environmental chemicals. Increasingly, other environment contaminants were taken into consideration; one example of this research was a study of mercury pollution (48). The original motive of the committee had been the assessment of indirect risks of pesticide exposure to humans; but in the end, the effects on nature in itself became a subject of research. The CNB thus functioned as a group that catalyzed the development of a new area in toxicology.

The CNB played an important role in a variety of ways, not least by providing the necessary organizational structure to establish links between scientists working for diverse government agencies, government laboratories, and nature conservation. The research of these groups was discussed in this forum; scientists involved in regulation brought proposed measures forward for discussion; biologists from the Central Laboratory of the TNO involved in OECD and EEC programmes on the environmental toxicity of chemicals reported back about these programmes. The participants stress that the strength of the committee was that it provided a forum for informal discussions of useful research and probable policy measures. Its informal character also meant that industrial scientists could participate and talk about the industry side of the issue without fear that the government would use their information. The CNB organized working parties to facilitate these discussions. The themes covered included water, pollution, and wildlife.

In the working groups the themes provided a structuring influence on discussions. For example the soil working group was strongly interested in pesticides. In this group more scientists related to agricultural interests were active, and pesticides are clearly an agricultural concern. Also the soil is the means of production of agriculture. The working group on water problems was directed towards polychlorobiphenyls and organochlorine compounds. Generally scientists in this area had a broader interest as a consequence of the broader spectrum of water pollution. In recent years the interests of soil scientists are changing in the same direction because of increased soil pollution (49).

One result of this structure was an easier adjustment of scientific work to relevant problems. In this respect the function of the CNB can be compared to that of Dutch Sector Councils in providing a meeting place between policy makers and scientists. Scientists, however, organized the CNB without explicit science policy aims in mind. Rather, the emergence of the CNB can be seen as a type of bottom-up process of adjustment of science to social needs. Its main effect on research was that it provided a structured forum for discussions of emerging environmental research. The perspective within these discussions was decidedly a scientific one; with regard to policy the CNB took a middle position between specific interests (like agriculture) and nature conservation. The strategy of the CNB was to use scientific evidence to convince the government that it should take action to control the use of pesticides (50). The participation of industry and government scientists provided the toxicologists with the possibility of becoming visible as scientific experts in the area.

Another effect was that the new groups involved could obtain financial assistance from the CNB. During this period contract research was not readily available from the ministries. The aim of stimulating research could be served through a small budget for research made available by the CNB. Proposals for research could be drafted and put forward to the working groups. This financing also played a role in the development of research in the Utrecht group (51). One of the changes facilitated by these funds was the introduction of Gas Liquid Chromatography in the field of environmental toxicology. Koeman, for example, worked in the Institute of Veterinary Pharmacology and Toxicology on a variety of environmental issues with these instruments. The largely informal network of the CNB also served to inform public opinion of the scientists' concerns. By providing information on the sources of contamination the toxicologists provided environmentalists with arguments for their case.

The CNB can be seen as a formal expression of the role which communication and collaboration with an external audience can play in the formation of a new specialty.

7. Conclusions

I have shown how Dutch toxicologists sought to develop a new area of toxicological research, no longer simply in reaction to outside problems. While linking their research to the needs of government and industry, the toxicologists spent a lot of effort in interesting these groups in their own perspective. I am suggesting that this can be seen as the structuring of an audience. The first phase of this effort created some room for research related to safety evaluation inside the government laboratory RIV. The relation with industry built up in this development was only a transitory phase, whereas in the CIVO establishing a relationship of close collaboration with industry was an end in itself. The contrast between the developments in the RIV and CIVO shows how differences in institutional structure can shape the relations between scientists and non-scientific groups.

Changes in legislation were actively pursued as a means of stabilizing the RIV's toxicological work. In contrast to Johnston's observation quoted in the introduction, it is not so much the case that pressures shape toxicology as that close interaction between scientists and government creates an area for scientific research. In such a new area, scientists try to establish relations with audiences that are significant for their purposes. In a sense, they may use their relations with one audience (in the case of the RIV, industry) to enlist the interest of a more relevant audience (government). In the case of CIVO, the relation with the audience was shaped as a market relation in which attention to the "customer's" questions became central. The location of toxicology within the RIV also provided constraints hindering the toxicologists from developing a fundamental toxicology.

In the second phase toxicologists emerging from this background sought to broaden the area of their expertise and research. In order to do so they explored a socially relevant issue: the safety problems of pesticides. Informed by the environmental concern of a variety of audiences, the problem shifted to environmental toxicology in general. In this case, use was made of a variety of audiences to draw societal attention to the issue. The main groups involved in this process were the nature conservationists, government scientists, and the public at large. Fundamental research was oriented to the support of environmental questions. This development has been aided by a lack of institutional pressures in the University of Utrecht. The attention of this group to fundamental problems, as well as their attention to environmental toxicology, can be seen as a reaction to constraints in the RIV.

The stabilization of research on environmental problems found expression in the CNB, which provided an organized forum for the adjustment of scientific research to deal with these problems. This largely informal network organized around the issue of pesticides acted as a forum for reputations in a local and international context. The CNB also engaged in arousing public attention and the interest of the government in such problems. Regulatory laws have been developed by the personnel of agencies and government officials who have maintained close contact with CNB researchers. The organization of interests surrounding the establishment of research facilities and new research areas can thus be seen as a result of the actions of scientists influencing the public debate and organizing attention for the problems on which they want to do research. With the development of Dutch toxicological research, I hope to have shown how variations in institutional location influenced the collaborative strategies of toxicologists with outside audiences.

Notes and References

1. Bruno Latour, "Give me a laboratory and I will raise the world", in K. D. Knorr-Cetina and M. Mulkay (eds.), *Science Observed*. London: Sage, 1983, pp. 141—170.

2. R. M. Friedman, "Constituting the polar front, 1919—1920", *Isis*, 1982, 343—362.
3. R. Whitley, *The Intellectual and Social Organization of the Sciences*. Oxford: Clarendon Press, 1984.
4. W. Shrum, R. Wuthnow and J. Beniger, "The organization of technology in advanced industrial society: A hypothesis on technical systems," *Social Forces* **64** (1), pp. 46—63.
5. *Ibid.*
6. R. Johnston, "The characteristics of risk assessment research", in J. Conrad (ed.), *Society, Technology and Risk Assessment*. London: Academic Press, 1980, 104—122.
7. B. Gillespie, D. C. Eva and R. Johnston, "Carcinogenic risk assessment in the United States and Great Britain," *Social Studies of Science* **9**, 1979, 265—301.
8. H. van Genderen and G. J. van Esch, "25 jaar preventief toxicologisch onderzoek in het Rijksinstituut voor de Volksgezondheid," *Berichten uit het RIV, 1974*, Verslagen 33/34, 1975.
9. Interview with Prof. Dr. H. van Genderen, 8 August 1985.
10. G. J. van Esch, *De Kunst van de Toxicologie* (*The Art of Toxicology*) Toespraak van Van Esch RIV, maart 1985. See also Note 8.
11. Note 9. See also J. Whorton, *Before Silent Spring, Pesticides and Public Health in pre-DDT America*. Princeton: Princeton University Press, 1974.
12. Note 9.
13. P. Flipse, "Taak en werkwijze van de Commissie voor Fytofarmacie," *TNO-Nieuws*, 1969. 511.
14. Interview with Prof. Dr. H. Van Genderen, 2 December 1985.
15. Van Genderen and others participated in international organizations involved in regulation like the World Health Organization. One of the examples is participation in the Joint FAO/WHO Expert Committee on Food Additives in the fifties.
16. Note 14.
17. Note 9.
18. Currently the name of the institute is the TNO-CIVO Toxicology and Nutrition Institute.
19. *TNO-CIVO Toxicology and Nutrition Institute, Department of Biologic Toxicology* Zeist, 1985. 3—4.
20. A. P. De Groot, *Relaties van de afdeling Biologische Toxicologie met overheid en bedrijfsleven* (relations of the division of biological toxicology with government and industry) Voordracht Voedingsorganisatie TNO, 25 oktober 1974.
21. Interview with Dr. D. C. Leegwater, CIVO, 8 januari 1986.
22. He worked in that period for the Quartermaster Food and Container Institute of the US Army in Chicago. The research was directed to the nutritive value of dried food. A. P. De Groot, "Controle van nieuwe produkten," *TNO-project*, 1974, pp. 174—180.
23. One of the reasons BP chose CIVO was that they wanted to avoid the leaking of information to its English competitors.
24. This Council is paid by the cigarette industry; research supported by these funds is judged by the Council, membership in which is based upon scientific credentials.
25. Note 20.

26. Emphasis has been on the work for industry; publishing was seen as an extra for motivated scientists.
27. Changes in the amount of funding available from industry recently forced CIVO to become more innovative and to compete for grants for pure research.
28. Only within two pharmaceutical firms has toxicological research come into a separate existence: Philips Duphar and Organon.
29. A. M. Coles, *Social and political factors in the development of toxicology*, unpublished Ph.D. dissertation, University of Aston, Birmingham, 1984.
30. The Dutch university system was organized along the lines of the "chair-based system" as described by B. R. Clark in "Academic power: Concepts, modes, and Perspectives", in J. H. Van de Graaf (ed.), *Academic Power, Patterns of Authority in Seven National Systems of Higher Education*. New York: Praeger Publishers, 1978.
31. One of the scientists interviewed characterized this nicely: "When you want to do rewarding immunological research, you certainly should not pick immunotoxicology. Immunology is the science which provides insight into the immune system of the mouse, while to work in toxicology means experimentally to work on rats. So to obtain scientific credit it is not a good area to work in; it is simply too small." Prof. Dr. W. Seinen, 13 August 1983.
32. R. Hohlfeld, "Two scientific establishments which shape the pattern of cancer research in Germany: Basic science and medicine," in N. Elias, H. Martins and R. Whitley (eds.), *Scientific Establishments and Hierarchies*. Sociology of the Sciences, Volume VI. Dordrecht: Reidel, 1982, pp. 145—168.
33. This attitude is still in existence, as illustrated by reactions to a recent report by the Dutch Toxicological Society on education. While academics tend to argue for a place for toxicology next to their own territory, people working in toxicity testing laboratories and the government stress the need for specialists in specified areas who are not necessarily *a priori* extensively trained in toxicology. *Toxicologenopleiding in Nederland*, Rapport van de Werkgroep Toxicologen-opleiding, Nederlandse Vereniging voor Toxicologie, 1983.
34. Biological toxicology was coined as a term to distinguish research interests in the biological mechanisms underlying toxicity from merely analyzing toxic chemical substances in biological material, as had been done previously.
35. H. Van Genderen, *De Toxicologische Gesteldheid van het Milieu*, Inaugurele Rede Rijksuniversiteit Utrecht, Utrecht 1963. While the overall tone of the address is very practice-oriented, it is also stressed that fundamental biological and biochemical research is necessary to solve problems in the long run.
36. For example, Van Genderen recalls having a lot of trouble with the biology faculty to get them paying scientists for the department, while the university was unable to organize a system in which the Faculty of Veterinary Science got paid for the education of biology students working for the department. This partly is a consequence of Dutch university financing.
37. *Publikaties van het Instituut voor Veterinaire Farmacologie en Toxicologie 1961—1980.*
38. Interview with Dr. B. Blaauboer, Utrecht, 20 March 1984.
39. Interview with Prof. Dr. H. J. Koeman, Wageningen, 2 August 1985.

40. H. Van Genderen, "Nevenwerkingen in soorten," *TNO-Nieuws*, 1972, 524–527.
41. J. H. Koeman et al., "Insecticides as a factor in the mortality of the Sandwich Tern," *Mededelingen Rijksfaculteit Landbouwetenschappen Gent* **22** (3/4), 1967, 841–854. The spread of other chemicals was also studied, for example in J. H. Koeman et al., "Chlorinated biphenyls in fish, mussels and birds from the river Rhine and the Netherlands coastal areas," *Nature* 1969, 1126–1128.
42. See Note 39. The impact, by number of citations, of the publication in *Nature* by Koeman (1969) – 150 citations in 8 years – indicates that interest and possibly publication opportunities for this type of work were available.
43. See Note 37.
44. NJN, the Dutch Youth Nature Conservancy Organization, where membership was possible up to 23 years of age.
45. M. F. Mörzer Bruyns, "Nature preservation and chemical pollution," *TNO-Nieuws*, 1972, pp. 528–531. The committee was made up of biologists from the Research Institute for Nature Management (RIN) at Arnhem and Leersum, the Institute of Ecological Research (IOO) at Arnhem, Central Veterinary Institute (CDI) at Lelystad, Zoological Museum, University of Amsterdam, Biological Laboratory of the Free University at Amsterdam, Netherlands Society for the Protection of Birds at Utrecht, Commission for Faunapreservation of the Nature Conservation Council (part of the ministry of culture recreation and welfare) at Rijswijk, Institute of Veterinary Pharmacology and Toxicology of the State University of Utrecht, Research Institute of Public Health (RIV) at Bilthoven, Plant Protection Service at Wageningen, Department for Nature Management of the Agricultural University of Wageningen, research Dept. of Shell Netherlands and CIVO.
46. Interview with Drs. E. H. Hueck-Van der Plas, 10 November 1985.
47. H. Van Genderen (1972), *op. cit.*, 527.
48. Note 46.
49. *Ibid.*
50. "The publication of the data of research (. . .) contributed much to the publicity of the most interesting facts. This has been important and more especially to the benefit of the responsible authorities." M. F. Mörzer Bruyns, *op. cit.*, 531.
51. Note 39.

EPILOGUE

THE CAUSES AND CONSEQUENCES OF COLLABORATIONS BETWEEN SCIENTISTS AND NON-SCIENTIFIC GROUPS

JOSKE BUNDERS AND LOET LEYDESDORFF

In the summer of 1983, when we started to formulate our ideas about a comparative study of the "bottom up" initiatives of scientists who wanted their research to be more socially relevant and hence had sought collaboration with non-scientific groups, we acted primarily from our concern with social reform, having both been active members of the science shops movement in the Netherlands. We knew from these experiences that the process of translating social demand from public interest groups into the science system, the feedback of expertise on social action, and the resistance and receptivity of the science system to such demands were aspects of a much more complex and multiform process than had previously been described in the literature. The available models had either taken a linear perspective (1) — allowing sometimes for feedbacks as well — or had then been elaborated into an inter-actor perspective (2). In other cases the relations of the science system with its environment had been conceptualized as resource-relations among which the researchers involved had to make selections (3). The specific historical and cultural conditions under which these "bottom up" collaborations occur, which are so evidently tied to cultural developments — as we were aware from our backgrounds in the aftermath of the student revolts of the late 1960s — were only seldom brought to the fore, let alone analyzed in terms of the sociology of science. Reflection on these experiences in different settings appeared a promising domain for further analysis, but we did not have at our disposal a solid conceptual framework to assess these experiments. To begin to develop such a framework, in the theme paper for the volume we asked potential contributors to distinguish between the socio-political conditions in their specific national history and culture,

S. Blume, J. Bunders, L. Leydesdorff and R. Whitley (eds.), The Social Direction of the Public Sciences. Sociology of the Sciences Yearbook, Vol. XI, 1987, 331–347.

on the one hand, and, on the other hand, the systematic significance of the processes which were brought about in the science system.

Now that we have come to the end of the book, we have again to raise the question of what can be learned from the experiences in the case studies described in this volume of relevance to the sociology of the sciences. Can we describe these experiences better than before as instances of general patterns of collaboration between scientists and non-scientific groups? If so, are we able to catalogue relevant dimensions for comparison, raise new questions from a more analytic point of view, or perhaps even set an agenda for future research on such questions?

The studies gathered in this volume show an overwhelming variety of patterns. However, we think that we can proceed to order them in terms of the socio-political conditions of the cooperative processes in a specific national history and culture, and in terms of the substantial significance of these processes for the knowledge production process. The first aspect largely determines the degree of the external group's integration into the science system; the second aspect is largely determined by the compatibility between the cognitive goals of the cooperation and developments in the field(s) involved. The scientists involved in cooperations have to match their options and their resources also within this dimension if they are to be successful in science. In our opinion, therefore, the relative success of cooperations like those described in this volume can be assessed in terms of these two types of outcome: that is, increased integration of the external group into the discourse about science and science policy, or at the cognitive level the emergence of new scientific activities or lines of research, or even scientific fields.

We recognize that these two outcomes are not wholly independent: that both can develop over time, and that they may interact. By becoming more acquainted with the mechanisms of the knowledge production process, clients can build up a greater involvement in the development of the sciences, and hence the translation process can be better managed from their side. On the other side, the need of scientists to mobilize resources from within their local environments may motivate them to perform translations of external demands and to build up expertise in this respect, including improving their ability to translate external questions into questions which can be dealt with from within a specific scientific tradition.

However, for analytical purposes it may be worth maintaining the distinction between these two elements since they enable us to order the various cooperations along two dimensions: the degree of *integration* of the external group into the science system, and the degree of *cognitive compatibility* of the demands put forward by these external groups. We will now discuss the first dimension with regard to the contributions to this volume. The second dimension is dealt with in section 2. In the last section we will combine the two dimensions and look at the initiatives aimed at stabilizing cooperation processes.

1. Dimensions of Cooperation

Cooperations with a high degree of cognitive compatibility allow the scientists involved greater opportunities to publish their results in line with the intellectual standards and values of their fields. Thackray, cited by *Blume*, presents an example of such cooperation in chemistry, when he describes Adams as having been able to find "a harmonious way of combining the growing interdependence of academic research and chemical manufacturing in the 1920s and 1930s." When there is such a correspondence between external (e.g. industrial) interest and academic standing, the cooperation may succeed without much conflict. For instances in which such a balanced mutual interest is extended beyond the individual scientist to include groups of scientists and non-scientists, Blume introduces the concept of "common social projects." An example of such strong interaction in a completely different domain is elaborated and somewhat modified in *Wagner*'s contribution to this volume, which describes reform coalitions between social scientists and policy-makers in France, Italy and West Germany. The "common social projects" can be contrasted with other types of cooperation in which socially conflicting interests emerge, and where the cognitive purpose of the common enterprise is not easily identifiable in a single way or within an existing disciplinary framework. As we know from the study of coalitions, one of the partners may have a "hidden agenda", for example, the long-term research programme of biologists (*Bunders*), or else through the process of integration it may become clear that what was originally believed to be a common social project gave rise in the end to new differentiations and to conflicts of interest. *Wagner* speaks in this respect of a "honeymoon syndrome" which is at the root of the demise of such coalitions. In *Fridjonsdottir*'s paper the focus is on the role of unions in developments similar to those Wagner describes, but now in Sweden, where the unions introduced a new element into the political forces around the emergence of the sociology of work. Here we see socially conflicting interests entering the emergence of this branch of sociology in a national context, while to some extent this conflict is neutralized through the further internal differentiation of the discipline.

However, social problems are not always easily translated into or accommodated to fit the rules of the game in the scientific system. Conflicting interests may disturb the process of mutual adjustment because of differences in priorities, concepts, or norms, etc. An extreme example of such non-compatibility on the cognitive side is elaborated by *Benguigui.* He describes French fishermen who continued to ask scientists to develop methods to restock the sea with fish. But extensive aquaculture — which is what they asked for — was not on the priority list of the marine biologists, and so restocking experiments did not fit into their goals. From the perspective of the marine biologists the reason for this is primarily technical: experiments under laboratory conditions can be more easily monitored than the results of restocking experiments in extensive aquaculture. Therefore, the demand of the fishermen was percieved as cognitively incompatible with the research agenda.

The demand for other research methods, such as more holistic approaches, is widespread among proponents of "alternative" agriculture, "alternative" medicines, etc. Although demands for what is essentially an alternative scientific method may seem at first sight to have *per se* a low degree of cognitive compatibility, under certain historical and cultural conditions such demands can challenge the existing science system and its disciplinary structure. An example of this is given by *Ramasubban and Singh,* who describe the situation of a third-world country in which groups of scientists have lost their faith in the traditional ways scientists have dealt with problems concerning health, food production, and poverty. Moreover, demands for a multi-disciplinary approach to problems usually do not fit easily into the dynamics of scientific fields. *Cramer, Eyerman & Jamison* argue in their case, however, that the underlying shared ideologies may appeal to the institutional development of new environmental sciences, at least during the "honeymoon period"; *Von Gizycki* describes a case in which the articulation of demand by a group of patients who suffer from an eye disease functioned at the cognitive level as an integrating device for the specialists involved. Although multidisciplinary demands may by definition have a low degree of cognitive compatibility when they are raised, in the history of science under specific conditions they can be among the most powerful incentives for innovation (4). The specificity of these conditions should, in our opinion, be related to the balance of risks which scientists have to assess when they feel motivated to go ahead without having from their traditional training the skills necessary to solve the problems they are facing and without having the certainty of the reward systems of their fields (5). *Bunders* suggests that in biology the discrepancy between the demands of the reputational reward system of the field and the aims of the research group can be understood by looking at the conflicting demands of a particular reputational reward system. Scientific groups which have to deal with this problem can seek non-scientists' resources to be able to work on aims which are very poorly rewarded but which might prove to be extremely important in contributing to the attainment of a long-term main goal in the field.

An interesting situation arises when scientists see opportunities to change the cognitive structure of their field in relation to the demands of external groups. In this context, *Blume* cites Friedman's account of the Norwegian scientist Bjerknes, who in the 1920s saw the possibility of exploiting the demand of aviators and farmers for better weather forecastings in order to enhance the conceptual development of meteorology (6).

The second dimension along which we will attempt to order the different cases is the degree of integration of the external partner into the science system. One conceptual problem with the notion of integration is the question of *what* precisely these different clients are integrated into. The integration can be shaped at different levels which again can interact with one another. Are the clients integrated directly into the group of scientists involved, into the local or national organization of the scientific enterprise, or into the development of the field? We think that this is an empirical question which may vary among fields, societies, and clients: integration, when successful, gears the inflow of resources from outside both to social interests and to the core of the knowledge production process. In most cases the relations between "integration" and "the social conditions for integration" can be made a research topic in itself.

Broadly speaking in the context of this book, however, if we consider the different forms this integration may take from the perspective of participation in the organized knowledge production process, it seems that partners who are accustomed in one way or another to cooperating with scientists will meet with fewer barriers to the collaboration, will have developed mechanisms within their organization to overcome such barriers, and will eventually know how to handle the conflicts of interests which may be involved. This dimension, which from now on we will call "historical integration" to emphasize its socially contingent character, may be compared with the notion of participation in the knowledge society as it was introduced in Vol. X of this series (7).

Examples of a high degree of historical integration can be found in university-industry relations as reviewed by *Blume*. Particularly in the "science based" industries, such relations have now existed for many decades, and sometimes go back to the 19th century. The traditional examples are the scientification of the chemical and the electrotechnical industries, but nowadays the same relations can be found in many more branches, including the so-called "new technologies" (8).

The social sciences, however, can also have important ties to industry. *Fridjonsdottir* and *Wagner* independently mention the role of industrialists initiating and financing empirical research in the social sciences. Here, however, the civil servants from the growing state bureaucracies seem to have been the most important clients, especially in the post-war period.

We may see something of the mechanisms of the integration process in cases in which the degree of integration is low but the cognitive compatibility is high, or vice versa. For example, *Leydesdorff and Van den Besselaar* focus in their contribution on the relations between the trade unions and the natural sciences in a Western European country. Here the integration has historically been blocked, while the cognitive compatibility of the questions involved may still be high. In the subfield of the sociology of work central to *Fridjondottir*'s article, unions played an important role in the development of the field despite earlier political arrangements (in the 1950s) which prevented scientists at that time from becoming involved in industrial conflicts. The particular historical circumstances may lead to a low degree of integration of an important social partner for specific sciences in a specific period. *Amsterdamska*'s analysis of the relations between intellectuals and the Polish "Solidarity" movement is also an interesting case in this respect.

One should, however, be aware that the degree of integration of the external group is primarily a historical factor which will be important when we want to explain the occurrence of collaborations between scientists and non-scientific groups in contingent social terms. The other dimension — the cognitive compatibility *of the demand* — is more important for the substantial dynamics of cooperation at the level of knowledge production. Of course, the major task which we will turn to in the last section is to combine the two dimensions in the analysis. It is there, in our opinion, that the main research questions for the future will be located.

Analytically, we can already anticipate that we will be able to distinguish between cooperative processes in which both the degree of integration and the degree of cognitive compatibility is high — as for example in some university-industry relations — and other forms of cooperation which are lacking in one or both of these respects. We may call these "alternative" cooperations.

The first type of alternative cooperation is characterized by a high degree of integration and a low degree of cognitive compatibility. In this type of cooperation, traditional partners cooperate, although for some reason the demands of the external groups do not fit into the cognitive developments in the scientific field. One example of such a cooperation is described by *Bunders* in her contribution to this volume: the cooperation between firms and biologists in biotechnology in the Netherlands. The main goal of this cooperation has been to develop new biotechnological techniques. However, as *Bunders* argues, the development of new techniques is a badly rewarded activity with a low research priority in biology.

The second type of alternative cooperation can be characterized by a low degree of historical integration and a high degree of cognitive compatibility. This is illustrated in the paper by *Leydesdorff and Van den Besselaar* about the relations between trade unions and natural scientists. Although alternative options can sometimes be perceived on the side of science and technology, and although the unions are eager to stimulate developments compatible with their own interests, the institutional organization in the wider society is such that serious barriers have to be taken into account.

Finally, we have alternative cooperations which can be characterized by both a low degree of historical integration and a low degree of cognitive compatibility. For instances of such cooperations we can look at the cooperation between scientists and underprivileged groups in the Third World as described by *Ramasubban and Singh.*

2. The Dynamics of Integration

In this section we want to focus more specifically on the process of integration of a client group in the collaborative effort with a scientific group.

Such cooperations, as far as we can see now, involve a variety of elements. In the first place there is the communication of the external group's demands to the scientists. Secondly, there is the development of knowledge by the scientist through research to meet the demands. Thirdly, there is the communication of the results back to the external group. The major contextual factors in the cooperation include the social and economic position of the external group, the disciplinary contexts of the scientific groups involved, and the political and institutional environment in which the cooperation develops.

The position of the client group in the wider society is important in several respects because it shapes the political and cognitive orientation of the questions put forward and the answers which are expected. When the demands of an external group are considered legitimate by the general public and by influential groups in society, major barriers such as a lack of integration and a lack of cognitive compatibility may be overcome. An example of this is provided by the cooperation with the environmental movement in the 1970s (*Cramer et al.*). However, the major reason why scientists cooperate with external groups, as we assess it from the different papers, concerns their access to new resources: not necessarily financial, but including also apparatuses, access to materials and new domains, information, or legitimation. In the marine biology example given by *Benguigui*, the fishermen seem to be able even to influence the choices in the research of the scientists, because they enjoy certain legal and traditional rights — about pumping seawater, etc. — which they can use in negotiation. Another example of this mechanism — one not dealt with extensively in this volume — is the introduction of the Plant Variety Protection Act in the United States in 1970. This patent-like protection of newly developed plant varieties gave American firms a device for gaining the cooperation of scientists, because the plants and micro-organisms they would like to use in their research were no longer in the public domain (9).

The possession of such resources is particularly important for the cooperative efforts of groups whose demands reflect a low degree of cognitive compatibility. The chance for scientists to gain prestige among their colleagues through the results of a cooperation then decreases, and additional resources can sometimes compensate for this problem. Finally, another factor of importance seems to be the ability of the external group to specify the type of knowledge which it would find useful, which of course focuses the research to be done on that question. In order to specify their demand in this respect, the group needs to know about the ways in which it can influence the solution of the problem with which it is confronted. When the external group does not have the relevant information, this situation can itself block the process of integration (cf. *Leydesdorff and Van den Besselaar*).

When we now consider how the nature of the scientific group affects types of cooperation, we have once again to distinguish between the scientific field and the scientists involved. The orientation of the main developments with respect to social problems varies widely over the different fields. Of course, this variety influences the scientists' attitudes and particularly their inclination to engage in concrete cooperations, because their disciplinary background is the main resource which they can exchange in the collaboration. Other factors, such as their ability to popularise scientific results, or local, institutional, and political motives, may modify their receptiveness to a call for cooperation. The issue of popularisation, which was dealt with extensively in Vol. IX of this series (10) returns here particularly account of the cooperation between ecologists and the environmental movement in *Cramer et al.*

Wagner, Fridjonsdottir, and *Amsterdamska* each in a different way demonstrate the high receptiveness of sociologists to new demands from social actors. Perhaps the cognitive structure of sociology is such as not to exclude new problem definitions, methods, techniques, and concepts whenever they seem necessary (11). However, almost all the other sciences are less receptive to external demands. The issue of whether or not a *particular* social demand formulated by an external group can be met by scientists within a specific (sub)field raises questions about the cognitive and social structure of that (sub)field. *Bunders* argues in this volume that cooperation based on the development of techniques can have a low degree of cognitive compatibility in one field (biology) and a high degree of cognitive compatibility in another (physics).

The relationships between the external groups and the scientists involved are heavily influenced by the degree to which both groups share common values, whether these be substantive (e.g. what *Blume* refers to as "a heightened sense of shared social purpose"), or more procedural, about the rules of the game. Perhaps substantial shared values become critical when the degree of cognitive compatibility is not very high, as in the case of cultural emancipation. In such cases these values may compensate the scientists for a lack of scientific rewards. *Cramer et al.* analyze the cooperation between certain biologists and the environmental movement as sharing not only certain values but essentially a cosmology from which specific knowledge interests can be derived.

Communication between the scientists and the external group may be problematic, particularly when the non-scientist is not familiar with the scientific jargon and the scientist is not very skilled in popularizing his work. However, when there is a scientist in the external group, this problem is generally absent. The cooperation between the Retina Pigmentosa Society and the medical world, according to *Von Gizycki*, was notably improved by the dual role of the "Forschungsreferent" who was both a patient and a scientist. We know also from evaluations from the Amsterdam science shop that the presence of someone with higher education in the client group was statistically the determining factor to ensure client satisfaction (12).

Finally, we should mention the influence of government policies and legislation on the emergence and eventual success of cooperations between scientists and non-scientific groups. Science policies, the funding policies of universities, etc., can influence the room for manoeuvre at lower levels in a variety of ways by stimulating market forces to intervene or by shielding the science system from external pressures. Other forms of government policy can also facilitate or prevent certain types of cooperation, as shown by *Groenewegen* in the case of the emergence of toxicology as a practical science, and by *Fridjonsdottir* when she mentions the arrangements that prevented Swedish trade unions from interfering in employer policies in the 1950s as an explanation of the delay in cooperation between scientists and trade unions.

In summary: the cooperations most likely to occur are those with a high degree of integration and a high degree of cognitive compatibility. Sometimes, however, changes in socio-economic perspectives can be the cause of unusual demands posed by the external partners traditionally involved in cooperations with scientists. Resources then play a very important role in further cooperation. A lack of resources can, in some situations, be compensated by shared values. This may be the case particularly when these shared values are linked to the wider social environment of the cooperation, for example in cooperations with important social movements. Resources and shared values become critical when there is not only a low degree of historical integration but also a low degree of cognitive compatibility.

3. Processes to Stabilize Research Based on Demands of External Groups

In this section we will address the question of what mechanisms can be distinguished in the various contributions for the stabilization of research based on the demands of external groups. Under what conditions do stabilization processes occur, and what problems are involved in these processes? Can we link the types of stabilization processes to the two dimensions we introduced above, and to the typology derived from this scheme?

It will be clear that cooperation is less likely to emerge as a stable process under circumstances which imply low degrees of cognitive compatibility and historical integration. However, sometimes other circumstances can compensate for the problems which arise from these hindrances. This leads to the conclusion that cooperation can be enhanced by changing the circumstances which lead to the low degrees of compatibility and integration, or by creating circumstances which compensate for these situations.

The degree of integration of an external group can be enhanced by giving this group institutionalized access to the scientific system. It may be that the degree of cognitive compatibility of the external demands, when low, can be enhanced in the long term only through the emergence and eventually the institutionalization of a new subfield. The establishment of a specific institute seems to be the means par excellence to compensate for a lack in one or both of the noted dimensions if one wishes to stabilize the cooperation for other reasons.

From these considerations we would like to distinguish three types of stabilization process:

1. the institutionalization of the external group's access to the science system;

2. the institutionalization of a new field or subfield; and
3. the establishment of a new institute outside the disciplinary framework of the academic sciences.

Although all three developments may stimulate cooperative efforts between scientists and non-scientific groups — and in practice one always finds some combination of several of these elements — policy instruments should be developed in terms of their ability to overcome the specific barriers involved in the cooperation process.

Institutionalization of the external group's access to the science system is most explicitly described in *Fridjondottir*'s account of the Swedish "research circles". These "circles", which include both local union members and social science researchers, operate under the favourable conditions of the sector-wise funding system in Sweden, in which the various interest groups play an important role. Fridjonsdottir suggests that the origins of these circles can be traced to the more general re-orientation of university research and education in Sweden. (However, we think such tendencies can be found in other Western European countries as well, although perhaps they are weaker.)

One specific university policy instrument to promote the access of underprivileged groups is the science shops, which are mentioned occasionally throughout this volume. *Leydesdorff and Van den Besselaar* deal briefly with their social and cultural background in the Netherlands. Even though science shops have only limited funds available for scientists, they are sometimes very successful because they assume cognitive compatibility between questions in society and questions which interest researchers, and hence concentrate primarily on lowering thresholds. The cognitive distance which makes the development of cooperation unlikely can be decreased considerably through the intermediary role of the shopkeeper, particularly when the latter shares values with the external group.

The institutionalization of new fields or subfields of science is central to several contributions in this volume: the introduction of toxicology as a field (*Groenewegen*), the redefinition of ecology as a field (*Cramer et al.*), and substantial changes and extensions in sociology (*Wagner, Fridjonsdottir*). In all these cases, the development of the field was preceded by cooperation between scientists and external groups which initially had cognitively different problems. The translation of these problems into a field perspective is the central question in these contributions.

In their study of the emergence of ecology as a subfield of biology, *Cramer et al.* distinguish several phases in this cooperation. In the first phase, biologists popularized biological research results that demonstrated the damage being done to nature by pollution. In the second phase, biologists and environmental movements in several countries undertook political action to stimulate governments to formulate environmental policies. The general awareness of environmental problems created conditions for institutionalizing environmental studies and subsequently for creating the structures of a scientific community. Once these were established, in a third phase the interactions between ecologists and the environmental movement again became less intense.

Likewise, *Wagner* defines a problem definition phase in which policy makers and social scientists became aware of a common interest in instruments for social engineering, a second "honeymoon" phase, and a demise of the coalition once the field was institutionalized. From *Groenewegen*'s contribution to this volume, however, we learn that in the case of toxicology the lack of strong relations with social movements capable of influencing the general public retarded developments, at least at the institutional level in the Netherlands. Moreover, for a long time the direction in which the field should be developed was unclear, and the leading researchers were hesitant to decide whether the new research area should be linked to zoology, ecology, or veterinary pharmacology. This constellation resulted in a considerable institutional build-up, with thresholds not too high for external partners, before a clear research programme emerged.

The job expectations of students trained in new areas seem to have been a major concern in the institutionalization of new fields. In the cases presented in this volume, jobs often seem to have been found in governmental institutes which were established at the same time as the scientific (sub)fields. Independently of this side effect, it seems apparent that after the stabilization a "translation" (13) had actually taken place: the relations between external groups and scientists were redefined, and cognitive exchanges in the new network always differed considerably from previous arrangements.

Finally, we come to the last form of stabilization process which we have distinguished above: the establishment of a new institute. Institutes for problem-oriented and applied research outside academia are common in all modern societies. In some countries specialized research agencies cover the sectoral needs of scientific knowledge, while in others there are more direct links between the university system, the research system, and the sectoral needs of society. We have already noted the sectoral organization of university research in Sweden.

In France, there exist national agencies with specific missions not directly connected to the university system. In this volume *Benguigui* focuses on the establishment of the "Centre National pour l'Exploitation des Oceans" (CNEXO) by General de Gaulle's government. One of the tasks of CNEXO was to introduce *intensive* aquaculture. However, the French fishermen wanted *extensive* aquaculture, but as they had no resources they eventually had to support competitive governmental institutes whose goals were more in line with their own.

In most countries the establishment of a new institute is a matter of public policy, but disappointment with governmental policies can also lead to attempts to establish institutes. *Ramasubban and Singh* analyse the establishment of grassroots organizations from this perspective. Here we find an extreme case: traditional scientific institutions are neither accessible for the needs and demands of the population nor, in most areas, capable of keeping up with the pace of international scientific developments. University scientists become alienated from the public science system, and turn to new coalitions to fight poverty and famine, and to stimulate developments at the regional level.

The different institutional forms which we have discussed also generate different effects and side-effects. The emergence of a new subfield, and the inclusion in that subfield of the scientists who had formerly been active in the cooperation, changes their career perspectives, their potential audiences, and the theoretical framework of the substance of their work. The establishment of new institutions may lead, on the one hand, to problems such as those the Bolsheviks had to face once they had created scientific cadres which were "both red and expert" (*Balzer*), and whose critiques could no longer be dismissed as "counter-revolutionary". On the other hand, the organization of scientific results along bureaucratic lines may result in stifled and distorted scientific activities.

A lack of resources, which is usually the major problem facing new cooperations in the early stages, is of course most easily overcome by the establishment of a specific institute. When the emphasis is on improving access to existing organizations or deflecting the direction of existing lines of research, the resource issue becomes a more hidden factor in the development.

4. Conclusions

In the previous sections we have tried to order some elements from the different contributions to this volume according to a rather simple framework of two dimensions. We think that this effort has led to some results gained from comparison among the different cases. Of course, many other perspectives for comparing the different cases are possible, such as for example the questions of whether and how the external group profited from the cooperation. Pursuing these questions would, however, lead us away from the perspective of the sociology of science.

Our analysis in the two dimensions of cognitive compatibility and historical integration has made us aware that the distinction between a cognitive unit of analysis and a more institutional unit of analysis, which is common in the sociology of science, returns once again when we deal with the external relations of science. The traditional picture of internal cognitive criteria versus external non-cognitive criteria fades away, and a two-dimensionality of *all* the problems related to the science system emerges: between, on the one hand, the social actors intervening in the scientific enterprise, forming coalitions, inter-organizational structures, and arrangements, and the scientists stepping into these relations to mobilize resources or from more idealistic motives; while on the other hand, the substance of the demand is a cognitive reflection of what we might call the proto-scientific mode of a social problem (14), which probably because of its cognitive character appeals to scientists in different degrees, depending on their receptiveness to the external demands of the cognitive structure of their disciplines.

Of course, in real life the two dimensions interact, and the question for the external actor becomes that of finding "the precise mix of cognition and organizational power which is necessary to act upon the science and technology system" (*Leydesdorff and Van den Besselaar*). When the scientists are "over-powered" in such relations, they ought to resist; when they succumb in such a situation, for example because of resource dependency, they may be distracted from the cognitive goals of their fields, and in the long term their knowledge may be reduced to obsolete expertise. However, when there is no countervailing power in the cooperation, there are no resources to be mobilized to counter-balance the rigidities of the institutions and cultural traditions which feed the scientists with resources institutionally. At this level, the sociology of interorganizations, in which cultural traditions, cognitions, and skills are more fully elaborated as exchange media, may provide us in the long run with an intellectual perspective.

The other direction to go is that of the sociology of knowledge. What makes a demand of non-scientists a "proto-scientific" insight, i.e., a question which is prestructured to be scientified? As we have argued, one element is the cognitive compatibility of the core of a demand with ongoing developments in the given discipline. However, this is not enough, because we occasionally find successful cooperations which rely on shared values, on more normative and expressive elements of interaction than mere cognition. *Cramer et al.* in this volume used the concept of "knowledge interest" as a common denominator for social groups and scientists guiding their cognitions at the epistemological level. Others have proposed concepts such as "transepistemic arenas" which essentially interact with scientific discourse (15). At the same time, we should learn from the older traditions to which a term like "knowledge interests" refers, and remember that these cooperations are not related to transcendental subjects but to concrete social actors who have discovered that in modern times the sciences and their related technologies are also a realm of social conflict.

Notes and References

1. W. v.d. Daele, W. Krohn and P. Weingart (eds.), *Geplante Forschung*. Frankfurt a.M.: Suhrkamp, 1979, pp. 22 f. See for a critique of linear models in innovation studies: D. Mowery and N. Rosenberg, "The influence of market demand upon innovation. A critical review of some recent empirical studies", *Research Policy* **8**, 1979, 102—153.
2. Among others: E. von Hippel, "A customer-active paradigm for industrial product idea generation", *Research Policy* **7**, 1978, 240—266; G. Dosi, "Technological paradigms and technological trajectories", *Research Policy* **11**, 1982, 147—162. See also: M. Callon, "Some elements of a sociology of translation: domestication of the scallops and the fishermen in St Brieuc Bay", in J. Law (ed.), *Power, Action and Belief. A New Sociology of Knowledge*. Sociological Monograph Series, London/Boston: Routledge & Kegan Paul, 1986.
3. K. Knorr, "Scientific communities or transepistemic arenas of research? A critique of quasi-economic models of science", *Social Studies of Science* **12**, 1982, 101—130.
4. M. J. Mulkay, *The Social Process of Innovation*. London etc.: Macmillan, 1972. See also: G. Lemaine, R. McLeod, M. Mulkay and P. Weingart (eds.), *Perspectives on the Emergence of Scientific Disciplines*. The Hague/Paris: Mouton, 1976.
5. M. Douglas (ed.), *Essays in the Sociology of Perception*. London: Routledge & Kegan Paul, 1982. See also: R. Hagendijk and J. Cramer, "Intellectual traditions as cognitive constraints", *Social Science Information*, 1987 (forthcoming).

6. R. M. Friedman, "Constituting the polar front 1919—1920", *Isis* **73**, 1982, 343—362. See also: R. M. Friedman, *Appropriating the Weather: Vilhelm Bjerknes, Aeronautics, and the Construction of Modern Meteorology, 1898—1924.* Cornell University Press (forthcoming).
7. G. Böhme and N. Stehr (eds.), *Science and Society: The Impact of Scientific Knowledge on Social Structures.* Sociology of the Sciences Yearbooks, Vol. X, Dordrecht/Boston: Reidel, 1986.
8. B. J. Culliton, "Monsanto gives Washington U $23.5 million", *Science* **216**, 1982; F. H. Buttel, J. T. Cowan, M. Kennedy and J. Kloppenburg Jr., "Biotechnology in agriculture: The political economy of agribusiness reorganization and industry-university relationships", *Research in Rural Sociology and Development* **1**, 1984, 315—348.
9. *Ibid.*, p. 326.
10. T. Shinn and R. D. Whitley (eds.), *Expository Science: Forms and Functions of Popularisation.* Sociology of the Sciences Yearbooks, Vol. IX, Dordrecht/Boston: Reidel, 1985.
11. R. D. Whitley, *The Intellectual and Social Organization of the Sciences.* Oxford: Oxford University Press, 1984.
12. J. Jacobse, *Is de klant koning?* Amsterdam: Wetenschapswinkel, Universiteit van Amsterdam, 1980.
13. Cf. M. Callon, *op. cit.*, 1986, Note 2.
14. L. Leydesdorff, A. Teulings and P. Ulenbelt, "Trade union participation in university research policies", *International Journal of Institutional Management in Higher Education* **8**, 1984, 143 f.
15. K. Knorr, *op. cit.*, Note 3.

INDEX